MHC Protocols

METHODS IN MOLECULAR BIOLOGY™

John M. Walker, SERIES EDITOR

224. **Functional Genomics:** *Methods and Protocols,* edited by *Michael J. Brownstein and Arkady Khodursky, 2003*
223. **Tumor Suppressor Genes:** *Volume 2: Regulation, Function, and Medicinal Applications,* edited by *Wafik S. El-Deiry, 2003*
222. **Tumor Suppressor Genes:** *Volume 1: Pathways and Isolation Strategies,* edited by *Wafik S. El-Deiry, 2003*
221. **Generation of cDNA Libraries:** *Methods and Protocols,* edited by *Shao-Yao Ying, 2003*
220. **Cancer Cytogenetics:** *Methods and Protocols,* edited by *John Swansbury, 2003*
219. **Cardiac Cell and Gene Transfer:** *Principles, Protocols, and Applications,* edited by *Joseph M. Metzger, 2003*
218. **Cancer Cell Signaling:** *Methods and Protocols,* edited by *David M. Terrian, 2003*
217. **Neurogenetics:** *Methods and Protocols,* edited by *Nicholas T. Potter, 2003*
216. **PCR Detection of Microbial Pathogens:** *Methods and Protocols,* edited by *Konrad Sachse and Joachim Frey, 2003*
215. **Cytokines and Colony Stimulating Factors:** *Methods and Protocols,* edited by *Dieter Körholz and Wieland Kiess, 2003*
214. **Superantigen Protocols,** edited by *Teresa Krakauer, 2003*
213. **Capillary Electrophoresis of Carbohydrates,** edited by *Pierre Thibault and Susumu Honda, 2003*
212. **Single Nucleotide Polymorphisms:** *Methods and Protocols,* edited by *Piu-Yan Kwok, 2003*
211. **Protein Sequencing Protocols, 2nd ed.,** edited by *Bryan John Smith, 2003*
210. **MHC Protocols,** edited by *Stephen H. Powis and Robert W. Vaughan, 2003*
209. **Transgenic Mouse Methods and Protocols,** edited by *Marten Hofker and Jan van Deursen, 2002*
208. **Peptide Nucleic Acids:** *Methods and Protocols,* edited by *Peter E. Nielsen, 2002*
207. **Recombinant Antibodies for Cancer Therapy:** *Methods and Protocols.* edited by *Martin Welschof and Jürgen Krauss, 2002*
206. **Endothelin Protocols,** edited by *Janet J. Maguire and Anthony P. Davenport, 2002*
205. ***E. coli* Gene Expression Protocols,** edited by *Peter E. Vaillancourt, 2002*
204. **Molecular Cytogenetics:** *Protocols and Applications,* edited by *Yao-Shan Fan, 2002*
203. ***In Situ* Detection of DNA Damage:** *Methods and Protocols,* edited by *Vladimir V. Didenko, 2002*
202. **Thyroid Hormone Receptors:** *Methods and Protocols,* edited by *Aria Baniahmad, 2002*
201. **Combinatorial Library Methods and Protocols,** edited by *Lisa B. English, 2002*
200. **DNA Methylation Protocols,** edited by *Ken I. Mills and Bernie H, Ramsahoye, 2002*
199. **Liposome Methods and Protocols,** edited by *Subhash C. Basu and Manju Basu, 2002*
198. **Neural Stem Cells:** *Methods and Protocols,* edited by *Tanja Zigova, Juan R. Sanchez-Ramos, and Paul R. Sanberg, 2002*
197. **Mitochondrial DNA:** *Methods and Protocols,* edited by *William C. Copeland, 2002*
196. **Oxidants and Antioxidants:** *Ultrastructure and Molecular Biology Protocols,* edited by *Donald Armstrong, 2002*
195. **Quantitative Trait Loci:** *Methods and Protocols,* edited by *Nicola J. Camp and Angela Cox, 2002*
194. **Posttranslational Modifications of Proteins:** *Tools for Functional Proteomics,* edited by *Christoph Kannicht, 2002*
193. **RT-PCR Protocols,** edited by *Joe O'Connell, 2002*
192. **PCR Cloning Protocols, 2nd ed.,** edited by *Bing-Yuan Chen and Harry W. Janes, 2002*
191. **Telomeres and Telomerase:** *Methods and Protocols,* edited by *John A. Double and Michael J. Thompson, 2002*
190. **High Throughput Screening:** *Methods and Protocols,* edited by *William P. Janzen, 2002*
189. **GTPase Protocols:** *The RAS Superfamily,* edited by *Edward J. Manser and Thomas Leung, 2002*
188. **Epithelial Cell Culture Protocols,** edited by *Clare Wise, 2002*
187. **PCR Mutation Detection Protocols,** edited by *Bimal D. M. Theophilus and Ralph Rapley, 2002*
186. **Oxidative Stress Biomarkers and Antioxidant Protocols,** edited by *Donald Armstrong, 2002*
185. **Embryonic Stem Cells:** *Methods and Protocols,* edited by *Kursad Turksen, 2002*
184. **Biostatistical Methods,** edited by *Stephen W. Looney, 2002*
183. **Green Fluorescent Protein:** *Applications and Protocols,* edited by *Barry W. Hicks, 2002*
182. **In Vitro Mutagenesis Protocols, 2nd ed.**, edited by *Jeff Braman, 2002*
181. **Genomic Imprinting:** *Methods and Protocols,* edited by *Andrew Ward, 2002*
180. **Transgenesis Techniques, 2nd ed.:** *Principles and Protocols,* edited by *Alan R. Clarke, 2002*
179. **Gene Probes:** *Principles and Protocols,* edited by *Marilena Aquino de Muro and Ralph Rapley, 2002*
178. **Antibody Phage Display:** *Methods and Protocols,* edited by *Philippa M. O'Brien and Robert Aitken, 2001*
177. **Two-Hybrid Systems:** *Methods and Protocols,* edited by *Paul N. MacDonald, 2001*
176. **Steroid Receptor Methods:** *Protocols and Assays,* edited by *Benjamin A. Lieberman, 2001*
175. **Genomics Protocols**, edited by *Michael P. Starkey and Ramnath Elaswarapu, 2001*
174. **Epstein-Barr Virus Protocols,** edited by *Joanna B. Wilson and Gerhard H. W. May, 2001*
173. **Calcium-Binding Protein Protocols, Volume 2:** *Methods and Techniques,* edited by *Hans J. Vogel, 2001*
172. **Calcium-Binding Protein Protocols, Volume 1:** *Reviews and Case Histories,* edited by *Hans J. Vogel, 2001*
171. **Proteoglycan Protocols,** edited by *Renato V. Iozzo, 2001*
170. **DNA Arrays:** *Methods and Protocols,* edited by *Jang B. Rampal, 2001*
169. **Neurotrophin Protocols,** edited by *Robert A. Rush, 2001*
168. **Protein Structure, Stability, and Folding,** edited by *Kenneth P. Murphy, 2001*
167. **DNA Sequencing Protocols,** *Second Edition,* edited by *Colin A. Graham and Alison J. M. Hill, 2001*

METHODS IN MOLECULAR BIOLOGY™

MHC Protocols

Edited by

Stephen H. Powis

Centre for Nephrology
Royal Free and University College Medical School
University College London, London, UK

Robert W. Vaughan

South Thames Tissue Typing, Division of Medicine
Guy's Campus, King's College, London, UK

Humana Press **Totowa, New Jersey**

999 Riverview Drive, Suite 208
Totowa, New Jersey 07512

www.humanapress.com

This publication is printed on acid-free paper. ∞

ANSI Z39.48-1984 (American Standards Institute)

Permanence of Paper for Printed Library Materials.

Cover design by Patricia F. Cleary.

For additional copies, pricing for bulk purchases, and/or information about other Humana titles, contact Humana at the above address or at any of the following numbers: Tel.: 973-256-1699; Fax: 973-256-8341; E-mail: humana@humanapr.com; or visit our Website: www.humanapress.com

Printed in the United States of America. 10 9 8 7 6 5 4 3 2 1

Library of Congress Cataloging in Publication Data

MHC protocols / edited by Stephen H. Powis, Robert W. Vaughan.
p. ; m. -- (Methods in molecular biology ; 210)
Includes bibliographical references and index.
ISBN 0-89603-548-4 (alk. paper)
1. Major histocompatibility complex--Genetic aspects--Laboratory manuals. 2. Immunogenetics--Laboratory manuals. 3. Genetic polymorphisms--Laboratory manuals. I. Powis, Stephen H. II. Vaughan, Robert W. III. Methods in molecular biology (Totowa, N.J.); v. 210
[DNLM: 1. Major Histocompatibility Complex. 2. HLA Antigens. 3. Polymorphism (Genetics) WO 680 M6177 2002]
QR184.315.M463 2002
616.07'96--dc21

20022190243

Preface

The aim of *MHC Protocols* is to document protocols that can be used for the analysis of genetic variation within the human major histocompatibility complex (MHC; HLA region). The human MHC encompasses approximately 4 million base pairs on the short arm of chromosome 6 at cytogenetic location 6p21.3. The region is divided into three subregions. The telomeric class I region contains the genes that encode the HLA class I molecules HLA-A, -B, and -C. The centromeric class II region contains the genes encoding the HLA class II molecules HLA-DR, -DQ, and -DP. In between is the class III region, originally identified because it contains genes encoding components of the complement pathway. The entire human MHC has recently been sequenced ***(1)*** and each subregion is now known to contain many other genes, a number of which have immunological functions.

The study of polymorphism within the MHC is well established, because the region contains the highly polymorphic HLA genes. HLA polymorphism has been used extensively in solid organ and bone marrow transplantation to match donors and recipients. As a result, large numbers of HLA alleles have been identified, a process that has been further driven by recent interest in HLA gene diversity in ethnic populations. The extreme genetic variation in HLA genes is believed to have been driven by the evolutionary response to infectious agents, but relatively few studies have analyzed associations between HLA genetic variation and infectious disease, which has been difficult to demonstrate. In distinction, a huge number of studies have described associations between HLA alleles and human diseases, many of which are thought to have an immunological component (such as rheumatoid arthritis, diabetes mellitus, and multiple sclerosis). This enormous literature was initially collated by Tiwara and Terasaki *(2)* and has since continued to grow.

The study of MHC disease associations has also stimulated the analysis of genetic variation within other HLA region genes. Because of strong linkage disequilibrium, it has been reasoned that for some diseases the true susceptibility genes may not be the genes encoding the HLA molecules, but other nearby genes. Some of these, such as the *TAP* and *LMP* genes, are obvious candidate susceptibility genes for immunological disorders because of their role in the normal immune response. Similarly, linkage studies of MHC-associated disorders have stimulated the characterization of microsatellite markers and, more recently, single nucleotide polymorphisms (SNPs).

The study of human MHC polymorphism has also contributed to our understanding of many other areas of biology, including cellular immunobiology, virology, and genome evolution.

The first part of *MHC Protocols* describes electronic databases designed to catalog HLA region genes and their alleles. Chapter 1 describes the methods involved in accessing HLA sequence data from these databases and Chapter 2 describes one of these databases, 6ace, in more detail. Inevitably, these databases will continuously change as, for example, the number of identified HLA alleles increases and genomic information is better understood. Similarly, their underlying software and computer interfaces will become more refined. However, the underlying principles governing the use of these databases are likely to remain constant.

The second part describes DNA-based protocols for the study of polymorphism in HLA class I and II genes. The part begins with two chapters describing older methodology based on restriction fragment length polymorphism analysis. Although these techniques have now been superseded, they may still have a place in smaller research projects. Chapters 5 through 7 describe the SSO and SSP typing methodology that is now in use in most tissue typing laboratories. Chapter 8 describes a recently developed technique, reference strand-mediated conformation analysis, and Chapter 9 describes sequencing-based methodology. Because the number of HLA alleles is constantly increasing (as shown in Chapter 1), it is impossible for any of these chapters to comprehensively describe the analysis of all possible alleles. Nevertheless, the methodology is readily adaptable to new alleles.

Chapters 10 and 11 focus on the nonclassical HLA genes, HLA-E, HLA-G, and HLA-DM. Although these genes encode molecules that are structurally similar to class I and II HLA molecules, they differ in a number of key respects: their cell and tissue expression is unique, they exhibit only limited polymorphism, and they have developed specialized biological functions.

The third part of *MHC Protocols* describes methods for the study of polymorphism in non-HLA MHC genes. Although a large number of these genes have now been identified, we have chosen to concentrate on a small number that are of particular interest, either for functional reasons or because they have been associated with human disease. Chapter 12 describes polymorphism in the genes *TAP1* and *TAP2*. The products of these class II region genes bind noncovalently to form the TAP transporter, which pumps peptides from the cytoplasm into the lumen of the endoplasmic reticulum prior to binding by class I molecules. Chapters 13 and 14 concentrate on two class III encoded components of the complement pathway, C2 and C4, and Chapter 15 describes the TNF gene. Chapter 16 describes *MICA*, a recently described gene that has homology to HLA genes and appears to be highly polymorphic.

Finally, Chapter 17 details a number of microsatellites within the human MHC. Clearly, it is possible to identify many more MHC microsatellites, especially now that the entire human MHC genomic sequence is available. Furthermore, the development of MHC SNPs may eventually supersede the use of microsatellites. Nevertheless, the markers described in this chapter represent a well-characterized core set that can be used for medium resolution analysis of the region.

In a fast developing field such as MHC genetics, it is impossible to produce a written text that will remain up to date. However, we hope that the protocols described in *MHC Protocols* will provide a secure base for those who wish to study genetic variation in this fascinating part of the human genome.

Stephen H. Powis
Robert W. Vaughan

References

1. MHC Sequencing Consortium. (1999) Complete sequence map of a human major histocompatibility complex. *Nature* **401**, 921-923.
2. Tiwari J. L., Terasaki P. I. (1985) *HLA and Disease Associations.* Springer Verlag, New York, NY.

Contents

Contributors

J. Rafael Argüello • *Anthony Nolan Research Institute, London, United Kingdom*

Antonio Arnaiz-Villena • *Hospital Univeritario "12 de Octubre," Universidad Complutense, Madrid, Spain*

Stephan Beck • *The Sanger Centre, Hinxton, United Kingdom*

Ann B. Begovich • *Roche Molecular Systems, Alameda, CA; Children's Hospital Oakland Research Institute, Oakland, CA*

Teodorica L. Bugawan • *Roche Molecular Systems, Alameda, CA*

Mike Bunce • *Churchill Hospital, Oxford, United Kingdom*

Mary Carrington • *National Cancer Institute, Frederick, MD*

R. W. M. Collins • *Guy's and St. Thomas's Hospitals, London, United Kingdom*

Steven T. Cox • *Anthony Nolan Research Institute, London, United Kingdom*

Paul P. J. Dunn • *Southmead Hospital, Bristol, United Kingdom*

Gaby Fisher • *Anthony Nolan Research Institute, London, United Kingdom*

Eduardo Gomez-Casado • *Hospital Univeritario "12 de Octubre," Universidad Complutense, Madrid, Spain*

Roger Horton • *The Sanger Centre, Hinxton, United Kingdom*

Ann-Margaret Little • *Anthony Nolan Research Institute, London, United Kingdom*

J. Alejandro Madrigal • *Anthony Nolan Research Institute, London, United Kingdom*

Steven G. E. Marsh • *Anthony Nolan Research Institute, London, United Kingdom*

Jorge Martinez-Laso • *Hospital Univeritario "12 de Octubre," Universidad Complutense, Madrid, Spain*

GOTTFRIED MAUFF • *Laboratory of Dr. Kramer and Colleagues, Geesthacht, Germany*

ALASDAIR MCWHINNIE • *Anthony Nolan Research Institute, London, United Kingdom*

DEREK MIDDLETON • *Belfast City Hospital, Belfast, United Kingdom*

PRISCILLA V. MOONSAMY • *Roche Molecular Systems, Alameda, CA*

ANDREA PAY • *Anthony Nolan Research Institute, London, United Kingdom*

MARTHA PÉREZ-RODRÍGUEZ • *Anthony Nolan Research Institute, London, United Kingdom*

STEPHEN H. POWIS • *University College London, London, United Kingdom*

JAMES ROBINSON • *Anthony Nolan Research Institute, London, United Kingdom*

PETER M. SCHNEIDER • *Johannes Gütenberg-Universität Mainz, Mainz, Germany*

LORI L. STEINER • *Roche Molecular Systems, Alameda, CA; Children's Hospital Oakland Research Institute, Oakland, CA*

HENRY A. F. STEPHENS • *University College, London, United Kingdom*

HÉLÈNE TEISSERENC • *Hôpital Saint-Louis, Paris, France*

ROBERT W. VAUGHAN • *South Thames Tissue Typing, Division of Medicine, Guy's Campus, Kings College, London, United Kingdom*

JOHN E. VOLANAKIS • *Biomedical Sciences Research Center "A Fleming," Vari, Greece*

F. WILLIAMS • *Belfast City Hospital, Belfast, United Kingdom*

ANTHONY GERARD WILSON • *University of Sheffield, Sheffield, United Kingdom*

ZENG-BIAN ZHU • *University of Alabama at Birmingham, Birmingham, AL*

I

Databases

1

HLA Informatics

Accessing HLA Sequences from Sequence Databases

James Robinson and Steven G. E. Marsh

1. Introduction

Scientists working in the human leukocyte antigen (HLA) field have access to a number of different informatics resources for the analysis and interpretation of HLA sequences. Recent advances in bioinformatics have resulted in an increase in the number of tools and facilities available for sequence analysis. Researchers can now utilize these tools to analyze the genomic information within their own field; in addition, the HLA field also has a number of groups developing software specifically for this area. Within this chapter, we will be discussing both the more general tools and the specialist HLA informatics tools available.

There are a number of different sequence databases available and the type of database can influence the information retrieved. Databases range from the large international data repositories like the European Molecular Biology Laboratory Nucleotide Sequence Database (EMBL), GenBank®, and the DNA Databank of Japan (DDBJ) through to specialist systems like the ImMunoGeneTics project IMGT/HLA Database. It is therefore imperative that the user understands the merits and flaws of the various systems. This

From: *Methods in Molecular Biology, vol. 210: MHC Protocols*
Edited by: S. H. Powis and Robert W. Vaughan © Humana Press Inc., Totowa, NJ

chapter will provide information on the various options available to the user, to permit an informed decision when selecting a database. The main databases discussed in this article are the large public sequence databases, protein structure databases and databases specializing in the human major histocompatibility complex (MHC).

All of the databases mentioned in this chapter are accessed via the Internet. The advent of the Internet has revolutionized the dissemination of sequence information, allowing scientists from all over the world to log on and access the many databases that relate to gene sequences using the World Wide Web (WWW). The uniform resource locators (URLs) of all the databases.discussed will be included in **Section 6.** of this chapter.

2. Generalist Nucleotide Sequence Databases

2.1. EMBL/GenBank/DDBJ

The generalist databanks are not HLA-specific, but rather large international data repositories for all organisms. The three main general nucleotide sequence databases are the EMBL *(1)*, GenBank *(2)*, and the DDBJ *(3)*. These three databases form an international collaboration and exchange sequences daily, so that each contains identical data. Most published sequences can be found in these databases. The retrieval of HLA sequences from these systems is possible through a number of tools. The EMBL database will be used as the model system for all examples in this chapter.

A number of different methods can be used to retrieve HLA sequences. A conceptually simple approach would be to copy the entire database to the users system and then to remove all nonrelevant entries leaving a smaller HLA-related database. However, this method would actually take longer than most search tools, as only a small subset (around 0.001%) of the database represents HLA data. A more pragmatic approach is to use the search tools provided, although these may be time-consuming if large numbers of sequences are being retrieved. Search tools allow you to perform complex queries and download the retrieved entries in a specified format. The search engines for EMBL are maintained by the European Bioinformatics Institute (EBI). The search engines are also mirrored at other sites on the Internet, allowing the user to access a local server. The EMBL database is very large database containing over 17 million nucleotide sequences, and to this are linked other databases like SWISS-PROT and TREMBL (translations of EMBL) *(4)*. All these databases are available and can be searched at the EBI Web site. In this chapter, we will only discuss the retrieval of nucleotide sequences, although the same methods apply for the retrieval of protein sequences from the appropriate database.

The EMBL database receives around 14,000 sequences/12 megabases of sequence per day, and this data is processed and made available through flat files. These files include sequence features, which provide a large amount of additional data allowing for advanced queries. Sequence features are best described as a number of different motifs, references and data particular to the DNA sequence. The most common sequence features are the source of sequence (species, cell type, etc.), coding details, exon coordinates, and protein translation. Additional sequence features include promoter sequences and their coordinates and complimentary sequences. There are two main approaches for sequence searching. One is to search on particular keywords or features contained within the sequence documentation, and the other is to search for the actual nucleotide or protein sequence.

2.2. Sequence Similarity Searches

Sequence similarity searches look for matches to the actual sequence from larger databases. These matches are based on a number of similarity measures and, in general, retrieve identical or highly similar sequences. The most recognized sequence similarity search tool is the Basic Local Alignment Search Tool (BLAST) algorithm *(5)*. This widely distributed algorithm has been refined over many years *(6)* and is now the premier sequence similarity search tool. There are number of tools within the BLAST family, BLASTN for nucleotides, BLASTP for proteins, and BLASTX for translated sequences. BLAST works by searching for areas of local similarity between sequences (*see* **Fig. 1**). These regions are then linked to form a score for a particular sequence. The higher the score, the more accurate the match. The BLAST algorithm does suffer from a particular problem when using it to identify some HLA sequences. The first general problem with BLAST is that very short sequences cannot be used, the minimum sequence length is 20 bps. The second problem associated with HLA sequences is due to the high similarity between HLA alleles. BLAST can accurately retrieve HLA entries from a database where there are a huge range of sequences, i.e., EMBL. However, when used on a more specialist subset of data, care is needed when analyzing the results. BLAST has its own scoring system and uses this rather than sequence identity. In this system, a 546-bp sequence of approx 95% identity may score higher than a shorter (e.g., 270 bps) sequence of 100% identity. Users should therefore study the large output files carefully, as identical matches may not appear first in the output.

2.3. Sequence Search Tools

The EBI provides one of the most advanced flat file search tools, the Sequence Retrieval System (SRS) *(7)*. This tool allows the user to search on any of the sequence features, accession numbers, keywords, or sequence description and is probably the best method of retrieving HLA sequences from a general nucleotide database. The disadvantage of the SRS system is that it does require some familiarity with the flat file format (*see* **Fig. 2**). Once users are accustomed to the way data is presented, they can quickly build up very complex queries. An advantage of the SRS tool is that it can also be

---

Fig. 1. *(opposite page)* SRS search tool showing flat file output. The figure shows the standard output of a flat file, which is a text-based file that identifies different sections by the line headings, e.g., AC, accession number; DE, description; KW, keywords; DR, cross references to other databases; and SQ, sequence. Lines also begin with O, phylogeny; R, references; and F, features. The output file shown is taken from the EMBL Nucleotide Sequence Database.

```
ID   HSMHCA1A   standard; DNA; HUM; 1098 BP.
XX
AC   M24043; M24029;
XX
SV   M24043.1
XX
DT   23-APR-1990 (Rel. 23, Created)
DT   04-MAR-2000 (Rel. 63, Last updated, Version 12)
XX
DE   Human MHC class I HLA-A1 chain gene (A1,2; B8,5), complete cds.
XX
KW   cell surface antigen; cell surface glycoprotein; class I gene;
KW   integral membrane protein; major histocompatibility complex.
XX
OS   Homo sapiens (human)
OC   Eukaryota; Metazoa; Chordata; Craniata; Vertebrata; Euteleostomi; Mammalia;
OC   Eutheria; Primates; Catarrhini; Hominidae; Homo.
XX
RN   [1]
RP   1-1098
RX   MEDLINE; 89235215.
RA   Parham P., Lawlor D.A., Lomen C.E., Ennis P.D.;
RT   "Diversity and diversification of HLA-A,B,C alleles";
RL   J. Immunol. 142(11):3937-3950(1989).
XX
DR   GDB; 119310; HLA-A.
DR   IMGT/HLA; HLA00001; A*01011.
DR   SWISS-PROT; P30443; 1A01_HUMAN.
XX
FH   Key             Location/Qualifiers
FH
FT   source          1..1098
FT                   /db_xref="taxon:9606"
FT                   /organism="Homo sapiens"
FT                   /map="6p21.3"
FT   CDS             1..1098
FT                   /codon_start=1
FT                   /db_xref="SWISS-PROT:P30443"
FT                   /note="MHC HLA-A1 chain"
FT                   /gene="HLA-A"
FT                   /protein_id="AAA59652.1"
FT                   /translation="MAVMAPRTLLLLLSGALALTQTWAGSHSMRYFFTSVSRPGRGEPR
FT                   FIAVGYVDDTQFVRFDSDAASQKMEPRAPWIEQEGPEYWDQETRNMKAHSQTDRANLGT
FT                   LRGYYNQSEDGSHTIQIMYGCDVGPDGRFLRGYRQDAYDGKDYIALNEDLRSWTAADMA
FT                   AQITKRKWEAVHAAEQRRVYLEGRCVDGLRRYLENGKETLQRTDPPKTHMTHHPISDHE
FT                   ATLRCWALGFYPAEITLTWQRDGEDQTQDTELVETRPAGDGTFQKWAAVVVPSGEEQRY
FT                   TCHVQHEGLPKPLTLRWELSSQPTIPIVGIIAGLVLLGAVITGAVVAAVMWRRKSSDRK
FT                   GGSYTQAASSDSAQGSDVSLTACKV"
XX
SQ   Sequence 1098 BP; 221 A; 332 C; 368 G; 177 T; 0 other;
     atggccgtca tggcgccccg aaccctcctc ctgctactct cgggggccct ggccctgacc        60
     cagacctggg cgggctccca ctccatgagg tatttcttca catccgtgtc ccggcccggc       120
     cgcggggagc cccgcttcat cgccgtgggc tacgtggacg acacgcagtt cgtgcggttc       180
     gacagcgacg ccgcgagcca gaagatggag ccgcgggcgc cgtggataga gcaggagggg       240
     ccggagtatt gggaccagga gacacggaat atgaaggccc actcacagac tgaccgagcg       300
     aacctgggga ccctgcgcgg ctactacaac cagagcgagg acggttctca caccatccag       360
     ataatgtatg gctgcgacgt ggggccggac gggcgcttcc tccgcgggta ccggcaggac       420
     gcctacgacg gcaaggatta catcgccctg aacgaggacc tgcgctcttg gaccgcggcg       480
     gacatggcag ctcagatcac caagcgcaag tgggaggcgg tccatgcggc ggagcagcgg       540
     agagtctacc tggagggccg gtgcgtggac gggctccgca gatacctgga gaacgggaag       600
     gagacgctgc agcgcacgga cccccccaag acacatatga cccaccaccc catctctgac       660
     catgaggcca ccctgaggtg ctgggccctg ggcttctacc ctgcggagat cacactgacc       720
     tggcagcggg atggggagga ccagacccag gacacggagc tcgtggagac caggcctgca       780
     ggggatggaa ccttccagaa gtgggcggct gtggtggtgc cttctggaga ggagcagaga       840
     tacacctgcc atgtgcagca tgagggtctg cccaagcccc tcaccctgag atgggagctg       900
     tcttcccagc ccaccatccc catcgtgggc atcattgctg gcctggttct ccttggagct       960
     gtgatcactg gagctgtggt cgctgccgtg atgtggagga ggaagagctc agatagaaaa      1020
     ggagggagtt acactcaggc tgcaagcagt gacagtgccc agggctctga tgtgtctctc      1080
     acagcttgta aagtgtga                                                    1098
//
```

```
BLASTN 2.0MP-WashU [27-Aug-2000] [irix6.5p-r10k 11:16:53 22-Sep-2000]

Copyright (C) 1996-2000 Washington University, Saint Louis, Missouri USA.
All Rights Reserved.

Reference:  Gish, W. (1996-2000) http://blast.wustl.edu

Notice:  this program and its default parameter settings are optimized to find
nearly identical sequences rapidly.  To identify weak similarities encoded in
nucleic acid, use BLASTX, TBLASTN or TBLASTX.

Query=  EMBOSS_001
        (1098 letters)

Database:  embl
           867,002 sequences; 2,191,064,830 total letters.
Searching....10....20....30....40....50....60....70....80....90....100% done

                                                                 Smallest
                                                                   Sum
                                                          High  Probability
Sequences producing High-scoring Segment Pairs:          Score  P(N)      N

EM_HUM:HSMHCA1A M24043 HUMAN MHC CLASS I HLA-A1 CHAIN GEN...  5490  8.7e-242  1
EM_HUM:HSHLAAW36 X61700 HUMAN RNA FOR HLA-AW36 ANTIGEN        5454  3.7e-240  1
EM_HUM:HS07161 U07161 HUMAN MHC CLASS I HLA-A (HLA-A-0102...  5445  9.3e-240  1
EM_HUM:HOSA17224 Y17224 HOMO SAPIENS MRNA FOR MHC CLASS I...  4964  4.5e-239  2
EM_HUM:HSHLA11E X13111 HUMAN MRNA FOR HLA-A11E, A MHC CLA...  5382  5.1e-237  1
EM_HUM:HSMURA1 D16841 HUMAN MRNA FOR HLA-A11 ANTIGEN (A11...  5382  6.5e-237  1
EM_HUM:AF165065 AF165065 HOMO SAPIENS HLA CLASS I ANTIGEN...  5373  1.7e-236  1
EM_HUM:HSOKATAK2 D16842 HUMAN MRNA FOR HLA-A11 ANTIGEN (A...  5373  1.7e-236  1
EM_HUM:HS505741 U50574 HUMAN MHC CLASS I HLA-A ALLELE (HL...  5364  4.2e-236  1
EM_OM:PPMHC01A L39093 PAN PANISCUS (CLONES KIA192, KIA193...  5328  1.6e-234  1

WARNING:  Descriptions of 4088 database sequences were not reported due to the
          limiting value of parameter V = 100.

>EM_HUM:HSMHCA1A M24043 HUMAN MHC CLASS I HLA-A1 CHAIN GENE (A1,2; B8,5),
          COMPLETE CDS.
        Length = 1098

  Plus Strand HSPs:

 Score = 5490 (829.8 bits), Expect = 8.7e-242, P = 8.7e-242
 Identities = 1098/1098 (100%), Positives = 1098/1098 (100%), Strand = Plus /
Plus

Query:     1 ATGGCCGTCATGGCGCCCCGAACCCTCCTCCTGCTACTCTCGGGGGCCCTGGCCCTGACC 60
             ||||||||||||||||||||||||||||||||||||||||||||||||||||||||||||
Sbjct:     1 ATGGCCGTCATGGCGCCCCGAACCCTCCTCCTGCTACTCTCGGGGGCCCTGGCCCTGACC 60

Query:    61 CAGACCTGGGCGGGCTCCCACTCCATGAGGTATTTCTTCACATCCGTGTCCCGGCCCGGC 120
             ||||||||||||||||||||||||||||||||||||||||||||||||||||||||||||
Sbjct:    61 CAGACCTGGGCGGGCTCCCACTCCATGAGGTATTTCTTCACATCCGTGTCCCGGCCCGGC 120

Query:   121 CGCGGGGAGCCCCGCTTCATCGCCGTGGGCTACGTGGACGACACGCAGTTCGTGCGGTTC 180
             ||||||||||||||||||||||||||||||||||||||||||||||||||||||||||||
Sbjct:   121 CGCGGGGAGCCCCGCTTCATCGCCGTGGGCTACGTGGACGACACGCAGTTCGTGCGGTTC 180
```

used to launch other applications, e.g., BLAST. The SRS tool also allows the users to customize the output of searches, meaning that you can quickly see how relevant entries are to the search criteria. SRS can be found at the EBI Web site and can be used to search a number of different databases.

The GenBank search engine, Entrez *(8)* also works on accession numbers, but provides several advanced options. These include bulk retrievals of entries in a preformatted manner. This is very useful for retrieving sequences, once a list of known accession numbers is available. The search engine for DDBJ is restricted to searching via accession numbers.

2.4. Other Generalist Sequence Databases

There are other databases that are not included in the EMBL/GenBank/DDBJ collaboration. These systems also have their own search engines and retrieval facilities, and similar problems with regard to data integrity and sequence retrieval also apply, which will be discussed later (*see* **Subheading 2.5.**).

The Genome Sequence DataBase (GSDB) *(9)* is run by the National Center for Genome Resources, USA. The GSDB database is included here, as there are a number of HLA sequences unique to this database and GSDB does not automatically forward these to the other main databases. The GSDB search engine retrieves entries and sequence features, but only by accession number. This means that an accession number must be known before a sequence can be retrieved.

Fig. 2. *(opposite page)* Example BLAST outputs. Selected sections of a standard BLAST output for a search using the A*01011 sequence against the EMBL database. The first part of the figure shows the sequences and descriptions retrieved by the EMBL accession numbers. The variation in Keywords and Description can be seen here; also note the lack of official nomenclature. Please note inclusion of Pygmy Chimpanzee (*Pan paniscus*) sequence which is high in the scoring due to their high sequence similarity. The second part of this figure shows how BLAST displays the sequence identities. Identity is shown by a pipe (|) between the bases, and mismatches have no join.

The Human Genome Project's Genome Database (GDB) *(10)* is the official repository for genomic mapping data from the human genome project. It includes data on genes, clones, and polymerase chain reaction (PCR) markers and details on the variation within the human genome for polymorphisms and mutations, including allele frequency data. The main obvious advantage to this system is that this site is only for human data, and the user does not need to immediately restrict the search options.

2.5. Advantages and Disadvantages of Generalist Databases

The advantage of using the generalist sequence databases is the large number of sequences available, covering a wide range of data relating to HLA. Also, the generalist databases contain sequences from other species, which may not be included in the specialist databases. However, there are also problems associated with accessing sequences from the generalist databanks. The main problem lies in the definition of the sequence. Upon submission to EMBL, a number of steps are taken to ensure that the sequence is accurately described and up-to-date. The generalist databanks provide checks on all entries, but the author assigns keywords and description, and these can vary. Therefore, despite the work of the World Health Organization (WHO) Nomenclature Committee for Factors of the HLA System in monitoring the HLA allele designations and maintaining the sequences, they have no control of how sequences are defined in the generalist databanks. Simply searching for the HLA keyword will not retrieve all relevant entries. Alleles may be incorrectly named, i.e., A*01011 may not be the sequence found under all entries described as A*01011. Some entries contain outdated or unofficial designations, such as HLA-DS now renamed HLA-DQ, causing further confusion, as the exact nature of the sequence is not always apparent. Additional complications occur as sequences are updated or become out of date. Updating a sequence can only be done by the author, so if a sequence is shown to contain errors or is incorrectly named, it is down to the individual author to update these sequences. Other common sequence errors are the inclusion of primer sequence in an entry and the addition of bases presumed to exist at the start of an exon to ensure the first codon is complete. These problems are currently being addressed by EMBL, but it can be seen that it can be difficult to accurately retrieve the sequences required.

The general databanks provide a good initial source of data for the HLA user, although the accuracy of some sequences cannot be guaranteed. Indeed, for some work, the sequences provided may be suitable. The facilities provided by the EBI through SRS give a very efficient search engine capable of highly complex queries. The feature qualifiers contained within the EMBL database provide the most detailed source of additional data currently available. For detailed work on noncoding sequences, the EMBL database is currently the best source of information. For other types of HLA-based research, this database may not be suitable due to the problems discussed.

3. Specialist HLA Sequence Databases

Recently, several specialist databases for the human MHC have been developed. In each case, the curators of these databases work within the field of histocompatibility and immunogentics, and the data contained is of a greater quality and is usually annotated in much more detail. However, with all databases maintained by institutes not directly involved with the WHO Nomenclature Committee for Factors of the HLA System *(11)*, there can be problems with consistency in naming and the accuracy of sequences used.

The Chromosome 6 (6ace) database previously known as MHC-DB *(12,13)* holds data on the complete genomic structure of the human MHC, and those wanting to access a genomic view of the HLA region should use this database. The 6ace database is covered in detail in Chapter 2, "Accessing HLA Sequencing Data Through the 6ace Database."

The MHCPEP *(14)* database maintained at the Walter and Eliza Hall Institute, Australia, is a specialist database of peptide binding sequences available for the MHC. This a curated database and contains over 13,000 peptide sequences. The database can be searched using the SRS search engine and is currently linked to other databases like SWISS-PROT. Although not directly used for retrieving HLA alleles, this database provides additional data on the specific area of MHC-bound peptides.

During 1999, the human mutations database was developed at the EBI. This provides an alternative database that can be used for searching for mutations in polymorphic systems. The human mutation database *(15)* works with the IMGT/HLA database to provide flat files for the SRS server, which indicates mutations. The flat file format is based on the standard reference sequence for each allele and details all mutations compared to the reference sequence. In the case of HLA alleles, the flat file features detail the sequence position and the change, i.e., G->T. This provides the user with a list of mutations, but restricts how the actual sequences can be accessed and used in various applications. This database is very specialized and may therefore only be suited to certain types of research.

3.1. IMGT/HLA Sequence Database

This IMGT/HLA database was released in 1998 and was designed to provide the HLA community with a specialist HLA sequence database *(16)*. The database was designed in collaboration with the WHO Nomenclature Committee for Factors of the HLA System and the EBI and can be considered as the primary source for HLA data. The database is part of the international IMGT database project *(17,18)*. This project includes the IMGT/HLA database and IMGT/LIGM database, which contains immunoglobulin (Ig) and T-cell receptor (TcR) sequences. The IMGT/LIGM database will not be discussed in this chapter, but readers should be aware of its existence.

The IMGT/HLA database provides the user with a database containing only sequences officially recognized and named by the WHO Nomenclature Committee for Factors of the HLA System. The sequences originally derived from flat file entries in the EMBL database have been expertly annotated to remove known errors and to assign standardized keywords and descriptions, thus ensuring a

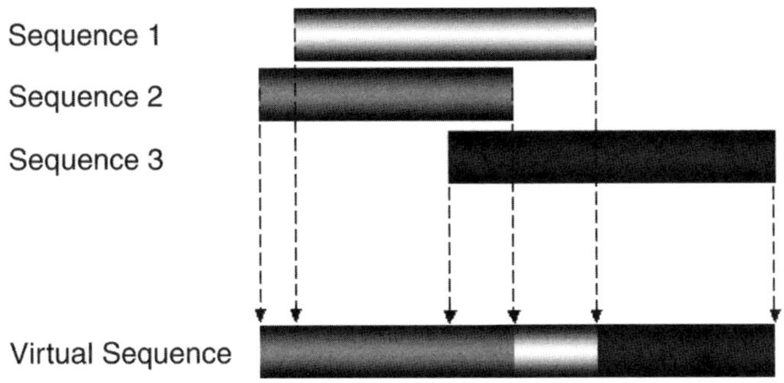

Fig. 3. Construction of a virtual sequence. A virtual sequence is constructed for each allele, which represents the longest available sequence for each allele. This is generated by combining the component entries from EMBL/GenBank/DDBJ to create a single entry. The diagram shows three component entries aligned and merged to form a virtual sequence.

consistency between entries that was not previously available. The database also contains multiple EMBL-derived sequences for single alleles, allowing the component sequences to be combined to create a virtual sequence for each allele (*see* **Fig. 3**). A standard entry from the IMGT/HLA database contains information on the individual alleles together with information on the source cell or individual from which the allele was sequenced.

The IMGT/HLA database is available in a number of different formats including via the Web site, flat files, EBI SRS server, EMBL CD-ROM and FTP. The flat files available on the EMBL CD-ROM, SRS server, and FTP are specific to the IMGT/HLA database and contain standardized keywords and description to aid in retrieval. The FTP site allows users to download the database in a variety of formats (FASTA, Protein Information Resource [PIR], multiple sequence format [MSF], text alignments, and flat files), which are suitable for the majority of sequence analysis tools.

3.1.1. IMGT/HLA Web Site and On-Line Tools

The IMGT/HLA Web site is the main access point for the HLA community. The Web site provides a centralized resource with tools to access and analyze the HLA sequences. The database also includes the background information and documentation to support the use of these tools. The main tools are for allele queries, the generation of sequence alignments, and cell queries, and each of these tools is unique to the IMGT/HLA database.

The allele query tool is designed to allow the user to retrieve the full sequence and nomenclature for any allele. The user-friendly interface prompts the user to enter the locus and allele name, i.e., Locus "A" and allele name "0101" or a wildcard to permit the user to do a more general search. The output includes data on the source individual or cell, ethnic origin, references, and the official nucleotide and protein sequence. Other features include hypertext links to Medline and cross references to other sequence databases, as well as local links to the cell database and help pages. The sequences are provided in a standard alignment format. The IMGT/HLA flat files replicate this output using the standardized EMBL flat file format. Essentially, the Web tool should be used for the basic enquiries on any specific allele. If a more complex query is required, then the SRS search tool can be used to selectively search the IMGT/HLA database. The output from the query tool is also linked to the SRS output for each flat file, meaning that users can easily navigate between the two tools.

Sequence similarity searches for unknown sequences can be performed using the BLAST search tool. The IMGT/HLA database provides the only BLAST engine containing the official WHO Nomenclature for Factors of the HLA System sequences. It is therefore recommended that for sequence similarity searches, this database should be used wherever possible. BLAST searches using EMBL or other databases cannot guarantee to match against official sequences.

The database provides an interactive tool for viewing the official HLA sequence alignments. The options available allow the user to customize the type of alignment and sequence selected as well as the output format. The alignment tool retains the same format as the text alignments, however users are no longer restricted to downloading an alignment of an entire locus just to view a single allele (*see* **Fig. 4**). The full locus text alignments are still available as before and can be downloaded from the Anthony Nolan Research Institute (ANRI) Web site (*see* **Subheading 4.**).

The database also provides access to a searchable database of the source cells or individuals for every allele. This data is used to provide background information on each allele and is unique to the IMGT/HLA database.

3.1.2. Submissions to the WHO Nomenclature for Factors of the HLA System

As well as providing HLA sequences for retrieval, the IMGT/HLA Web site also provides the submission tools for submitting new and confirmatory sequences to the WHO Nomenclature for Factors of the HLA System *(**11**)*. This is now the accepted method for submitting new sequences to the committee and includes sections for information on the source cell, references, and sequence features. Submissions are sent directly to the committee for processing, where they are assigned an official name before being loaded into the IMGT/HLA database. Details of new confirmatory and extended sequences submitted to the WHO Nomenclature for Factors of the HLA System are included in nomenclature update reports, which are published monthly in the journals *Tissue Antigens*, *Human Immunology*, and the *European Journal of Immunogenetics (**19–21**)* and also made available on the ANRI Web site.

3.1.3. Future Developments

At the present time, the database contains only HLA exon sequence data. The data contained within IMGT/HLA has to be expertly annotated before inclusion, and as such, it has taken longer to include all the feature information from other databases into the IMGT/HLA database. Therefore, at the time of writing, the EMBL database may contain additional, though largely unvalidated, data to that included in the IMGT/HLA database. In order to remedy this, a comprehensive review of all genomic HLA sequence data is currently underway, and it is expected that a future release of the

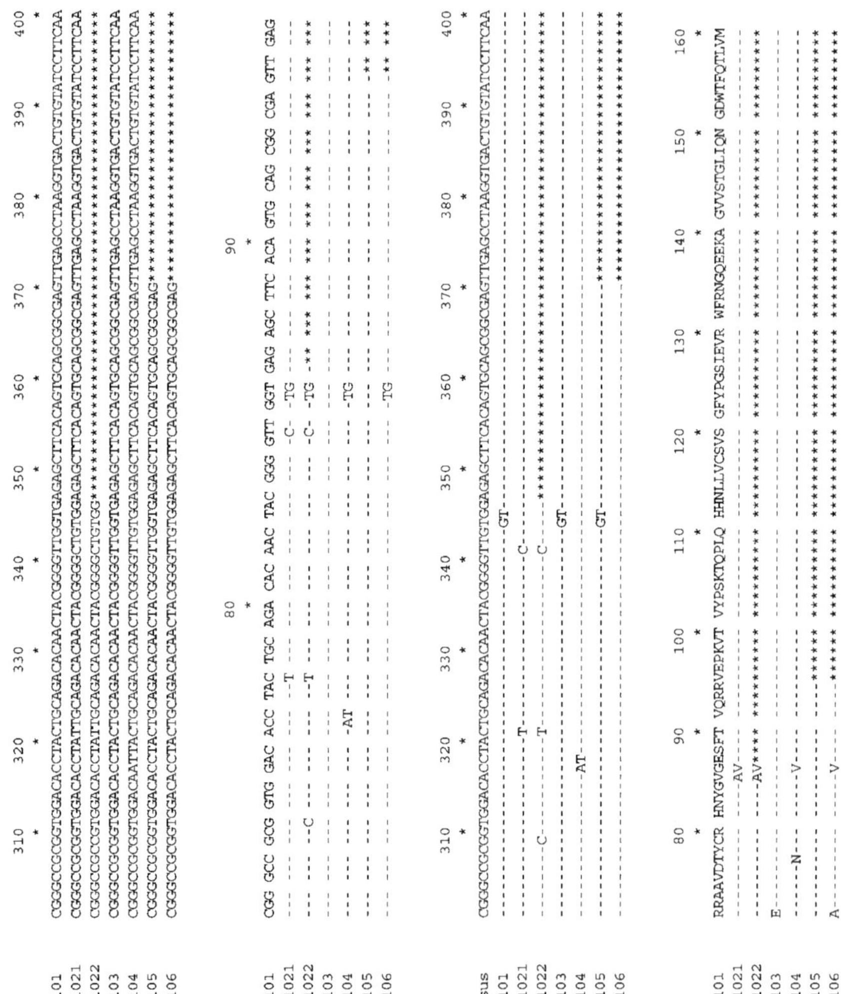

database will include both intron sequences and that derived from 5' and 3' untranslated regions. Other tools for HLA sequence manipulation and visualization are also being developed.

4. Sequence Alignments Web Sites

An alternative source of HLA sequences is the sequence alignment files that have been made available on a regular basis since 1989 as both publications in journals and also via the WWW. These sequences were originally produced and released in *Tissue Antigens*, *Human Immunology*, and the *European Journal of Immunogenetics* *(22,23)*. As the WWW became more available, these sequences were provided on a number of Web pages. The official sites were originally maintained by the Imperial Cancer Research Fund and then from 1996 by the ANRI. These Web sites were constructed from the data held in the HLA database *(24)*, maintained by Steven Marsh for the WHO Nomenclature Committee for factors of the HLA System, which, though not accessible to the public, was used to produce the sequence alignments. In addition to the official site, these alignments have also been made available by different institutes and societies. As part of the work of the IMGT/HLA database, these preformatted static sequence alignments continue to be made available on the ANRI Web site and are updated with each new release of the database (**Fig. 5**).

Fig. 4. *(opposite page)* Alignment formats available from the IMGT/HLA database. The examples shown are all based on alignment A, which shows seven DRB1*01 alleles. Alignment B shows these sequences with the mismatches highlighted and the sequence split into codons. In these alignments, a hyphen (-) indicates identity to the reference sequence, and an asterisk (*) denotes an unsequenced base. Alignment C shows how an alternative reference sequence can be used. Here for example, we have used a DRB1 consensus sequence. Alignment D represents a translation of the nucleotide sequence to produce a protein sequence alignment. In alignments A and C, the numbering is given for nucleotides, in alignments B and C, it is for codon or amino acid position.

```
                               Exon 1 | Exon 2
Nuc. Pos.           90                |                        120                                     150
Prot. Pos.           1                |                     10                                      20
DRB1*0101           GGG GAC ACC CGA C | CA CGT TTC TTG TGG CAG CTT AAG TTT GAA TGT CAT TTC TTC AAT GGG ACG GAG CGG GTG
DRB1*01021          --- --- --- --- - | -- --- --- --- --- --- --- --- --- --- --- --- --- --- --- --- --- --- --- ---
DRB1*01022          *** *** *** *** * | -- --- --- --- --- --- --- --- --- --- --- --- --- --- --- --- --- --- --- ---
DRB1*0103           --- --- --- --- - | -- --- --- --- --- --- --- --- --- --- --- --- --- --- --- --- --- --- --- ---
DRB1*0104           --- --- --- --- - | -- --- --- --- --- --- --- --- --- --- --- --- --- --- --- --- --- --- --- ---
DRB1*0105           *** *** *** *** * | -- --- --- --- --- --- --- --- --- --- --- --- --- --- --- --- --- --- --- ---
DRB1*0106           *** *** *** *** * | -- --- --- --- --- --- --- --- --- --- --- --- --- --- --- --- --- --- --- ---
DRB1*03011          --- --- --- A-- - | -- --- --- --- GA- T-C TC- -C- -C- --G --- --- --- --- --- --- --- --- --- ---
DRB1*03012          *** *** *** *** * | -- --- --- --- GA- T-C TC- -C- -C- --G --- --- --- --- --- --- --- --- --- ---
DRB1*03021          --- --- --- A-- - | -- --- --- --- GA- T-C TC- -C- -C- --G --- --- --- --- --- --- --- --- --- ---
DRB1*03022          --- --- --- A-- - | -- --- --- --- GA- T-C TC- -C- -C- --G --- --- --- --- --- --- --- --- --- ---
DRB1*0303           *** *** *** *** * | ** *** *** *** *** T-C TC- -C- -C- --G --- --- --- --- --- --- --- --- --- ---
DRB1*0304           *** *** *** *** * | ** *** *** --- GA- T-C TC- -C- -C- --G --- --- --- --- --- --- --- --- --- ---
DRB1*0305           *** *** *** *** * | -- --- --- --- GA- T-C TC- -C- -C- --G --- --- --- --- --- --- --- --- --- ---
DRB1*0306           *** *** *** *** * | ** *** --- --- GA- T-C TC- -C- -C- --G --- --- --- --- --- --- --- --- --- ---
DRB1*0307           --- --- --- A-- - | -- --- --- --- GA- T-C TC- -C- -C- --G --- --- --- --- --- --- --- --- --- ---
DRB1*0308           --- --- --- A-- - | -- --- --- --- GA- T-C TC- -C- -C- --G --- --- --- --- --- --- --- --- --- ---
```

Fig. 5. Detail of a static HLA sequence alignment format available from the ANRI Web site. In the alignments, a hyphen (-) indicates identity to the reference sequence, DRB1*0101, and an asterisk (*) denotes an unsequenced base. The exon/intron boundary is indicated by a pipe (|). The sequence is numbered by both nucleotide position, numbering from 1 the A of the ATG of the initiation methionine and the start of exon 1, and by the protein sequence, codons being numbered from 1, the first amino acid residue of the mature protein.

The advantage to this type of data source is that it is relatively quick and easy to access the basic sequence. The disadvantages are that there is no flexibility in searching, and the user is unable to perform any type of complex query. For that type of analysis, the IMGT/HLA database is recommended.

5. Conclusion

As can be seen, when accessing HLA sequences, the user has the option of a number of systems depending on the amount of information and accuracy in the data required. General requests for sequences can be made to the generalist databanks, and although these contain the greatest amount of sequence data information, the accuracy of these sequences can vary. Accessing sequences from approved sites can improve the quality of sequence, but maybe at the cost of additional information. The static HLA sequence alignment pages provide detailed source of sequences, but do not allow the user to specify any of the parameters used to produce the alignments. The IMGT/HLA database has overcome many of the problems discussed, but is still under development in some areas (intron sequences, protein structure). However, once this system is completed, it will provide the most accurate and up-to-date database for accessing HLA information. Used together, the specialist databases provide a good source of high quality data covering a range of areas important in the study of HLA. The 6ace and IMGT/HLA database illustrate this, with the IMGT/HLA database designed at the alleleic level and 6ace providing a genomic overview. Retrieving HLA sequences is becoming more straightforward, but the user should be prepared to spend time assessing all the databases before retrieving any sequences. This should ensure that data of the relevant type and quality is retrieved.

6. Electronic Resources

6.1. Sequence Databases

DDBJ—http://www.nig.ac.jp/
EMBL Nucleotide Sequence Database—http://www.ebi.ac.uk/embl/
GenBank—http://www.ncbi.nlm.nih.gov/Genbank/
Human Genome Mapping Project (GDB)—http://www.gdb.org/
Human Mutation Database—http://ebi.ac.uk/mutations/
SWISS-PROT Database—http://www.ebi.ac.uk/swissprot/

6.2. Search Tools

DDBJ Search Tools—http://www.nig.ac.uk.jp/
EBI BLAST Server—http://www.ebi.ac.uk/blast2/
EBI SRS Search Engine—http://srs.ebi.ac.uk/
GenBank Entrez Search Tool—http://www.ncbi.nlm.nih.gov/Entrez/

6.3. Specialist HLA Databases

HLA Informatics Group (ANRI)—http://www.anthonynolan.org.uk/HIG/
Human Chromosome 6 Database (6ace)—http://www.sanger.ac.uk/HGP/Chr6/
IMGT/HLA Sequence Database—http://www.ebi.ac.uk/imgt/hla/
IMGT/LIGM Database—http://imgt.cnusc.fr:8104/
MHCPEP—http://wehih.wehi.edu.au/mhcpep/

References

1. Baker, W., van den Broek, A., Camon, E., Hingamp, P., Sterk, P., Stoesser, G., and Tuli, M. A. (2000) The EMBL nucleotide sequence database. *Nucleic Acids Res.* **28**, 19–23.
2. Benson, D. A., Karsch-Mizrachi, I., Lipman, D. J., Ostell, J., Rapp, B. A., and Wheeler, D. L. (2000) GenBank. *Nucleic Acids Res.* **28**, 15–18.
3. Tateno, Y., Miyazaki, S., Ota, M., Sugawara, H., and Gojobori, T. (2000) DNA data bank of Japan (DDBJ) in collaboration with mass sequencing teams. *Nucleic Acids Res.* **28**, 24–26.
4. Bairoch, A. and Apweiler, R. (2000) The SWISS-PROT protein sequence database and its supplement TrEMBL in 2000. *Nucleic Acids Res.* **28**, 45–48.
5. Altschul, S. F., Gish, W., Miller, W., Myers, E. W., and Lipman, D. J. (1990) Basic local alignment search tool. *J. Mol. Biol.* **215**, 403–410.
6. Altschul, S. F., Madden, T. L., Schaffer, A. A., et al. (1997) Gapped BLAST and PSI-BLAST: a new generation of protein database search programs. *Nucleic Acids Res.* **25**, 3389–3402.
7. Etzold, T., Ulyanov, A., and Argos, P. (1996) SRS: information retrieval system for molecular biology data banks. *Methods Enzymol.* **266**, 114–128.
8. Schuler, G. D., Epstein, J. A., Ohkawa, H., and Kans, J. A. (1996) Entrez: molecular biology database and retrieval system. *Methods Enzymol.* **266**, 141–162.
9. Harger, C., Chen, G., Farmer, A., et al. (2000) The Genome Sequence DataBase. *Nucleic Acids Res.* **28**, 31–32.
10. Talbot, C. C. and Cutticchia, A. J. (1999) *Human Mapping Databases, Current Protocols in Human Genetics.* John Wiley & Sons, New York, pp. 1.13.1–1.13.12.
11. Bodmer, J. G., Marsh, S. G., Albert, E. D., et al. (1999) Nomenclature for factors of the HLA system, 1998. *Tissue Antigens* **53**, 407–446.
12. Newell, W. R., Trowsdale, J., and Beck, S. (1994) MHCDB—database of the human MHC. *Immunogenetics* **40**, 109–115.
13. Newell, W. R., Trowsdale, J., and Beck, S. (1996) MHCDB: database of the human MHC (release 2). *Immunogenetics* **45**, 6–8.
14. Brusic, V., Rudy, G. and Harrison, L. C. (1998) MHCPEP, a database of MHC-binding peptides: update 1997. *Nucleic Acids Res.* **26**, 368–371.
15. Lehvaslaiho, H., Stupka, E. and Ashburner, M. (2000) Sequence

variation database project at the European Bioinformatics Institute. *Hum. Mutat.* **15**, 52–56.
16. Robinson, J., Malik, A., Parham, P., Bodmer, J. G. and Marsh, S. G. (2000) IMGT/HLA database—a sequence database for the human major histocompatibility complex. *Tissue Antigens* **55**, 280–287.
17. Giudicelli, V., Chaume, D., Bodmer, J., et al. (1997) IMGT, the international ImMunoGeneTics database. *Nucleic Acids Res.* **25**, 206–211.
18. Ruiz, M., Giudicelli, V., Ginestoux, C., et al. (2000) IMGT, the international ImMunoGeneTics database. *Nucleic Acids Res.* **28**, 219–221.
19. Marsh, S. G. E. (2000) Nomenclature for factors of the HLA System, update March 2000. *Tissue Antigens* **27**, 163.
20. Marsh, S. G. E. (2000) Nomenclature for factors of the HLA System, update March 2000. *Hum. Immunol.* **61**, 718.
21. Marsh, S. G. E. (2000) Nomenclature for factors of the HLA System, update March 2000. *Eur. J. Immunogen.* **27**, 163.
22. Mason, P. M. and Parham, P. (1998) HLA class I region sequences, 1998. *Tissue Antigens* **51**, 417–466.
23. Marsh, S. G. E. (1998) HLA class II region sequences, 1998. *Tissue Antigens* **51**, 467–507.
24. Marsh, S. G. E. and Bodmer, J. G. (1993) HLA Class II Sequence Databank. *Hum. Immunol.* **36**, 44.

2

Accessing HLA Sequencing Data Through the 6ace Database

Roger Horton and Stephan Beck

1. Introduction

The chromosome 6 database (6ace) is one of a suite of databases available at the Sanger Centre, which serve the human genome sequencing communities of several chromosomes (1, 6, 20, 22, and X). Data may be retrieved in graphical or textural form using interactive windows and menus or simple command texts. The database and its management system are based on ACEDB and can be accessed via three main routes, a graphics interface, a text interface, and a Web interface. Here, we describe 6ace with particular emphasis on how to access major histocompatibility complex (MHC) and human leukocyte antigen (HLA) associated data.

2. Databases

2.1. ACEDB

Initially written as a *Caenorhabditis elegans* DataBase *(1)*, ACEDB is an object-based data management system specifically designed for use with genomic data. It is now being used for other genome projects as well as *C. elegans*, including *Drosophila*

From: *Methods in Molecular Biology, vol. 210: MHC Protocols*
Edited by: S. H. Powis and Robert W. Vaughan © Humana Press Inc., Totowa, NJ

melanogaster, *Arabidopsis thaliana*, *Schizosaccharomyces pombe*, and many more.

ACEDB can be used via an Xwindows graphical interface (XACE) (do not confuse with Xace, which is a chromosome-specific database), via TACE (a command line textural interface), or via WEBACE (T. Hubbard, personal communication), which is a Web browser interface. XACE versions allow data to be viewed not only as text fields but also as genetic and physical maps, or by using applications external to ACEDB to view images or sequence alignments. The maps are emulated in the WEBACE versions.

2.2. 6ace and MHCDB

The sequencing effort on human chromosome 6 was initially focused on the MHC, with the data being accumulated in the MHC database, MHCDB *(2)*. These data were later merged in the Chromosome 6 database *(3)*, 6ace, the use of which is described here. Although MHC peptides were formerly included in MHCDB, these are now catalogued in an independent database, MHCPEP, at the Walter and Eliza Hall Institute of Medical Research in Melbourne, Australia (http://wehih.wehi.edu.au/mhcpep/) *(4)*. The 6ace database incorporates data not just from the Sanger Centre Chromosome 6 Project *(5,6)* but from the whole chromosome 6 community.

2.3. 6ace Content

As there is rapid progress in the sequencing of the human genome, it is difficult to provide an up-to-date picture of the content of any genomic database. **Table 1** provides a snapshot of the situation in early 1998. Tools are available within the database for the user to gain information on the current status of the data at any time.

3. Access to 6ace

There are three means of getting access to 6ace. If you have your own UNIX system, you may copy the database from the Sanger Centre ftp site. If you are a registered user of the computing facili-

Table 1
Summary of the Content of the 6ace Database

		Database content (MHC data in parenthesis)	
Class	Subclass	Number	Megabase
Sequence	Genome_sequence	86 (32)	7.53 (1.19)
	- Sequenced_by = SC	59 (13)	6.20 (0.75 ≡ MHC class II)
Locus	Gene	402 (73)	
	Pseudogene	50 (29)	
Allele		662 (662)	
Image		27 (27)	
STS (RH mapped)		2532	
cosmid		499 (264)	
BAC		124 (0)	
PAC		6084 (11)	
YAC		69 (69)	

ties of the Human Genome Mapping Project Resource Centre (HGMP), you may access 6ace on-line. Finally, there is access via the Web to the Sanger Centre, where pages allow interactive use of 6ace. These means of access can be all reached from the Sanger Centre Chromosome 6 Web page (http://www.sanger.ac.uk/HGP/Chr6/) by looking under the Chromosome 6 Database entry.

3.1. Getting 6ace by ftp from the Sanger Centre

Releases of the Sanger Centre chromosome-specific databases take place at the beginning of each calendar month with weekly updates. The releases can be obtained from (ftp://ftp.sanger.ac.uk/pub/human/chr6/RELEASES/).

The weekly updates are found in (ftp://ftp.sanger.ac.uk/pub/human/chr6/weekly_release/).

There are README files in the directories, which give full instruction for installing the databases. Basically, it is required to copy a file of the form:

6ace_release_mm-yy.tar.Z

where mm and yy are month and year numbers, respectively. This file needs to be uncompressed (unzipped) and extracted using the UNIX tar command to give a 6ace directory.

The current release of the ACEDB code will also be required. This can be obtained from (ftp://ftp.sanger.ac.uk/pub/acedb/). The current version (4.5) of the code is in a directory ace4_5/.

3.2. Using 6ace at the HGMP

Details of registration and use of HGMP computing facilities can be obtained at (http://www.hgmp.mrc.ac.uk/). If suitable computer facilities are available, this enables the use of the graphical interface to 6ace. The HGMP database currently uses the monthly 6ace releases, but not the weekly updates. The HGMP also organizes training courses in the use of ACEDB.

3.3. Web Access to 6ace

The Sanger Web pages provide both graphical and textural access to 6ace. Major developments of this route of access are currently being made so as to provide better access to Sanger Centre data, in particular with WEBACE.

The WEBACE home page is at (http://webace.sanger.ac.uk/) and has a link to the full set of Sanger Centre ACEDB databases including 6ace. The user is then guided through the class structure of 6ace, and, if a suitable browser is being used, both graphics and textural objects can be displayed. This means of access will become increasingly popular with the expansion of applications available.

For those with a knowledge of TACE, the text version of ACEDB, another page (http://www.sanger.ac.uk/HGP/db_query/query.shtml) gives the user access to the QUERY THE CHROMOSOME SPECIFIC DATABASES page. Clicking on the "Open access" button leads to a TACE query form. Consult the help pages for a description of how to use TACE.

4. Abbreviations and Conventions

In describing the use of 6ace, the following abbreviations and conventions will be used here. In the majority of cases, they apply to the use of XACE.

CMW; Class Menu Window, the first menu window that appears in XACE.
MKS; Main KeySet window generated LMB DC on a Class in the CMW.
LMB, RMB; Left and Right Mouse Button, respectively.
SC, DC; Single and Double Click, respectively.
"text"; Text as it appears on the screen in 6ace.
'text'; Text to be entered from the keyboard by the user.
↵; The keyboard RETURN key.

The convention within ACEDB (apart from in the CMW and some of its menus) is that **bold text** is a link to further data and if clicked (LMB DC; LMB SC simply highlights the entry) will open up a further window.

Quit; This has two meanings in XACE. In all windows except the CMW, it may be present as a Quit button or may be selected from the RMB pull-down menu. When selected, it closes the window concerned.

In the CMW selected from the RMB menu, it closes and exits ACEDB after the warning: "Do you really want to quit acedb? Yes No".

In TACE, typing 'quit' closes 6ace immediately.

5. Structure of ACEDB Databases

The structure of data storage within an ACEDB database can be likened to a tree. The main menu access is via the trunk. This leads to classes that are the branches of the tree. Within classes, there are a number of objects like twigs on the branches. If the structure is followed through, eventually a tag field is reached, likened to a leaf, which contains an item of data. The skill in using the database is to know your way around this tree structure.

The structure in each class is contained within its Model. Looking at the Models can be of great help when trying to access and interpret data.

It should be remembered that many of the commonly encountered objects belong to subclasses. Thus, finished sequences are put in the Genome_sequence subclass, which is part of the Sequence class distinguished (or filtered) by having the Properties tag "Genomic_canonical". Similarly, Gene and Pseudogene are subclasses of Locus, filtered by the Type tag "Gene" and "Pseudogene", respectively.

In XACE, a list of classes and subclasses can be accessed by clicking (LMB DC) on the ▼ after "In class:" in the CMW. Model is itself a class that can be displayed and selected from. In TACE, typing "Classes" lists the classes and the number of objects they each contain. The command "Model", followed by a class name, will show the model structure.

WEBACE restricts the accessible classes to those on the menu lists. Browsing a class from the main Classes page generates a further page listing objects in the selected class. At the foot of this page is a "View...model" clickable button, leading to a display of the model structure.

At the UNIX level the model structures are all contained in the 6ace/wspec/ directory in the models.wrm file, whereas subclasses.wrm contains each subclass with its parent class on an "Is_a_subclass_of" tag together with the filter used. The appearance of data represented in a graphical form in XACE is determined by the Methods. These Methods can be viewed in XACE by selecting "Query" from the pull down menu found by holding down the RMB in the CMW. Entering 'find method' in the yellow "Query:" box lists the Methods. 'Find Method' can also be used in TACE even though TACE cannot display graphical objects.

6. 6ace in Use

6.1. XACE

Perhaps the most familiar version of the database, XACE contains a menu of commonly used classes, presented as a list in a window (CMW). List Items are clickable (LMB DC) and lead to windows containing KeySets (*sic*) of the selected class. The KeySets list entries are in turn clickable (if depicted in bold type) and can eventually lead to the display of the particular object of interest. Pull-down menus are available via the RMB, and windows buttons lead to other options. **Fig. 1** shows the 6ace CMW (first to appear when 6ace is opened) with "Genome_sequence" selected to give the MKS, and the object E1448 further selected in text mode to show full details of the entry for this cosmid. Note that, when there is a choice, XACE will try to display an object graphically. The text version of E1448 was selected by clicking on the toggle button at the top left of MKS "Show As…:" to select Text. An alternative for an object is to highlight the object name LMB SC and then select "Show as text" on the menu.

LMB SC on "Genome_sequence" and then entering 'E1448 ↵ ' in the yellow "Search:" box would generate a KeySet with just one entry and open the E1448 map directly.

6.2. TACE

In contrast to XACE, TACE provides textural access to ACEDB databases. It uses the query language that is also employed by the graphical version. On UNIX systems, TACE can be opened with the command 'tace' if paths have been set correctly. It may also be used via the Sanger Centre Web pages at (http://www.sanger.ac.uk/HGP/db_query/query.shtml) as described in **Subheading 3**.

TACE is not immediately user-friendly, in that the operator needs to know some commands to type in order to get started, whereas the XACE version has the advantage of menus and clickable entries in object lists, but, to the initiated, TACE provides a rapid tool for the interrogation of the database. Typing '?' at the TACE prompt (>)

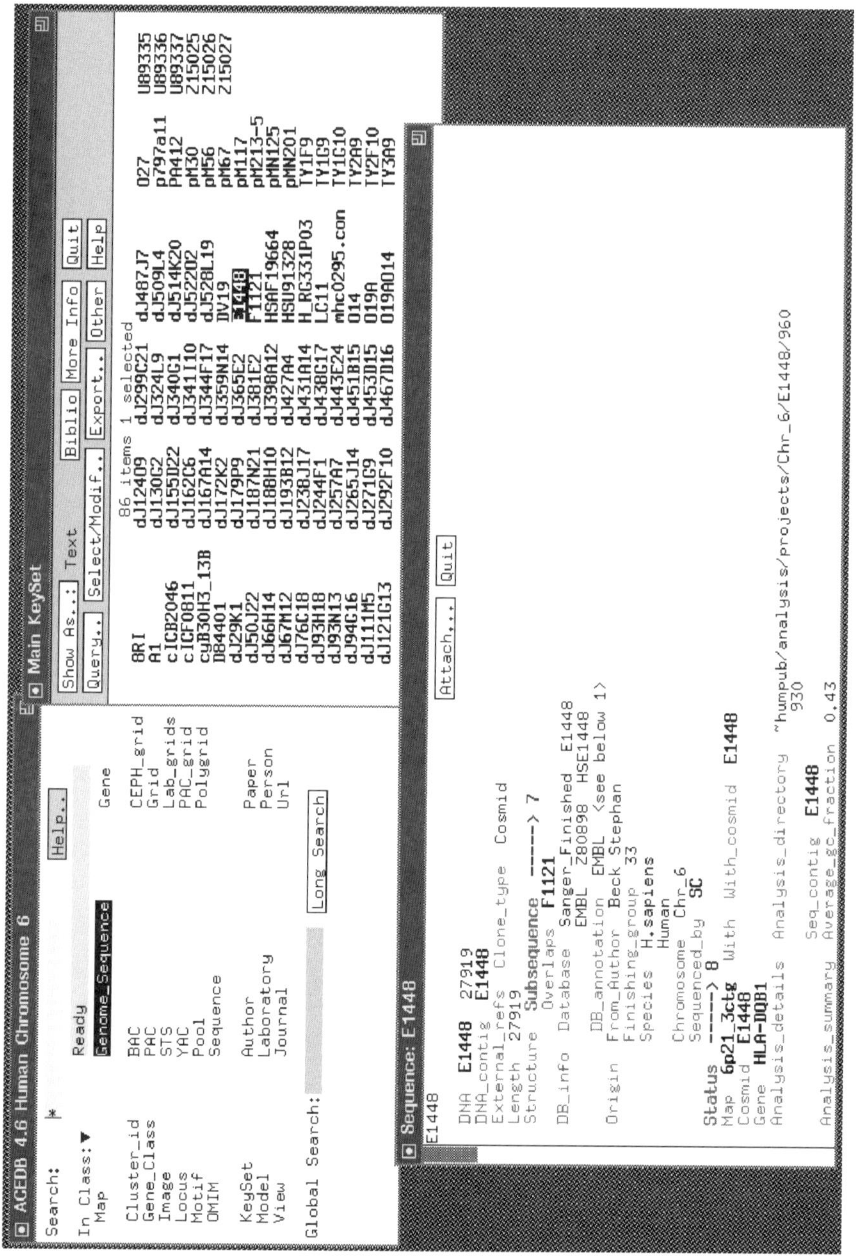
ACEDB 4.6 Human Chromosome 6
Search:
Help..
Ready
In Class:
Genome_Sequence
Long Search
Global Search:
Main KeySet
Show As... Text
Query.. Select/Modif.. Export.. Other Help
Biblio More Info Quit
86 items 1 selected
Sequence: E1448
Attach... Quit
E1448
DNA E1448 27919
DNA_contig E1448
Length 27919
Structure Subsequence -----> 7
Overlaps F1121
DB_info Database Sanger_Finished E1448
EMBL Z80898 HSE1448
DB_annotation EMBL <see below 1>
Origin From_Author Beck Stephan
Finishing_group 33
Species H.sapiens
Human
Chromosome Chr_6
Sequenced_by SC
Status -----> 8
Map 6p21_3ctg
Cosmid E1448
Gene HLA-DQB1
Analysis_details Analysis_directory ~humpub/analysis/projects/Chr_6/E1448/960930
Seq_contig E1448
Analysis_summary Average_gc_fraction 0.43

will list a help file of the commands and their function. Typing 'classes' lists the classes in the database and the number of objects each contains. To get the output obtained in XACE above for E1448 would require just the following two commands (at the > prompt, with output lines beginning //):

```
>find genome_sequence E1448
//Found 1 objects in this class
//1 Active Objects
>show
```

The data generated are of such a volume as to be virtually unreadable. To overcome this, write the data to a file using the command `>write E1448.ace` and examine your UNIX directory for the output.

6.3. WEBACE

Here, the window lists of XACE are replaced by selectable lists in windows within the Web pages. 6ace can be selected by clicking "acedb6" and then the "Browse DB" button. To obtain the E1448 details given above, it is necessary to make selections as follows (as illustrated in **Fig. 2**): "—>Genome_sequence" below "Sequence" and click the "Browse" button on the Classes page and then select "E1448" in the window generated. There is also the option of Text, Graphics (passive), and Graphics (active). Selecting the first followed by clicking the "show object(s)" button generates a "Sequence: E1448" page with the data as seen in XACE above. Links are shown in hypertext, and where the volume of data is large in a tag field, it may be in a compressed form beneath an "Expand" hypertext.

Fig. 1. *(opposite page)* Text display of Sequence E1448 in 6ace (XACE). Class Menu Window (*top left*) with the class "Genome_Sequence" selected; the Main KeySet for Genome_Sequence (*top right*) with the sequence name E1448 selected, and the "Show As...:" button toggled to Text; and the resulting Sequence: E1448 text window showing some of the database entries for the sequence.

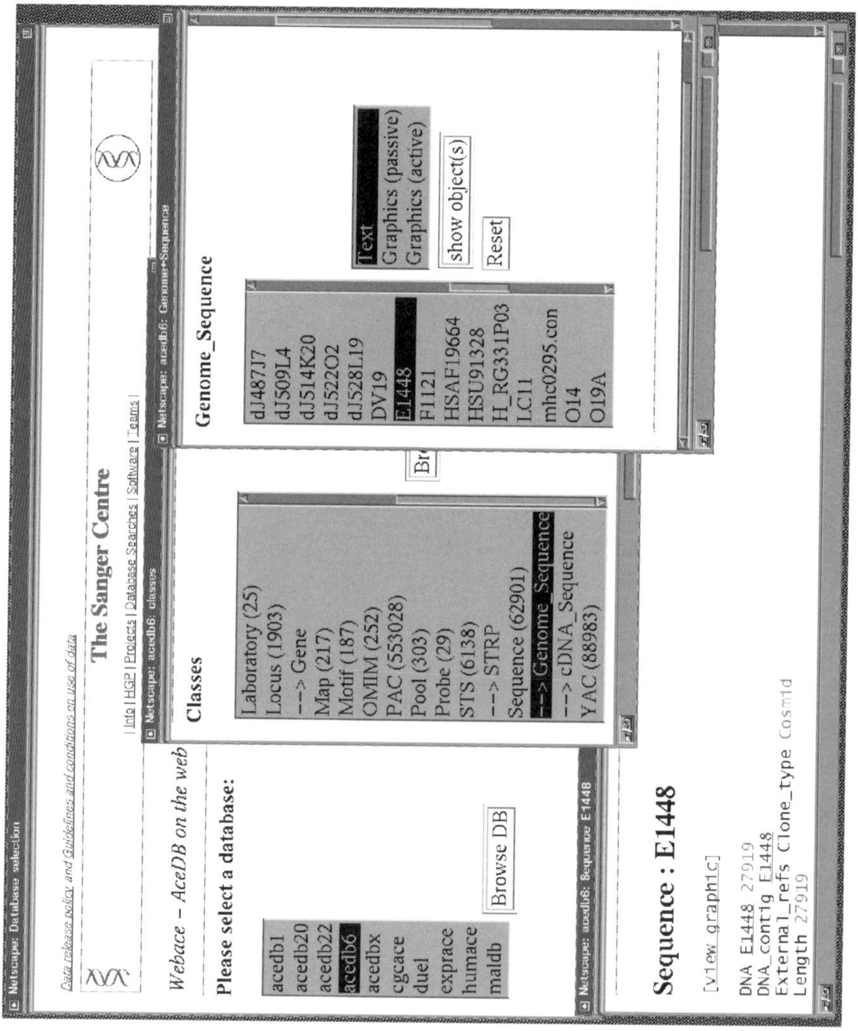
Netscape: Database selection
Data release policy and Guidelines and conditions on use of data
The Sanger Centre
| Info | HGP | Projects | Database Searches | Software | Teams |
Webace – AceDB on the web
Please select a database:
acedb1
acedb20
acedb22
acedb6
acedbx
cgcace
duel
exprace
humace
maldb
Browse DB
Netscape: acedb6: classes
Classes
Laboratory (25)
Locus (1903)
--> Gene
Map (217)
Motif (187)
OMIM (252)
PAC (553028)
Pool (303)
Probe (29)
STS (6138)
--> STRP
Sequence (62901)
--> Genome_Sequence
--> cDNA_Sequence
YAC (88983)
Netscape: acedb6: Genome+Sequence
Genome_Sequence
dJ487J7
dJ509L4
dJ514K20
dJ522O2
dJ528L19
DV19
E1448
F1121
HSAF19664
HSU91328
H_RG331P03
LC11
mhc0295.con
O14
O19A
Text
Graphics (passive)
Graphics (active)
show object(s)
Reset
Netscape: acedb6: Sequence E1448
Sequence : E1448
[view graphic]
DNA E1448 27919
DNA_contig E1448
External_refs Clone_type Cosmid
Length 27919

7. Examples

It is impossible here to give a complete set of examples of all the functions of 6ace, but the user is encouraged to explore the databases to gain experience. A novice user need not fear causing damage, as only those with write access can initiate changes. Only when users are running a local copy of the database and this is opened with TACE are they likely to have write access without specifically asking for it.

The MHC, also known as the HLA complex, is located on the short arm of human chromosome 6 (6p21.3). It is the most polymorphic and gene dense region of the human genome. The examples selected here, therefore, have been chosen to illustrate different types of variation and polymorphism present in HLA sequences.

7.1. Maps and Images in XACE

This example uses maps and images to illustrate the representation of variation in 6ace. A variation is either a substitution, insertion, or deletion of bases in the DNA sequence. XACE has the advantage over TACE of being able to display graphical objects such as the ACEDB genetic and physical maps. As WEBACE emulates XACE, it is also able to display these maps. Selecting "Maps" from the CMW lists all the genetic maps in 6ace. Of particular relevance to those with an interest in the HLA region will be "Chr_6" (the map of the entire chromosome) and "6p21_3ctg" (the map of the MHC region). These appear superimposed in **Fig. 3**. This shows, from left to right, the chromosome 6 ideogram flanked by a scale, in kb (kilobases), and the chromosome bands. The boxes (red) along the ideogram indicate regions selected for sequencing or where sequencing is already in progress. (Note: all colors given here are

---

Fig. 2. *(opposite page)* Text display of Sequence E1448 in 6ace (WEBACE). The three pages used to select the 6ace database (acedb6; top left), the Genome_Sequence class (top center), and the E1448 sequence text (top right), together with the resulting Sequence: E1448 text page (bottom).

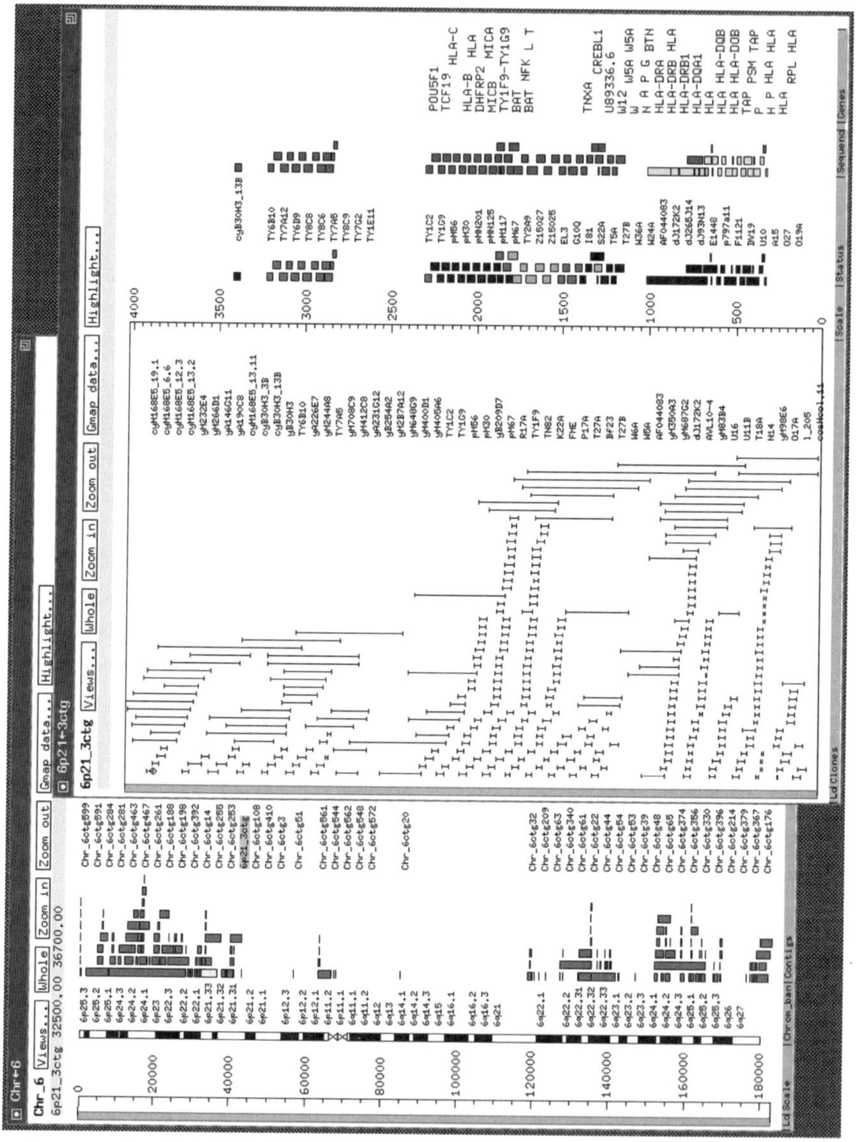

the default colors generally used in the XACE and WEBACE versions of 6ace. In XACE, it is possible for the user to alter these colors). The box associated with the MHC region (6p21_3ctg) has been selected using LMB SC. This selection changes the box color to pale blue, and the contig name appears highlighted in blue. The zoomed in view of 6p21_3ctg shows the status of the 4000 kb MHC region. To the left of the scale, the approximate positions and sizes of all yeast artificial chromosome (YAC), bacterial artificial chromosome (BAC), P1-based artificial chromosome (PAC), and cosmid clones are shown. A LMB SC on any clone or clone name will highlight the corresponding clone name or clone in pale blue. To the right of the scale the sequencing tiling path and corresponding clone names are shown. The key to the color-coded status of each clone can be viewed by LMB DC on the Status bar (green) at the foot of the map (grey, sequencing in progress; red, sequencing finished; black, analyzed and submitted to a public database). The mirrored tiling path to the right indicates where a particular clone is being or has been sequenced (pale blue, Sanger Centre; purple, others). LMB DC on any clone will bring up a text window with full details of its sequencing and with links to sequencing laboratory location and names of investigators involved. Finally, on the far right hand side, the names of genes are displayed alongside the clones in which they are encoded. The limitation of space in the magnification shown here results in the truncation or exclusion of some gene names. This may easily be overcome by repeated LMB SC on the "Zoom in" button until the required magnification is displayed. During zooming, the middle mouse button can be used to center on the object to be enlarged by SC on it or to scroll by SC at the top or bottom of the window.

Switching between different maps is very easy. The text window produced above (LMB DC on any clone) will contain a tag "DNA"

Fig. 3. *(opposite page)* The 6ace Chromosome 6 (*left*) and MHC (*right*) maps. The position of the latter (6p21_3ctg) is highlighted in the former. For further details see text.

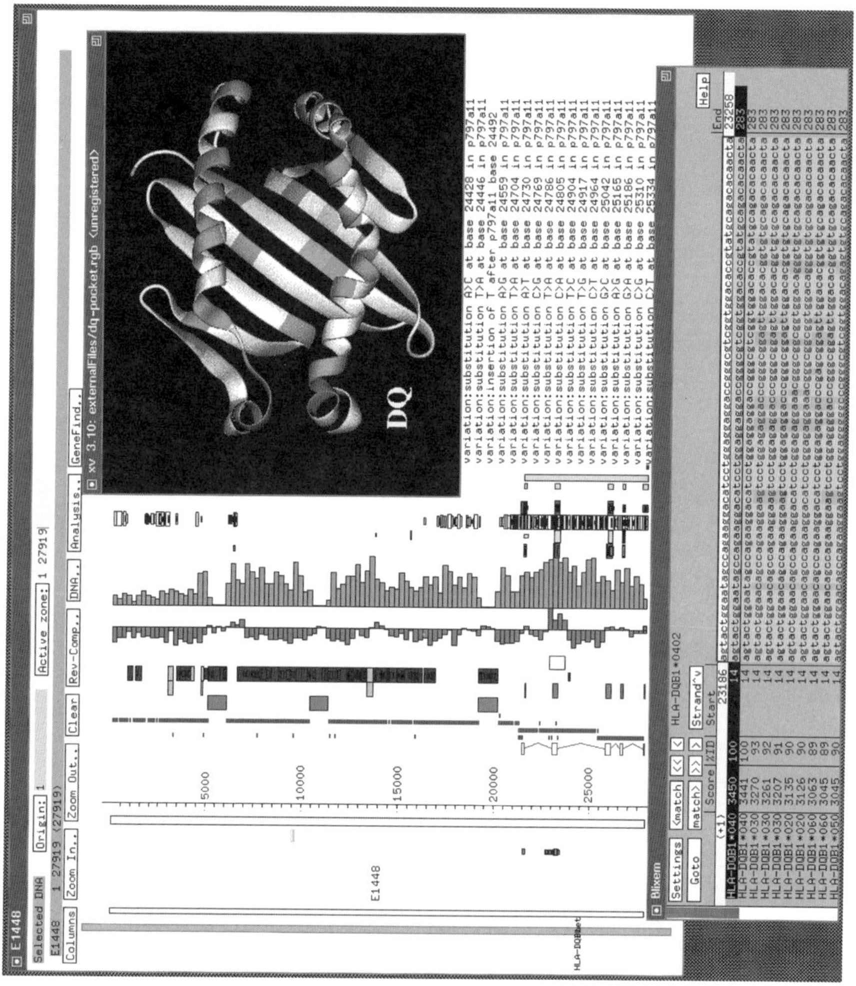
E1448
Selected DNA
Origin: 1
Active zone: 1 27919
Columns
Zoom In..
Zoom Out..
Clear
Rev-Comp..
DNA..
Analysis..
GeneFind..
xv 3.10: externalFiles/dq-pocket.rgb <unregistered>
DQ
Blixem
Settings
Goto
HLA-DQB1*0402
Help

followed by the clone name in bold text. This is a link to the physical map of the clone DNA sequence. LMB DC on this bold text brings up a map as shown in **Fig. 4** (main picture), in this case for clone E1448. On the left is a locator bar (light red when whole map is viewed, but changing to green when zoomed in) next to a box representing the clone (black outline; LMB DC on this box returns to the sequence text window). Names of genes within the clone are superimposed on the locator bar, as is that for *HLA-DQB1* here. To the right is a vertical bar (yellow) and scale in bases for the sequence in view. All the other details in the map relate to features of the DNA annotated using the HPREP analysis suite (G. Micklem and R. Durdin, personal communication). Those to the left of the yellow bar and scale represent features of the reverse complement, whereas those to its right represent features of the positive strand. The features include significant matches determined using Basic Local Alignment Search Tool (BLAST) *(7)* to entries in the protein databases (BLASTX; light blue boxes), to entries in the nucleotide databases (BLASTN; yellow boxes), and to vertebrate mRNA (brown boxes). At the far right, similarity to expressed sequence tag (EST) clusters are shown as pale violet boxes. Repeat sequences determined using RepeatMasker2 *(8)* (http://ftp.genome.washington.edu/cgi-bin/RepeatMasker) are also shown (Alus as green boxes, others in dark blue). Further information can be gained for all of the features, either by LMB SC on a box that brings up brief details in the banner line (light blue) at the top of the map, or by LMB DC that will call up a text window giving fuller details.

At the foot of the map, immediately to the right of the scale, is an exon and intron diagram for the *HLA-DQB1* gene (dark blue outline boxes and lines). The strength of ACEDB, in its ability to present a

---

Fig. 4. Representation of variation in 6ace. The sequence map of E1448 (*top left*) showing annotation of substitutions and of indels (for full description, *see* text); the DQ-pocket Image with residues color-coded according to variability (*inset, right*); a Blixem output of aligned *HLA-DQB1* allele sequences (*bottom*) compared with that of HLA-DQB1*0402.

combination of results visually, is illustrated by the concentration of features showing matches to the protein and DNA databases and the absence of repeat elements in the region of this gene. The presence of allele sequences is recorded as boxes (red) to the right of, and at the same horizontal level, as the gene exons. Holding down RMB on these allele-display boxes reveals a pull-down menu from which may be chosen "show multiple DNA alignment of just this kind of homolgies". This selects an external application, Blixem *(9)* which enables the display of aligned DNA sequences of the gene's alleles (**Fig. 4**, foot). This application highlights the allelic variation in this gene. The allele sequences of HLA class I genes were taken from Arnett and Parham *(10)* and those of the class II region from Marsh and Bodmer *(11)*.

Central features of the map are two horizontal bar graphs representing GC content and variation rate. The former (red) shows GC content calculated for intervals of 250 bases on a scale arranged a round a midpoint of 50%. The latter shows the variation rate, using the same intervals, between E1448 and an overlapping sequence from a different haplotype (p797a11). The GC content is highest in the sequence adjacent to the second exon in DQB1, which coincides with the presence of a CpG island (yellow box) directly to the left of the bar graph. It is also of interest to note that the variation reaches local maxima of up to 10% (which is the highest level of sequence variation reported in the human genome so far) and that it extends far beyond the transcriptional unit of DQB1 *(12)*. The three gaps in the variation bar graph are due to major indels (cerise boxes), which, in this case, are all of retroviral origin (long terminal repeat [LTR] sequences). The relatively high GC content around the gene is also emphasized. The nature of individual variations are also shown as text, far right, and as small boxes (red) immediately to the left of the gene diagram, where they often form a continuous bar because they are so numerous.

Although all these features can be seen in text form in TACE, only the maps in XACE and WEBACE allow their spacial visualization.

6ace via XACE also provides a means to view the structures and structural variability of HLA molecules using either xv (John Bradley; ftp://ftp.cis.upenn.edu) or RasMol *(13)* if these applications are installed on the computer system in use. With the former, images constructed using Prepi suite *(14)*, in which residues are color-coded according to their variability from constant (white) to most variable (red) according to their Shannon entropy, *(15)* can be displayed. With the latter, a 3D interactive display enables the rotation of images of protein tertiary structure.

To access these, LMB DC on the class Image in the CMW and then select from the MKS in which external image files are listed. Select (LMB DC) from the MKS and in the Image object window LMB DC on "Pick_me_to_call". Selecting, thus, for the "DQ_pocket" image calls the structure shown as an inset in **Fig. 4**, in which the positions within the structure of the DQ antigen, which are most variable, are banded in red.

In summary, **Fig. 4** illustrates the different ways in which HLA variation and polymorphism data can be accessed and viewed in 6ace.

7.2. Obtaining the DNA Sequences of All HLA-DQB1 Alleles

Obtaining DNA sequences in a format compatible with analysis software is a common problem in bioinformatics. The 6ace database contains allele sequences for genes in the MHC region imported from external databases *(10,11)*. This example shows how to obtain the sequences for all the alleles of an HLA gene in fasta format.

7.2.1. XACE

In the CMW enter HLA-DQB1* in the yellow "Search:" box and LMB DC on ▼ after "In class:". From the menu window, which appears LMB DC on Allele, to bring up a MKS of all *HLA-DQB1* alleles. These are, of course, objects in the Allele class. What we want are the Sequence objects associated with them. To get a KeySet of these, LMB SC on "Query…" in the MKS, and in the resulting "Query:" yellow box type 'follow sequence ↵'. A further KeySet window will appear with the Sequences listed. RMB on the "Export..." button in this window gives a menu from which select "DNA in FASTA format". Enter a file name in the yellow "File:" box (note that ACEDB will add the extention .dna to this), check that the green "Directory:" box contains an entry you are happy to write to, and press ↵. You should get a message similar to "# I wrote 28 sequences Continue". Click "Continue". Examine your UNIX directory for a file containing all the allele DNA sequences in FASTA format beginning as follows:

```
>HLA-DQB1*0201
agagactctcccgaggatttcgtgtaccagtttaagggcatgtgctactt
caccaacgggacagagcgcgtgcgtcttgtgagcagaagcatctataacc
gagaagagatcgtgcgcttcgacagcgacgt............
```

7.2.2. TACE

In TACE, the following session leads to the writing of a file, dqb1.dna, containing all the allele sequences:

```
> find allele HLA-DQB1*
// Found 34 objects in this class
// 34 Active Objects
> follow sequence
// Autocompleting sequence to Sequence
// Found 28 objects
// 28 Active Objects
> dna dqb1.dna
// 28 object dumped// 28 Active Objects
>
```

Note that although 34 alleles are listed, only 28 appear to have a DNA sequence attached.

7.2.3. WEBACE

This facility is not yet available in WEBACE.

8. Conclusion

The examples above give an idea of the variety of HLA data available in the 6ace database, but there is much that has not been described, such as the ability to query the database and to use analysis and gene finding tools, to which the user may progress with experience. Those working on the HLA region may also want to take advantage of other databases, such as IMGT (ImMunoGeneTics; http://www.ebi.ac.uk/imgt/) a database of nucleotide sequences of important genes of the immune system, or the HLA Peptide Binding Prediction database *(16)* (http://bimas.dcrt.nih.gov/molbio/hla_bind/). There is a wealth of information available to those willing to explore.

Acknowledgments

The writers would like to acknowledge the contribution made to the 6ace database by their colleagues at the Sanger Centre and, in particular, to the members of the Chromosome 6 Project group (http://www.sanger.ac.uk/HGP/Chr6/) and also to the wider community of those working on the chromosome and in the MHC region for free access to the data now included in the database. The curation of the 6ace database is funded by the Wellcome Trust.

References

1. Durbin, R. and Thierry-Mieg, J. (1994) The ACEDB genome database, in *Computational Methods in Genome Research* (Suhai, S., ed.), Plenum Press, New York, pp. 45–55.
2. Newell, W. R., Trowsdale, J., and Beck, S. (1996) MHCDB database of the human MHC (release 2). *Immunogenetics* **45**, 6–8.
3. Theaker, A. J., Maslen, G. L., Scott, C. E., et al. (1997) The chromosome 6 database at the Sanger Centre. *DNA Seq.* **8**, 167–172.
4. Brusic, V., Rudy, G., Kyne, A. P., and Harrison, L. C. (1998)

MHCPEP, a database of MHC-binding peptides: update 1997. *Nucleic Acids Res.* **26**, 368–371.

5. Avis, T., Clark, E. K., Flack, T. L., et al. (1997) The chromosome 6 sequencing project at the Sanger Centre. *DNA Seq.* **8**, 131–136.
6. Mungall, A. J., Humphray, S. J., Randby, S. A., et al. (1997) From long range mapping to sequence-ready contigs on human chromosome 6. *DNA Seq.* **8**, 113–154.
7. Altschul, S. F., Gish, W., Miller, W., Myers, E. W., and Lipman D. J. (1990) Basic local alignment search tool. *J. Mol. Biol.* **215**, 403–410.
8. Smit, A. F. A. and Green, P., unpublished results.
9. Sonnhammer, E. L. L., and Durbin, R. (1994) A workbench for large scale sequence homology analysis. *Comput. Appl. Biosci.* **10**, 301–307.
10. Arnett, K. L., and Parham, P. (1995) HLA Class I nucleotide sequences. *Tissue Antigens* **46**, 217–257.
11. Marsh, S. G. E. and Bodmer, J. G. (1995) HLA Class II region nucleotide sequences. *Tissue Antigens* **46**, 258–280.
12. Horton, R., Niblett, D., Milne, S., et al. (1998) Large-scale sequence comparisons reveal unusually high levels of variation in the *HLA-DQB1* locus in the class II region of the human MHC. *J. Mol. Biol.* **282**, 71–97.
13. Sayle, R. A. and Milnerwhite, E. J. (1995) Rasmol-biomolecular graphics for all. *Trends Biochem. Sci.* **20**, 374–376.
14. Islam, S. and Sternberg, M. J. E., unpublished results (http://www/sbg.ic.ac.uk/prepi/).
15. Shannon, C. E. and Weaver, W. (1949) The Mathematical Theory of Communication. University of Illinois Press, Urbana.
16. Parker, K. C., Bednarek, M. A., and Coligan, J. E. (1994) Scheme for ranking potential HLA-A2 binding peptides based on independent binding of individual peptide side-chains. *J. Immunol.* **152**, 163–175.

II

Polymorphism in Classical and Nonclassical HLA Genes

3

HLA Typing by Restriction Fragment Length Polymorphism Analysis

Robert W. Vaughan

1. Introduction

As a practical method for the identification of human leukocyte antigen (HLA) class I or II alleles, the restriction fragment length polymorphism (RFLP) approach has almost entirely been superceded by other DNA-based methods, in particular those based on polymerase chain reaction (PCR). Historically, RFLP identification of *HLA class II* polymorphism was important at a time when class II antisera were relatively poorly defined and B cell separation for serological typing from peripheral blood was sometimes problematic. The close correlation found between serology and RFLP pointed to the conserved organization of introns within the major histocompatibility complex (MHC), and the RFLP method quickly established that HLA and disease associations were not due to gross genomic reorganizations (insertions or deletions) in the areas examined. Some relatively recent publications have used RFLP for the identification of *HLA class II* polymorphism in large population studies *(1)*. In this chapter, the basic method for Southern blot and the identification of RFLP will be outlined. This approach may also find application for the identification of polymorphism in new genes within the MHC and perhaps for the nonclassical *HLA* genes.

From: *Methods in Molecular Biology, vol. 210: MHC Protocols*
Edited by: S. H. Powis and Robert W. Vaughan © Humana Press Inc., Totowa, NJ

2. Materials

1. Swelling buffer: 10 m*M* Tris-HCl, pH 7.5, 20 m*M* NaCl, 5 m*M* $MgCl_2$.
2. TE: 10 m*M* Tris-HCl, pH 7.5, 1 m*M* EDTA, pH 8.0.
3. Nuclei lysis buffer: 10 m*M* Tris-HCl, pH 8.2, 400 m*M* NaCl, 2 m*M* EDTA, pH 8.0.
4. Phenol. Phenol must be buffered to a pH above 7.8 to ensure the DNA remains in the aqueous phase. Solid phenol is melted at 37°C, and an equal volume of 0.5 *M* Tris-HCl, pH 8.0, is added. The mixture is stirred and left to separate into two phases. The upper aqueous phase is aspirated off and the process is repeated until pH paper gives a reading of pH 8.0 for the phenol. An equal vol of 0.1 *M* Tris-HCl, pH 8.0, is added, and the solution is stored at 4°C. A pinch of 8-hydroxyquinoline may be added to slow oxidation (and color the phenol phase pale yellow). The phenol should be discarded if any pink coloration is visible, as this indicates oxidation.
5. Chloroform. **Warning:** Chloroform is toxic and volatile. Care should be taken to minimize exposure, and it should be used in a frame with a hood.
6. UV transilluminator. Protective goggles should be worn at all times and exposure to skin should be minimized.
7. Horizontal submarine gel apparatus and power pack.
8. 5X TBE electrophoresis buffer: 108 g Trizma base (Sigma, St. Louis, MO, USA), 27.5 g boric acid, 20 mL. 0.5 *M* EDTA, pH 8.0, made up to 1 L with water. Use at 0.5X TBE for agarose gel electrophoresis
9. 20X SSC: 350.6 g NaCl, 175.4 g sodium citrate, made up to 2 L with water.
10. Denaturing solution: 175.4 g NaCl, 40 g NaOH, made up to 2 L with water.
11. Neutralizing solution: 175.4 g NaCl, 242.2 g Tris base, 1500 mL water. Adjust to pH 7.0 with concentrated hydrochloric acid (about 50 mL), and make solution up to 2 L.
12. Ethidium bromide. Ethidium bromide is a powerful mutagen, and extreme care should be exercised in its use. Gloves should be worn when handling solutions containing it, and, when weighing out the powder, a mask should also be worn. Solutions and spills may be decontaminated with activated charcoal, and the residue may then be sealed and incinerated.
13. 6X Gel loading buffer (GLB): 0.25% bromophenol blue, 0.25% xylene cynaol FF, 15% Ficoll (Type 400, Pharmacia) in water.

14. Oligolabeling buffer (OLB). Solution O: 6.25 mL 2 *M* Tris-HCl, pH 8.0, 1.25 mL 1 *M* $MgCl_2$, 2.5 mL water (filtered and stored at 4°C). Solution A: 1 mL solution O, 5 mL of 0.1 *M* dTTP, dATP, and dGTP (store at –20°C). Solution B: 2 *M* Hepes, pH 7.0 (store at 4°C). Solution C: Hexanucleotide primers at 90 OD U/mL in TE (store at –20°C). To make 5X OLB, solutions A, B, and C are mixed together in the ratio of 100:250:150. Aliquots can be stored at –20 C.
15. Hybridization fluid: 25 mL 20X SSC, 10 mL 100X Denhardt's, 10 mL 20% sodium dodecyl sulfate (SDS), 55 mL water.
16. 50X Denhardt's Solution: 5 g Ficoll (Type 400; Amersham Pharmacia Biotech, Piscataway, NJ, USA), 5 g polyvinylpyrrolidone, 5 g bovine serum albumin (Fraction V, Sigma), made up to 500 mL with water, filtered and stored at –20°C.
17. Sonicated salmon sperm. A solution of 10 mg/mL of salmon sperm DNA (Sigma) is made up in water and sonicated for 10 min. The solution is then boiled and aliquoted and stored at –20°C.
18. LB medium: 10 g NaCl, 5 g Bacto yeast extract (Difco, Detroit, MI, USA), 10 g Bacto tryptone (Difco), make to 1 L with water and autoclave.

3. Methods

3.1. Separation of Genomic DNA from Peripheral Blood (*see* Notes 1 *and* 2)

3.1.1. Phenol/Chloroform Extraction

1. Twenty milliliters of peripheral blood is taken and anticoagulated with 0.5 mL. of 0.5 *M* EDTA buffered to pH 7.5 with sodium hydroxide. (NB heparin may interfere with restriction enzymes.) The blood may be stored at –20°C prior to extraction of the DNA or used immediately.
2. The thawed or fresh blood is mixed with an equal vol of swelling buffer and centrifuged 1000 *g* for 10 min. The supernatant is removed, and the pellet is resuspended in 40 mL of swelling buffer, and the process is repeated. On resuspension for the second time, 0.5 mL of 10% Nonidel P-40® (NP40) is added to the swelling buffer and the suspension is kept on ice for 15 min before centrifugation at 1000 *g* for 10 min.
3. The pellet should now consist of a rather slimy pale mass of cells and nuclei, with most of the red cells lysed. If this is not the case, a fur-

ther swelling buffer wash may be included. The pellet is resuspended in 5 mL of nuclei lysis buffer, and 200 µL of 20% SDS is added together with 150 µL proteinase K solution. The mixture is resuspended using a wide bore plastic pipet, and left to incubate at 37°C overnight. If there is some urgency to the separation, then 2 h incubation at 55°C is sufficient.

4. After the proteinase K digestion, the suspension should be glutinous, as most of the DNA will be in solution. To separate the protein from the DNA, 5 mL of buffered phenol is added, and the solution is roller-mixed for 15 min. Chloroform (1 mL) is then added, and the solution is centrifuged for 10 min at 1000*g*. The upper aqueous phase is removed to a clean tube using a cut-off plastic pipet, and the process repeated.
5. The aqueous solution is then washed twice with chloroform to remove any traces of phenol. Chloroform (5 mL) is added to the aqueous phase, and the mixture is centrifuged at 1000 *g* for 5 min. The upper aqueous phase is removed carefully, using a wide bore plastic pipet to a clean tube and this stage is repeated.
6. The aqueous phase containing the DNA is then dialyzed against TE overnight at 4°C in dialysis tubing bags.
7. The final stage is to precipitate the purified DNA from solution. The DNA solution is cut from the dialysis bags and poured into a 50-mL Falcon® tube. One milliliter of 5 *M* sodium chloride is added and mixed gently. Double the vol of absolute alcohol is added carefully, and the DNA forms a white fibrous precipitate, which can be spooled-out on a plastic rod. The DNA is resuspended in TE and left to dissolve at 4°C. This process can be encouraged by a 30-min incubation at 65°C, but may still take about 1 week.

3.1.2. Salt Extraction (see **ref. 2***)*

As in **Subheading 3.1.** for **steps 1–3**, but:

3. Particular care should be taken to ensure a pale pellet before the Proteinase K digestion.

Then:

4. Add 2 mL of saturated sodium chloride solution (6 *M* NaCl) to the DNA solution and shake vigorously for 30 s. Centrifuge the solution at 1250 *g* for 20 min to precipitate the protein. The upper phase

should be a clear fluid containing the DNA. This fluid is transferred to a clean tube avoiding any strands of protein.

5. The DNA is precipitated by adding 2X the vol of absolute alcohol and spooling-out the white fibrils of DNA using a plastic rod.

3.2. Spectrophotometric Determination of DNA Concentration

1. DNA solution (20 µL) is diluted in 980 µL TE, and the optical density of the solution is read on a spectrophotometer at 260 and 280 nm.
2. TE alone is the reference negative control.
3. A solution of double-stranded DNA at 50 µg/mL has an optical density at 260 nm of 1. Therefore, the concentration of the unknown is given by the formula: OD at 260 nm × 50 (dilution factor) × 50 = Concentration (µg/mL). The purity of the DNA solution can also be estimated, as the ratio of the OD at 260/280 is greater than 1.8 for pure DNA. A lower ratio indicates protein contamination.

3.3. Restriction Enzyme Digestion

1. The selection of the appropriate restriction enzyme will depend on the gene to be examined (*see* **Table 1**). For example *Taq*I recognizes and cuts double-stranded DNA at the sequence 5'-TCGA-3' and has been found to be most useful for the analysis of HLA-DRB, -DQA, and -DQB genes, as it gives informative fragments that correlate well with serologically detected polymorphism.
2. Genomic DNA (5–10 µg) is mixed with 5 µL of 10X restriction buffer and 50 U of the restriction enzyme. Distilled water is added to make a final vol of 50 mL.
3. *Taq*I (Life Technologies, Rockville, MD, USA) is incubated at 65°C for 4–6 h.
4. To ensure digestion is complete, 5mL aliquots are electrophoresed through a 0.6% agarose minigel (*see* **Subheading 3.4.1.**).
5. If streaking is obvious when viewed under UV light, further enzyme is added and the incubation is continued.
6. The fully digested DNA is electrophoresed for 16 h. at 3 V/cm through 0.6% agarose gels containing 1 µg/mL ethidium bromide in 0.5X TBE buffer (*see* **Subheading 3.4.2.**).
7. DNA digested with *Hind*III is used as a size marker for each gel.

Table 1
The 10th International Histocompatibility Workshop, Restriction Enzymes and Probes Used

Restriction enzymes used	Probes used	Reference
*Taq*I	class I	*(8)*
*Eco*RI	*DRB*	*(9)*
*Msp*I	*DRA*	*(10)*
*Bam*HI	*DQB*	*(11)*
*Pst*I	*DQA*	*(12)*
*Bgl*II	*DOB*	*(13)*
*Hind*III	*DNA*	*(14)*
*Sst*I	*DPB*	*(15)*
*Eco*RV	*DPA*	*(16)*
*Pvu*II	*C4*	*(17)*
*Kpn*I	*21-OH*	*(18)*
*Hinc*II	*C2*	*(19)*
	BF	*(20)*

The method was carefully standardized and 58 laboratories contributed information. Those interested should refer to the chapters in **ref. *21***.

3.4. Submarine Gel Electrophoresis

3.4.1. Minigel

1. Add 0.6 g agarose (BRL 5510UA) to 100 mL 1X TAE buffer. Boil and cool in a 65°C water bath.
2. Add 5 μL ethidium bromide (10 mg/mL), stir gently, and pour into the gel plate with tape sealed ends. Insert comb to form wells and ensure no air bubbles have formed. Allow to set.
3. Remove tape and comb, and fully immerse gel in tank containing 1X TAE to which 0.5 μg/L ethidium bromide has been added.
4. Add 5 μL of enzyme-digested DNA mixed with 2.5-μL GLB to each well and electophorese at 100 V for 2 h.
5. Place gel on a UV transilluminator, check for enzyme digestion (absence of streaking), and equal DNA quantities.

3.4.2. Transfer Gel

1. The transfer gel is prepared and loaded in a similar way to the minigel, however 300 mL of agarose is made up for a 20 × 20 cm gel in 0.5X TBE.
2. Depending on well size, 25–45 µL of digested DNA is loaded with 5 µL of GLB.
3. It is important to include two lanes of λ/*Hind*III digested DNA in marker lanes.
4. The gel is run at 50 V for 16 h (overnight).
5. Check at the end of the electrophoresis run that the 2.0 Kb λ/*Hind*III marker has run at least 15 cm into the gel.

3.5. Southern Blotting

1. After electrophoresis, the gels are photographed under UV light with a ruler placed along the marker track to measure the migration of the known band sizes.
2. The gels are then soaked in denaturing solution for 30 min, followed by two 15-min immersions in neutralizing solution. This process renders the DNA single-stranded.
3. The DNA is then transferred to a nylon membrane (Hybond® N, Amersham International) by capillary blotting *(3)* in 5X SSC. The transfer is complete after 16 h, whereupon the filter is washed briefly in 5X SSC to remove any adherent agarose and air-dried. The DNA is then fixed to the membrane by UV irradiation, 5 min on the transilluminator or automatically in the Stratalinker. The filters are then sealed in a plastic bag and kept at 4°C before use.

3.6. Preparation of cDNA Probes

Two methods may be used to prepare cDNA probes. The conventional method is to transform *Escherichia coli* with a plasmid containing the desired probe as an insert and use the in vivo amplification of bacterium and plasmid. Purification of the plasmid DNA from the bacterial DNA is followed by excision of the probe and the separation of insert from plasmid DNA on an agarose gel. The PCR method (**Subheading 3.6.6.**) is much quicker and cheaper.

In vitro amplification of the plasmid insert is achieved directly using amplimers on either side of the plasmid insert site. The insert is purified on a sephacryl spin column.

3.6.1. Preparation of Competent Cells (4)

1. 10 milliliters of sterile LB broth is inoculated with a single *E. coli* colony (JM101 or HB 101 strains have been used) from an agar plate or 10 µL of a glycerol stock.
2. The culture is incubated overnight at 37°C with agitation. The overnight culture (250 µL) is added to 20 mL of fresh sterile LB broth at 37°C and incubated until the cells have reached log phase growth. This can be estimated from the OD reading at 600 nm when a reading of 0.5 is reached.
3. The cells are pelleted by spinning at 2000 *g* for 5 min at 4°C.
4. The pellet is gently resuspended in 20 mL of ice-cold 80 m*M* $CaCl_2$ and left on ice for 1 h.
5. The cells are repelleted and resuspended in 2 mL of ice-cold 80 m*M* $CaCl_2$ and left on ice for an additional 1 h.

3.6.2. Transformation

1. Approximately 50 ng of plasmid containing the desired cDNA insert in 100 µL of sterile TE is added to 250 µL competent cells in a sterile microfuge tube.
2. The cells are gently mixed to avoid breaking the pili, and left on ice for 1 h.
3. The tube is then heat-shocked at 42°C for 2 min, and the tube is returned to the ice bath.
4. One milliliter of LB at 37°C is added, and the tube is incubated at 37°C for 1 h.
5. The transformed cells, having taken up the plasmid, are then added to 1 L of LB at 37°C in a flask, and the appropriate antibiotic is added. The plasmid contains an intact antibiotic resistance gene, which varies with the plasmid and insert site, to enable only transformants to grow in the antibiotic-containing medium.
6. The 1-L culture is incubated at 37°C for 24 h with shaking.

3.6.3. Separation of Plasmid DNA by Alkaline Lysis ***(5)***

1. The bacterial cells are pelleted at 2000 *g* for 15 min at 4°C.
2. The cells are resuspended in 50 mL ice-cold Solution I, to which 100 mL Solution II and 50 mL Solution III is added. The solutions are mixed and spun at 2000 *g* for 5 min at 4°C.
3. The supernatant is drained through a muslin cloth into a sterile bottle and the white glutinous retentate is discarded.
4. An equal vol of propan-2-ol is added to the filtered supernatant and mixed thoroughly. The DNA is precipitated and collected as a pellet by spinning the supernatant at 2000 *g* for 15 min at 4°C.
5. The pellet is resuspended in TE.

3.6.4. Purification of Plasmid DNA by Cesium Chloride Density Gradient Centrifugation

This method depends on density differences between closed circular plasmid DNA and the larger linear bacterial DNA.

1. 5 milliliters of plasmid solution is made up to 9 mL, and 10 g of CsCl is added and mixed until dissolved.
2. Ethidium bromide is added (250 µL of a 10-mg/mL solution), and the solution is pipeted into a sterile polyallomer ultracentrifuge tube and sealed.
3. The tubes are balanced (if necessary, a CsCl solution-only tube is used to balance) and placed opposite each other in a 70.1 Titanium fixed-angle rotor (Beckman Coulter, Brea, CA, USA). The rotor head is spun at 45,000 rpm (approx 10,000 *g*) for 16 h at 4°C.
4. The tubes are carefully extracted from the rotor and examined under UV light. The upper diffuse band is ignored, and the lower closed circular plasmid DNA band is extracted. This can be achieved by piercing the tube just below the plasmid band with a 21-gauge needle and aspirating off the band through a syringe.
5. Ethidium bromide is extracted from the solution by shaking with water-saturated butanol 3X or until the pink coloration has disappeared from the aqueous phase.
6. Two phenol extractions and one phenol/chloroform extraction (*see* **Subheading 3.1.**) are performed on the plasmid solution, and the

DNA is precipitated by adding 1/10 vol of sodium acetate and 2X vol of absolute ethanol.
7. The DNA is pelleted by spinning at 4°C in a microcentrifuge (12,500 rpm) for 15 min. The plasmid is resuspended in TE. Yields should approximate 5 mg plasmid/L of culture. The plasmid can be stored at –20°C.

3.6.5. Excision of cDNA Probe from Plasmid

The probe is cut from the plasmid using a restriction enzyme chosen to include as much of the coding sequence as possible, but excluding the vector (plasmid) DNA.

1. A 10-μg aliquot of plasmid is test-digested with the restriction enzymes at 37°C for 2 h. If two enzymes are necessary, then digestion conditions may have to be adjusted if their salt concentration requirements differ. The low salt enzyme can be used first and the concentrations raised for the higher salt enzyme.
2. The digest is tested by running on a 1% agarose gel with size markers. The fragment sizes are checked to ensure complete digestion.
3. A large-scale digest is then set up using identical conditions. The digest is then run on a 0.6% low-melting point agarose gel at 4°C. TAE is used as the running buffer, as borate ions may inhibit enzyme labeling reactions.
4. The probe fragment is cut out under UV light including as little agarose as possible and put in a weighed microfuge tube. Water (3 g) is added per gram of agarose, and the mixture is boiled for 7 min.
5. The probe is then aliquoted and stored at –20°C until use.

3.6.6. PCR Amplification of Probes

A simple method for the preparation of probes has been developed. The principal is to use the PCR technique to amplify the probe directly from the vector. Purification is accomplished by passing the reaction mixture down a G400 Sephacryl™ (Amersham Pharmacia Biotech) column. This removes the primers and unreacted nucleotide triphosphates, but allows the large probe fragment to pass through. As the initial concentration of vector containing insert can be kept extremely low, contamination with vector in the final preparation is negligible. Most of the class II probes are in pBR322 and cloned into the *Pst*I site, so amplimers on either side of this site are used.

1. One nanogram of vector (pBR322) containing the required cDNA insert is added to a sterile siliconized 0.5-mL microfuge tube.
2. PCR buffer (5 µL), 8 µL of dNTPs, 5 µL of each amplimer, and 0.3 µL of Amplitaq® (Applied Biosystems, Foster City, CA, USA) are added. Water is added to a vol of 50 mL, and one drop of light paraffin oil is placed on top to prevent fluid evaporation on heating.
3. The tube is put in a thermal cycler, and after 3 min at 95°C, the tube is given 30 cycles of: 10 s 95°C, 10 s 55°C, and 10 s 72°C.
4. The final 72°C incubation is continued for 1 min to ensure chain termination. After completion of the cycling the products can be stored at 4°C. The products are then passed through a G400 Sephacryl spin column, and the eluate is aliquoted and stored at –20°C.

3.7. Radiolabeling of cDNA Probes

The random priming method ***(6)*** can be used. Random hexamers of the four deoxynucleotide triphosphates are added to a single-stranded cDNA probe. The hexamers bind and act as initiators for DNA synthesis by DNA polymerase I large fragment (Klenow). This enzyme fragment incorporates deoxynucleotide triphosphates in the 5' to 3' direction without the exonuclease activity of the intact enzyme. One ^{32}P-labeled nucleotide triphosphate is added to the reaction mixture to give single-strands of probe labeled to a high specific activity.

1. cDNA probe (50 ng) is boiled for 5 min and cooled rapidly on ice.
2. OLB (10 µL), 5 L of ^{32}P-dCTP (3000 Ci/mM; Amersham Pharmacia Biotech) and 2 µL of Klenow (Gibco BRL) is added and the solution made up to 50ml.
3. The reaction can be carried out at room temperature overnight or for 4 h at 37°C. If the probe has been prepared from agarose it may be necessary to add the water before boiling the probe.

Separation of unincorporated hexanucleotides and nucleotides can be accomplished using a G50 Sephadex® (Amersham Pharmacia Biotech, Little Chalfont, England) spin column. G50 Sephadex retains small fragments in its pores but allows the larger cDNA probe fragments through.

1. The column is prepared in a 1-mL syringe with the barrel removed.
2. A small screw of siliconised glass wool is pushed in the tube as a plug, and G50 Sephadex slurry is poured in after.
3. The 1-mL tube is then spun at 500 *g* for 5 min, placed inside a plastic centrifuge tube.
4. The bed vol is adjusted to 1 mL, and the bed is washed 2X with 100 µL of TE at 750 *g* for 5 min.
5. A collecting tube is placed at the bottom of the centrifuge tube, and the labeling reaction is stopped with the addition of 50 µL of TE.
6. A 1-µL aliquot is kept for counting, and the 99 µL are pipeted onto the Sephadex column.
7. The column is spun at 750 *g* for 5 min, and 1 µL of the eluate is taken for counting.
8. The percentage incorporation of ^{32}P-CTP can be estimated from the Cerenkov counting on a β-counter of the precolumn and eluate aliquots. The incorporated counts are given by the equation: counts per min (cpm) × Z × 100 (dilution factor). Z is a correction factor for the β-counter detector, as the direct counting of γ rays underestimates the emission depending on the detector type and shape.

3.8. Hybridization of cDNA Probes

1. The filters are prehybridized for 1 h at 65°C with hybridization fluid containing 1 µg/mL of single-stranded sonicated salmon sperm DNA.
2. The radiolabeled probe is boiled in a pierced tube for 5 min and immediately placed on ice for 10 min.
3. The probe is then added to the filters at a concentration of 10–20 × 106 cpm/filter, and the incubation is continued for 16 h.
4. The filters are then washed 2x in 2X SSC containing 0.1% SDS at 65°C for 10 min. This is followed by 2 stringent washes in 0.2X SSC/0.1% SDS at 65°C for 20 min.
5. The filters are then left to dry for 10 min and wrapped in Saran Wrap (Dow Chemicals).

6. The filters are then exposed to Kodak XAR5 film in a daylight cassette, containing two intensifying screens, at –70°C. The film is developed after 24 h. If the bands are faint another film is inserted and left for up to 10 d at –70°C.

3.9. Interpretation of Autoradiograph Band Patterns

1. The Tenth International Histocompatibility Workshop 1987 remains as the most complete data source on the use of RFLP to define *HLA* and complement genes. This mammoth international collaborative effort used 13 probes, 93 cell lines, and 12 restriction enzymes (**Table 1**).
2. For HLA class II typing, perhaps the most comprehensive system for relating RFLP band patterns to recognized *class II DRB* and *DQ* alleles or haplotypes was developed by J. Bidwell and others using the restriction enzyme *Taq*1 *(7)*. This system became the standard for RFLP typing in the UK for many years.

4. Notes

1. In preparing genomic DNA, the simple cheap salt extraction method *(2)* is usually quite adequate. The use of phenol/chloroform is, perhaps, only worth attempting on old frozen-thawed samples. A quick chloroform rinse may also clean up samples that have been extracted using salt, but remain impure.
2. If phenol/chloroform extraction is used, remember that phenol is toxic and can cause severe skin burns. Gloves should be worn whenever handling phenol. Chloroform is toxic and volatile; care should be taken to minimize exposure and it should be used in a fume cupboard. Polycarbonate tubes (50-mL) are chloroform-resistant and of suitable volume. Because of the numerous changes of tubes, care should be observed in labeling of the samples, especially because chloroform washes off many labeling pen inks!

A comprehensive method of analyzing polymorphism of the C2 gene by RFLP can be found in Chapter 13.

References

1. Thursz, M. R., Kwiatowski, D., Allsop, C. E., Greenwood, B. M., Thomas, H. C., and Hill, A. V. (1995) Association between an MHC

class II allele and clearance of hepatitis B virus in the Gambia. *N. Eng. J. Med.* **332**, 1065–1069.

2. Miller, S. A., Dykes, D. D., and Polesky, H. F. (1988) A simple salting-out procedure for extracting DNA from human nucleated cells. *Nucleic Acid Res.* **16**, 1215.
3. Southern, E. M. (1975) Detection of specific sequences among DNA fragments separated by gel electrophoresis. *J. Mol. Biol.* **98**, 503–507.
4. Mandel, M. and Higa, A. (1970) Calcium dependant bacteriophage DNA infection. *J. Mol. Biol.* **53**, 159–162.
5. Birnboim, H. C. and Doly, J. (1979) A rapid alkaline extraction procedure for screening recombinant plasmid DNA. *Nucleic Acid Res.* **7**, 1513–1515.
6. Fineberg, A. P. and Vogelstein, B. (1984) A technique for radiolabelling DNA restriction endonuclease fragments to high specific activity. *Anal. Biochem.* **137**, 266–267.
7. Bidwell, J. L. (1988) DNA-RFLP analysis and genotyping of HLA-DR and DQ antigens. *Immunol. Today* **9**, 18–23.
8. Sood, A. K., Pereira, D., and Weissman, S. M. (1981) Isolation and partial nucleotide sequence of a cDNA clone for human histocompatibility antigen HLA-B by use of a oligodeoxynucleotide primer. *Proc. Natl. Acad. Sci. USA* **78**, 616–620.
9. Wake, C. T, Long, E. O., and Mach, B. (1982) Allelic polymorphism and complexity of the genes for HLA-DR beta-chains—direct analysis by DNA-DNA hybridization. *Nature* **300**, 372–374
10. Gustafsson, K., Wiman, K., Larhammer, D., Rask, L., and Peterson, P. A. (1984) Signal sequences distinguish class II histocompatibility antigen beta chains of different loci. *Scand. J. Immunol.* **19**, 91–97.
11. Sheldon, E. L., Kellogg, D. E., Watson, R., Levenson, C. H., and Erlich, H. A. (1986) Use of nonisotopic M13 probes for genetic analysis: application to HLA class II loci. *Proc. Natl. Acad. Sci. USA* **83**, 9085–9087.
12. Auffray, C., Lillie, J. W., Korman, A. J., et al. (1987) Structure and expression of HLA-DQ alpha and DX alpha genes: interallelic alternate splicing of the HLA DX alpha gene and functional splicing of the DQ alpha gene using a retroviral vector. *Immunogenetics* **26**, 63–73.
13. Tonnelle, C., DeMars, R., and Long, E. O. (1985) Do beta: a new beta chain gene in HLA-D with a distinct regulation of expression. *EMBO J.* **4**, 2839–2847.

14. Trowsdale, J. and Kelly, A. (1985) The human class II alpha chain gene DZ alpha is distinct from genes in the DP, DQ and DR subregions. *EMBO J.* **4**, 2231–2237.
15. Roux-Dosseto, M., Auffray, C., Lillie, J. W., et al. (1983) Genetic mapping of a human class II antigen beta-chain cDNA clone to the SB region of the HLA complex. *Proc. Natl. Acad. Sci. USA* **80**, 6036–6040.
16. Erlich, H., Stetler, D., Sheng-Dong, R., and Saiki R. (1984) Analysis by molecular cloning of the human class II genes. *Fed. Proc.* **43**, 3025–3030.
17. Belt, K. T., Carroll, M. C., and Porter, R. R. (1984) The structural basis of the multiple forms of the human complement component C4. *Cell* **36**, 907–914.
18. White, P. C., New, M. I., and Dupont, B. (1986) Structure of the human steroid 21-hydroxylase genes. *Proc. Natl. Acad. Sci USA* **83**, 5111–5115.
19. Bentley, D. R., Campbell, R. D., and Cross, S. J. (1985) DNA polymorhism of the C2 locus. *Immunogenetics* **22**, 377–390.
20. Morley, B. J. and Campbell, R. D. (1984) Internal homologies of the Ba fragment from human complement Factor B, a class III MHC antigen. *EMBO J.* **3**, 153–157.
21. Dupont, B., ed., *Immunobiology of HLA-Histocompatibility Testing 1987*; Vol. 1, Springer Verlag, NY, pp. 653–959.

4

PCR–Restriction Fragment Length Polymorphism Typing of Class I and II Alleles

Robert W. Vaughan

1. Introduction

Polymerase chain reaction restriction fragment length polymorphism (PCR-RFLP) as a method for the definition of human leukocyte antigen *(HLA)* alleles has most often been applied as a typing technique for *HLA class II*. The best and most comprehensive overview is provided by Inoko and Ota *(1)*. Methods have also been published for *HLA-C (2)* and a nested PCR-RFLP method for high resolution definition of *HLA-A (3)*. The technique consists of the generic amplification of an *HLA* locus or subregion by PCR amplification, followed by restriction enzyme digestion of the amplified products and gel electrophoresis with ethidium bromide to visualize the fragments. Acrylamide gels are preferred for the separation of small fragments. As the nucleotide sequence of alleles and therefore their restriction enzyme cut sites are known, careful selection of specific restriction enzymes that will cut certain alleles and not others can allow fine definition of *HLA* alleles.

The critical stage is the restriction enzyme digestion, as overdigestion can lead to cut sites other than those specified, and incomplete digestion can give false negative results. It is therefore important to use amplified DNA that will act as positive and negative

From: *Methods in Molecular Biology, vol. 210: MHC Protocols*
Edited by: S. H. Powis and Robert W. Vaughan © Humana Press Inc., Totowa, NJ

controls for the restriction enzyme digestion and to use the correct buffer, dilution, and temperatures.

PCR-RFLP is still widely used for the detection of single nucleotide substitutions in non-*HLA* genes, but the extreme polymorphism of HLA class II alleles has meant that a typing system for *HLA class II* can become complicated. The reduced price and easy synthesis of oligonucleotide probes has tended to favor amplification refractory mutation system (ARMS) or sequence-specific primer (SSP) methods. In comparison with PCR sequence-specific oligonucleotide (SSO) methods, however, PCR-RFLP would be the method of choice for small sample numbers, and it has been demonstrated to be accurate and reliable for *DRB1*, *DQA1*, *DQB1*, and *DPB1* typing in a comparison carried out at the time of the 11th International Workshop *(4)*.

2. Materials

2.1. PCR Amplification

1. Genomic DNA (200 ng).
2. PCR Buffer (*see* **Notes 1** and **2**): 50 m*M* KCl, 2.5 m*M* $MgCl_2$, 10 m*M* Tris-HCl, pH 8.4, 0.01% gelatin, 0.02% Nonidet® P-40 (NP40).
3. 200 μ*M* each of dATP, dTTP, dCTP, and dGTP.
4. 5' and 3' amplimer.
5. *Taq* DNA polymerase.
6. Thermal cycler.

2.2. Restriction Enzyme Digestion

1. Restriction enzyme.
2. Restriction enzyme buffer, often supplied with the enzyme.
3. Water bath.

2.3. Polyacrylamide Gel Electrophoresis

1. Gloves.
2. Detergent: (e.g., Alconox; Alconox, Inc., NY).
3. Ethanol.

4. Acrylamide: Scotlab (Shelton, CT), Easigel, cat. no. 5L-9223.
5. Suitable gel electrophoresis tank for running polyacrylamide gels (e.g., BRL vertical sequencing tank, model S2, cat. no. 580-1105).
6. 10X TBE: 890 m*M* Tris base, 890 m*M* borate, 20 m*M* EDTA, pH 8.0.
7. Urea.
8. Silane mixture: 3 mL ethanol, 50 µL 10% acetic acid, 5 µL methacryloxypropyltrimethoxysilane.
9. TEMED: N,N,N',N'-tetramethylethylenediamine.
10. Siliconizing solution: dimethyldichlorosilane.
11. APS: 10% Ammonium persulfate.
12. Ethidium bromide.
13. 4 Bulldog clips.
14. Yellow Scotch tape.

3. Methods

3.1. PCR Amplification

1. Amplimers appropriate for the locus of choice are chosen. *See* **Table 1** for amplimers to the major *HLA class II* region gene loci.
2. The reaction vol should be 50 µL to allow for repeat digestion if necessary.
3. Add 1 m*M* of each amplimer to the PCR buffer together with 200 µM each of dATP, dTTP, dCTP, and dGTP, 200 ng genomic DNA, and 2.5 U *Taq* DNA polymerase.
4. Denature for 3 min at 95°C and follow with 30 cycles of 30 s annealing (55°–62°C depending on amplimers), 30 s extension (72°C), and 30 s denaturing (94°C).

3.2. Restriction Enzyme Digestion

1. Take a 5-µL aliquot of the amplification mixture and digest with 1 to 2 U of the selected restriction enzyme. Use 1 µL of the manufacturer's supplied restriction buffer and make up the reaction vol to 10 µL.
2. In some cases it may be necessary to dilute the reaction to 20 µL. It is recommended to include a positive (amplimer with a known cutsite for the enzyme chosen) and negative control (amplimer with no cutsite for the enzyme chosen) for each enzyme digestion reaction used.

Table 1
References for PCR-RFLP Typing Schemes

Gene	References
HLA-DRB	5–7
HLA-DQA1	8,9
HLA-DQB1	5,10–12
HLA-DPA1	13
HLA-DPB1	9,14,15
HLA-A	3
HLA-C	2

3.3. Polyacrylamide Gel Electrophoresis

Polyacrylamide gels give a better size separation and will allow a more accurate size estimate than agarose gels and are preferable particularly if the expected fragment sizes fall below 100 bp.

1. The precise details will depend to some extent on the apparatus being used. Clean plates thoroughly with water and detergent and ensure they are free from dust and grease. Wipe with ethanol.
2. Use the silane mixture to siliconize the plates. Wipe over with the silane mixture in a fume cupboard, leave to dry, and wash with ethanol.
3. Clean spacers and combs with water and ethanol. Assemble plates around sealers and ensure there are no leaks by taping with yellow Scotch tape.
4. Prepare a 12% acrylamide gel: 12 mL acrylamide, 10 mL 10X TBE, 42 g urea made up to 100 mL with distilled water. Stir on a hot plate to dissolve urea.
5. Add 800 μL 10% APS and 80 μL TEMED, swirl to mix, and take up solution in a 50-mL syringe.
6. With plates at a 45° angle, add gel mixture steadily to one side of the plate, trying to ensure that no air bubbles appear. Any that do appear should be removed by tapping the plate.
7. Lean the plate down to approx 10° and place the comb between the plates. Add 2 bulldog clips to either side of the plates at the top, and another pair may be applied half way down.
8. Allow the gel to set (approx 1 h.). The remaining gel can be put in a beaker as a check for polymerization.

9. Once the gel has set, remove the tape from the bottom of the gel plate assembly, and clamp in place on the electrophoresis apparatus. The backplate should face outwards. A plate may be applied to the outside to ensure even heat distribution across the plate during the run.
10. Fill the upper and lower chambers with 1X TBE running buffer. Remove the gel comb. Using a needle and syringe, flush the unpolymerized acrylamide and excess urea from the wells immediately prior to loading.
11. Load the gel and run.

3.5. Allele Interpretation

Using this methodology, a number of authors have developed distinct typing schemes. These can be found in the publications referenced in **Table 1**.

4. Notes

1. Suppliers of *Taq* DNA polymerase often supply their own buffer, and it is often better to use this. Some *Taq* will work with suppliers buffer and not with generic buffer.
2. The amount of Mg may need to be adjusted.

A PCR-RFLP method for HLA-G is included in Chapter 10.

References

1. Inoko, H. and Ota, M. (1993) PCR-RFLP, in *Handbook of HLA Typing Techniques* (Hui, K. M. and Bidwell, J. L., eds.) CRC Press, Boca Raton, pp. 9–70.
2. Tatari, Z., Fortier, C., Bobrynina, V., Loiseau, P., and Charron, D. (1995) HLA-Cw allele analysis by PCR-restriction fragmant length polymorphism: study of known and additional alleles. *Proc. Natl. Acad. Sci. USA* **92**, 8803–8807.
3. Mitsunaga, S., Tokunaga, K., Kashiwase, K., Akaza, T., Tadokoro, K., and Juji, T. (1998) A nested PCR-RFLP method for high resolution typing of HLA-A alleles. *Eur. J. Immunogen.* **25**, 15–27.
4. Mizuki, N., Ohno, S., Sugimura, K., et al. (1992) PCR-RFLP is as sensitive and reliable as PCR-SSO in HLA class II genotyping. *Tissue Antigens* **40**, 100–103.

5. Uryu, N., Maeda, M., Ota, M., Tsuji, K., and Inoko, H. (1990) A simple and rapid method for HLA-DRB and -DQB typing by digestion of PCR-amplified DNA with allele specific restriction endonucleases. *Tissue Antigens* **35**, 20–31.
6. Ota, M., Seki, T., Fukushima, H., Tsuji, K., and Inoko, H. (1992) HLA-DRB1 genotyping by modified PCR-RFLP method combined with group specific primers. *Tissue Antigens* **39**, 187–202.
7. Yunis, I., Salazar, M., and Yunis, E. J.. (1991) HLA-DR generic typing by AFLP. *Tissue Antigens* **38**, 78–88.
8. Maeda, M., Murayama, N., Iishi, H., et al. (1989) A simple method for HLA-DQA1 genotyping by digestion of PCR-amplified DNA with allele specific restriction endonucleases. *Tissue Antigens* **34**, 290–298.
9. Ota, M., Seki, T., Nomura, N., et al. (1991) Modified PCR-RFLP method for HLA-DPB1 and -DQA1 genotyping. *Tissue Antigens* **38**, 60–71.
10. Nomura, N., Ota, M., Tsuji, K., and Inoko, H. (1991) HLA-DQB1 genotyping by a modified PCR-RFLP method combined with group-specific primers. *Tissue Antigens* **38**, 53–59.
11. Salazar, M., Yunis, J. J., Delgado, M. B., Bing, D., and Yunis, E. J. (1992) HLA-DQB1 allele typing by anew PCR-RFLP method: correlation with a PCR-SSO method. *Tissue Antigens* **40**, 116–123.
12. Mercier, B., Ferec, C., Dufosse, F., and Huart, J. J. (1992) Improvement in HLA-DQB typing by PCR-RFLP: introduction of a constant restriction site in one of the primers for digestion control. *Tissue Antigens* **40**, 86–89.
13. Al-Daccak, R., Wang, F. Q., Theophille, D., Lethielleux, P., Colombani, J., and Loiseau, P. (1991) Gene polymorphism of HLA-DPB1 and DPA1 loci in caucasoid population: frequencies and DPB1-DPA1 associations. *Hum. Immunol.* **31**, 277–285.
14. Salazar, M., Yunis, I., Alosco, S. M., Chopek, M., and Yunis, E. J. (1992) HLA-DPB1 allele mismatches between unrelated HLA-A,B,C,DR (generic) DQA1-identical unrelated individuals with unreactive MLC. *Tissue Antigens* **39**, 203–208.
15. Hviid, T. V., Madsen, H. O., and Morling, N. (1992) HLA-DPB1 typing with polymerase chain reaction and restriction fragment length polymorphism technique in Danes. *Tissue Antigens* **40**, 140–144.

5

PCR–Sequence-Specific Oligonucleotide Probe Typing for HLA-A, -B, and -DR

Derek Middleton and F. Williams

1. Introduction

The advantages of using molecular methods instead of serological typing to define the human leukocyte antigen (HLA) system have been well shown in transplantation and disease association studies *(1,2)*. There are many molecular methods available to define the HLA alleles. Described in this chapter is the sequence-specific oligonucleotide probe (SSOP) method. The basis of this method is the HLA locus-specific amplification by polymerase chain reaction (PCR) and the subsequent probing of this product by SSOP. Most of the vast polymorphism of the HLA system results from conversion events whereby small nucleotide sections of one allele (usually no more that 100 bases long) are transferred to another allele. Thus many of the sequences tend to be shared by alleles and not to be allele-specific. Therefore probes are used which are sequence-specific. In order to differentiate the alleles, a battery of probes is required, and it is the pattern of reaction of these probes that distinguishes the HLA alleles.

The detection system used in this laboratory consists of labeling the probes with digoxigenin (DIG) and detecting the presence of hybridization of these probes to an identical sequence present in the

From: *Methods in Molecular Biology, vol. 210: MHC Protocols*
Edited by: S. H. Powis and Robert W. Vaughan © Humana Press Inc., Totowa, NJ

PCR-amplified HLA allele of an individual by adding an antidigoxigen antibody conjugated with alkaline phosphatase (ALP). The ALP then uses CSPD as its chemiluminescent substrate, and the light emitted is detected by autoradiography.

To define all alleles at any specific locus at the same time would require a large number of probes (more than 100 for some loci), and the system would constantly need updating to take account of newly discovered alleles (whereas there were 25 HLA-A, 32 HLA-B, and 34 HLA-DR alleles in January 1991, the corresponding numbers at January 2001 were 209, 414, and 274). This is because although each allele has a specific probe pattern, the combined probe pattern of two alleles present in a heterozygous individual can be identical to the combined probe pattern of another heterozygous individual with two different alleles. In addition some applications of HLA typing, e.g., matching in renal transplantation, do not require resolution to the allele level.

In our laboratory, we use a 2-tier SSOP system. The first level of resolution (medium) is equivalent to very good serology, i.e., the allele group is defined, e.g., HLA-A*02. Thereafter, depending on the initial type, a second PCR specific for a group of alleles is performed, and a further set of probes is used to give definition to the allele level. Thus, the number of probes required is kept to a minimum and, except for exceptional circumstances, only the high resolution system needs alteration to take account of new alleles.

2. Materials

2.1. General Reagents

1. Buffer 1 (4X): 0.4 *M* maleic acid, 0.6 *M* NaCl, pH 7.5. Add 300 mL 4 *M* NaCl and 400 mL 2 *M* maleic acid followed by 200 mL 4 *M* NaOH to approx 800 mL distilled water (dH_2O). Add 27 g NaOH pellets. (A white precipitate forms when all reagents are added—this will disappear as the pH approaches 7.0). Cool to room temperature and adjust pH to 7.5 by adding 4 *M* NaOH by drops. Adjust vol to 2 L with dH_2O and sterilize by autoclaving.

2. Buffer 2: 2% blocking reagent in buffer 1. Combine 768 mL 5% blocking reagent (in buffer 1), 288 mL 4X buffer 1 and 864 mL dH_2O. Leave 5% blocking reagent at room temperature for 10 min before use.
3. Buffer 3: 0.1 *M* Tris-HC1, 0.1 *M* NaCl, 0.05 *M* $MgCl_2$, pH 9.5. Add approx 1400 mL dH_2O to 200 mL 1 *M* Tris-HCl, pH 9.5, and 50 mL 4 *M* NaCl. Add 100 mL of filter-sterilized 1 *M* $MgCl_2$ and mix. Adjust pH to 9.5 and make up to 2 L with dH_2O. Do not autoclave as precipitates tend to form. Store at room temperature for up to 1 wk.
4. Buffer hybridization: 192 mL 2% blocking reagent, 144 mL 6X sodium chloride sodium phosphate EDTA (SSPE), 48 mL 5X Denhardts solution, 48 mL 0.1% N-laurylsarcosine, 0.96 mL 0.02% sodium dodecyl sulfate (SDS) and make up to 480 mL with dH_2O.
5. Buffer washing: 0.3% Tween® 20 in buffer 1. Add 14.4 mL Tween 20 to 1200 mL 4X buffer 1 and make up to 4800 mL by adding dH_2O.
6. Blocking reagent: 5% in buffer 1 (Boehringer, Lewes, England; 1096176). Prepare 2 L of 1X buffer 1 by combining 500 mL 4X buffer 1 with 1500 mL dH_2O. Add 100 g blocking reagent in parts, with vigorous mixing using a magnetic stirrer, to approx 1600 mL 1X buffer 1. As blocking reagent is supplied in 50-g tubs, there is no need to weigh out. Heat to 65°C until blocking reagent is dissolved. Allow to cool to room temperature and make up to 2 L with buffer 1. Sterlize by autoclaving and store at 4°C.
7. Cresol red: 10 mg/mL, sodium salt (Sigma, St. Louis, MO, USA; C9877). Add 200 mg to some dH_2O taken from measured 20 mL dH_2O in a sterile universal. Resuspend in remaining vol. Filter-sterilize and dispense into 1-mL aliquots and freeze at –20°C.
8. CSPD (Boehringer, 1655884). Vortex mix and centrifuge CSPD in microcentrifuge for 1 min before use. Dilute CSPD stock solution (25 m*M*, 11.6 mg/mL) 1:100 in buffer 3.
9. Denhardts Solution (50X): 1% polyvinyl pyrrolidone (PVP), 1% Ficoll®, 1% bovine serum albumin (BSA). Prepare 200 mL of 2% PVP and 2% Ficoll by adding 4 g of each to 180 mL dH_2O. (Prepare this solution in a fume cupboard. PVP is harmful if inhaled.) Dissolve with gentle mixing and make up to 200 mL with dH_2O. Sterilize by autoclaving and cool to room temperature. Add 4 g of BSA to 200 mL of above solution slowly with gentle mixing. When the BSA has dissolved, make up to 400 mL with dH_2O, mixing well. Filter

solution through 0.45-μm filter and aliquot. Do not autoclave. Store at –20°C. Leave to thaw at 4°C the evening before it is to be used.

10. anti-DIG-ALP conjugate (Boehringer; 1093274). Just prior to use, remove anti-DIG-ALP stock conjugate (0.75 U/μL) from the refrigerator, vortex mix for 15 s, and centrifuge for 1 min in a microcentrifuge. Make a 1:10,000 dilution of the conjugate in buffer 2 (i.e., 192 μL of anti-DIG-ALP conjugate in 1920 mL of buffer 2).
11. 0.5 *M* EDTA, pH 8.0. Add 186.1 g of EDTA Na_22H_2O in parts to 800 mL dH_2O. Adjust the pH to 8.0 using 4 *M* NaOH. Make up to 1 L with dH_2O and sterilize by autoclaving.
12. Ethidium bromide 10 mg/mL (Sigma; E-1510).
13. Gel loading buffer (GLB): add 8 g sucrose (slowly) to 10 mL dH_2O and mix by inversion until dissolved. Then add 1 mL 1 *M* Tris, pH 7.6, 2 mL 0.5 *M* EDTA, 1 mL 10% SDS, 0.02 g cresol red. Make up to 20 mL with dH_2O. Do not autoclave.
14. dH_2O: Double-distilled H_2O or equivalent. Note that dH_2O used to set up PCR is of ultra-high purity quality.
15. $MgCl_2$: supplied with *Taq* enzyme.
16. dNTP's (Amersham Pharmacia Biotech, St. Albans, England; 27-2094).
17. NH_4 buffer: supplied with *Taq* enzyme.
18. N-Laurolysarcosine (1%): barrier face mask should be worn when weighing N-laurylsarcosine. Dissolve 10 g N-laurlysarcosine in approx 800 mL dH_2O. Adjust vol to 1 L with dH_2O and autoclave.
19. Nylon membrane (Boehringer; 1417 240).
20. PCR plates 96-well (Advanced Biotechnologies, Epsom, England; AB-0366).
21. Size marker Φχ174/*Hae*III (0.1 mg/mL (Promega, Southampton, England; G1761). Add 450 μL dH_2O to vial of the size marker. Add 20 μL of GLB to 12 μL (1.2 μg) of the size marker and 8 μL of TE buffer. Store at 4°C.
22. 10% SDS: this reagent is extremely harmful if inhaled. Wear a mask when working with SDS powder. Also wear gloves. Wash skin thoroughly if in contact with SDS. Wipe down work area after use. Preferably, add SDS to dH_2O in fume cupboard. SDS sometimes comes out of solution, but will go back on heating. Add 100 g of SDS in parts to approx 800 mL dH_2O. As SDS is supplied in 100-g tubs,

there is no need to measure. Apply heat (up to 68°C) if necessary to assist dissolution. Allow to cool to room temperature and adjust the vol to 1 L. Do not autoclave.

23. 20X SSPE: 3 *M* NaCl, 0.2 *M* NaH_2PO_4, 0.02 *M* EDTA, pH 7.4. Add 350.6 g NaCl followed by 48 g NaH_2PO_4 to approx 1600 mL dH_2O. Then add 80 mL 0.5 *M* EDTA, pH 8.0. Adjust the pH to 7.4 using 4 *M* NaOH. Adjust vol to 2 L and sterlize by autoclaving.
24. Saran® Wrap (Genetic Research Instrumentation, Felsted, England; SW1).
25. 2X SSPE/0.1% SDS: Combine 240 mL 20X SSPE and 24mL 10% SDS. Make up to 2400 mL with dH_2O.
26. 5X SSPE/0.1% SDS: Combine 600 mL 20X SSPE and 24 mL 10% SDS. Make up to 2400 mL with dH_2O.
27. 2X Sodium saline citrate (SSC), pH7.0: 0.3 *M* NaCl plus 0.03 *M* tri-sodium citrate.
28. *Taq* enzyme (Bioline, London, England; M958013).
29. Thermofast plate (Advanced Biotechnologies; AB-0600).
30. 1 *M* Tris-HCl, pH 7.6: add 242.28 g Tris base to 1400 mL dH_2O. Adjust the pH to 7.6 by adding 100 mL concentrated HCl (wear a mask and goggles, and where possible, perform this in a fume cupboard). Allow the solution to cool to room temperature before making final adjustments to the pH. Make up to 2 L with dH_2O and sterilize by autoclaving. If the 1 *M* solution has a yellow color, discard it, and obtain better quality Tris. More than 100 mL concentrated HCl may be required.
31. Tris-borate EDTA (10X) (TBE): Add 216 g Tris, 110 g orthoboric acid and 80 mL 0.5 *M* EDTA to 1400 mL dH_2O. Adjust vol to 2 L with dH_2O and sterilize by autoclaving.
32. Tris-EDTA (TE) buffer: 10 m*M* Tris-HCl, 1 m*M* EDTA, pH 7.6. Combine 10 mL 1 *M* Tris-HCl, pH 7.6, with 2 mL 0.5 *M* EDTA and make up to 1 L with dH_2O. Sterilize by autoclaving and aliquot.
33. Enzyme boxes (Boehringer; 800058).
34. Gel sealer and casting tray (Merck, Ltd., Poole, England; 306/7252/12).
35. Robbins hybridization incubator (Robbins Scientific, Sunnyvale, CA, USA; 1040-60-2).
36. Robbins hydra (Robbins Scientific; 1029-60-1).

2.2. PCR Primers and Probes

1. **Tables 1–3** give the primers, the PCR amplification conditions, and the master mixes used for the medium and the allele resolution systems. The primers for HLA-A and -B loci give a locus-specific product covering exons 2 and 3, and the primer for HLA-DR gives a product from exon 2. This product is not specific for the HLA-DRB1 locus and amplifies alleles of other HLA-DR loci (e.g., HLA-DRB3 locus). Thus, it is necessary to include a further amplification for alleles of HLA-DRB1*03, -DRB1*11, -DRB1*13, and -DRB1*14. This is referred to as the HLA-DRB3/11/6 group.
2. The reason for two 3' end primers for HLA-B is because HLA-B*7301 differs in intron 3 from all other known alleles at this locus, and the extra primer is required to amplify this allele. In testing for HLA-B*27 alleles only, which many laboratories perform in individuals where ankylosing spondylitis is being queried, the extra primer is not required. The probe BL12 detects a sequence that is only found in HLA-B*27 alleles and HLA-B*7301. Thus, leaving out primer 3 BIN3-AC means that HLA-B*7301 is not amplified, and the BL12 probe is specific for alleles of HLA-B*27.

Table 1
HLA-A, -B, -DR Primers Used for SSOP Typing

	Primers	Sequence 5' ------------------------------------3'	Band Size
HLA-A		94 (intron 1) → 116	
GENERIC	A15	GAGGGTCGGGC(G/A)GGTCTCAGCCA	863
	AL#AW	TGGCCCCTGGTACCCGT	
		13 (intron 3) → 274 (exon 3)	
HLA-B		36 (Intron 1) → 57	
GENERIC	5 BINI-57M	GGGAGGAG(C/A)(G/A)AGGGGACCGCAG	970
		68 (Intron 3) → 37	
	3 BIN3-37M	AG(G/C)CCATCCCCG(G/C)CGACCTAT	
		68 (Intron 3) → 37	
	3 BIN3-AC	AGGCCATCCCGGGCGATCTAT	
HLA-B27	5 BINI-57M	see above	970
	3 BIN3-37M	see above	
HLA-DRB		(intron1) 15 (exon2) → 24	
GENERIC	AMP-A*	CCCCACAGCACGTTTC(T/C)TG	274
	AMP-B	CCGCTGCACTGTGAAGCTCT	
		279 (exon2) → 260	

Table 1 *(cont.)*
HLA-A, -B, -DR Primers Used for SSOP Typing

	Primers	Sequence 5' -----------------------------------3'	Band Size
		17 (exon2) → 38	
HLA-DRB 3/11/6 group	3/11/6 GF	GTTTCTTGGAGTACTCTACGTC	263
	AMP-B	CCGCTGCACTGTGAAGCTCT	
		279 (exon2) → 260	
		117 (Intron 1) → 2 (Exon2)	
HLA-A2 group	5'A2	CTCCTCGTCCCCAGGCTCT	1754
	Ex4ra	ATCTCAGGGTGAGGGGCT(T/C)G	
		274 (exon 4) → 255 (exon 4)	
		6 (exon 2) → 26 (exon 2)	
HLA-A24 group	A24F	CCACTCCATGAGGTATTTCTC	821
	AL#AW	TGGCCCCTGGTACCCGT	
		13 (intron 3) → 274 (exon 3)	
		6 (exon 2) → 26 (exon 2)	
HLA-A29 group	A29F	CCACTCCATGAGGTATTTCAC	1732
	Ex4ra	ATCTCAGGGTGAGGGGCT(T/C)G	
		274 (exon 4) → 255 (exon 4)	
		6 (exon 2) → 26 (exon 2)	
HLA-A26 group	A26F	CCACTCCATGAGGTATTTCTA	821
	AL#AW	TGGCCCCTGGTACCCGT	
		13 (intron 3) → 274 (exon 3)	
		26 (exon 1) → 44	
HLA-B7 group	B7F	TCCTCCTGCTGCTCTCGGC	1025
	BIn3-37m	AG(G/C)CCATCCCCG(G/C)CGACCTAT	
		68 (Intron 3) → 37	
		32 (exon 1) → 49	
HLA-B54 group	B54F	TGCTGCTCTGGGGGGCCC	1019
	BIn3-37m	AG(G/C)CCATCCCCG(G/C)CGACCTAT	
		68 (Intron 3) → 37	
		27 (exon 1) → 45	
HLA-B15 group	B15F	CCTCCTGCTGCTCTCGGGA	1026
	BIn3-37m	AG(G/C)CCATCCCCG(G/C)CGACCTAT	
		68 (Intron 3) → 37	
		32 (exon 1) → 49	
HLA-B13 group	B13F	TGCTGCTCTGGGGGGCAG	1592
	B13RC	TGGCCTCATGGTCAGAGAT	
		52 (exon 4) → 34	
		32 (exon 1) → 49	
HLA-B44 group	B13F	TGCTGCTCTGGGGGGCAG	1604
	B44R	CCAGCACCTCAGGGTGA	
		66 (exon 4) → 50	
		32 (exon 1) → 49	
HLA-B35 group	B13F	TGCTGCTCTGGGGGGCAG	1592
	B35R	TGGCCTCATGGTCAGAGAC	
		52 (exon 4) → 34	

( ) in primer indicates that at this position, two nucleotides are inserted when the primer is being made. The primer is referred to as being degenerate.

Table 2
PCR Master Mixes

	Master Mix	Stock Conc	End Conc
HLA-A Generic			
dH_2O	6200 µL		
Cresol Red/Suc	2000 µL	0.4/160 mg/mL	0.08/32 mg/mL
NH_4 Buffer	1000 µL	10X	1X
$MgCl_2$	300 µL	50 m*M*	1.5 m*M*
dNTPs	100 µL	20 m*M* each	200 µ*M* each
Each Primer (X3)	120 µL	25 µ*M*	0.3 µ*M*
Biotaq	40 µL	5U/ µL	2U/100 µL
HLA-B Generic			
dH_2O	9495 µL		
Cresol Red/Suc	3000 µL	0.4/160 mg/mL	0.08/32 mg/mL
NH_4 Buffer	1500 µL	10X	1X
$MgCl_2$	450 µL	50 m*M*	1.5 m*M*
dNTPs	150 µL	20 m*M* each	200 µ*M* each
Each Primer (X3)	120 µL	25 µ*M*	0.2 µ*M*
Biotaq	45 µL	5U/ µL	1.5U/100 µL
HLA-B27			
dH_2O	3205 µL		
Cresol Red/Suc	1000 µL	0.4/160 mg/mL	0.08/32 mg/mL
NH_4 Buffer	500 µL	10X	1X
$MgCl_2$	150 µL	50 m*M*	1.5Mm
dNTPs	50 µL	20 m*M*	200 µ*M*
Each Primer (X2)	40 µL	25 µ*M*	0.2 µ*M*
Biotaq	15 µL	5U/ µL	1.5U/100 µL
HLA-DRB Generic			
dH_2O	6380 µL		
Cresol Red/Suc	2000 µL	0.4/160 mg/mL	0.08/32 mg/mL
NH_4 Buffer	1000 µL	10X	1X
$MgCl_2$	300 µL	50 m*M*	1.5 m*M*
dNTPs	100 µL	20 m*M* each	200 µ*M* each
Each Primer (X2)	100 µL	25 µ*M*	0.25 µ*M*
Biotaq	20 µL	5U/ µL	1U/100 µL
HLA-DR 3/11/6 group			
dH_2O	6400 µL		
Cresol Red/Suc	2000 µL	0.4/160 mg/mL	0.08/32 mg/mL
NH_4 Buffer	1000 µL	10X	1X
$MgCl_2$	300 µL	50 m*M*	1.5 m*M*
dNTPs	100 µL	20 m*M* each	200um each
Each Primer (X2)	80 µL	25 µ*M*	0.2 µ*M*
Biotaq	40 µL	5U/ µL	2U/100 µL
HLA-A2 group			
dH_2O	6300 µL		
Cresol Red/Suc	2000 µL	0.4/160 mg/mL	0.08/32 mg/mL
NH_4 Buffer	1000 µL	10X	1X
$MgCl_2$	400 µL	50 m*M*	2.0 m*M*
dNTPs	100 µL	20 m*M* each	200um each
Each Primer (X2)	80 µL	25 µ*M*	0.2 µ*M*
Biotaq	40 µL	5U/ µL	2U/100 µL
HLA-A24 group			
dH_2O	6300 µL		
Cresol Red/Suc	2000 µL	0.4/160 mg/mL	0.08/32 mg/mL
NH_4 Buffer	1000 µL	10X	1X
$MgCl_2$	400 µL	50 m*M*	2.0 m*M*
dNTPs	100 µL	20 m*M* each	200um each
Each Primer (X2)	80 µL	25 µ*M*	0.2 µ*M*
Biotaq	40 µL	5U/ µL	2U/100 µL
HLA-A29 group			
dH_2O	6300 µL		
Cresol Red/Suc	2000 µL	0.4/160 mg/mL	0.08/32 mg/mL
NH_4 Buffer	1000 µL	10X	1X
$MgCl_2$	400 µL	50 m*M*	2.0 m*M*

(cont.)

Table 2 *(cont.)*
PCR Master Mixes

	Master Mix	Stock Conc	End Conc
HLA-A29 group (cont.)			
dNTPs	100 µL	20 m*M* each	200um each
Each Primer (X2)	80 µL	25 µ*M*	0.2 µ*M*
Biotaq	40 µL	5U/ µL	2U/100 µL
HLA-A26 group			
dH_2O	6300 µL		
Cresol Red/Suc	2000 µL	0.4/160 mg/mL	0.08/32 mg/mL
NH_4 Buffer	1000 µL	10X	1X
$MgCl_2$	400 µL	50 m*M*	2.0 m*M*
dNTPs	100 µL	20 m*M* each	200um each
Each Primer (X2)	80 µL	25 µ*M*	0.2 µ*M*
Biotaq	40 µL	5U/ µL	2U/100 µL
HLA-B7 group			
dH_2O	6400 µL		
Cresol Red/Suc	2000 µL	0.4/160 mg/mL	0.08/32 mg/mL
NH_4 Buffer	1000 µL	10X	1X
$MgCl_2$	300 µL	50 m*M*	1.5 m*M*
dNTPs	100 µL	20 m*M* each	200 µ*M* each
Each Primer (X2)	80 µL	25 µ*M*	0.25 µ*M*
Biotaq	40 µL	5U/ µL	2U/100 µL
HLA-B54 group			
dH_2O	6300 µL		
Cresol Red/Suc	2000 µL	0.4/160 mg/mL	0.08/32 mg/mL
NH_4 Buffer	1000 µL	10X	1X
$MgCl_2$	400 µL	50 m*M*	2.0 m*M*
dNTPs	100 µL	20 m*M* each	200 µ*M* each
Each Primer (X2)	80 µL	25 µ*M*	0.2 µ*M*
Biotaq	40 µL	5U/ µL	2U/100 µL
HLA-B15a & B15b groups			
dH_2O	6400 µL		
Cresol Red/Suc	2000 µL	0.4/160 mg/mL	0.08/32 mg/mL
NH_4 Buffer	1000 µL	10X	1X
$MgCl_2$	300 µL	50 m*M*	1.5 m*M*
dNTPs	100 µL	20 m*M* each	200 µ*M* each
Each Primer (X2)	80 µL	25 µ*M*	0.2 µ*M*
Biotaq	40 µL	5U/ µL	2U/100 µL
HLA-B13 group			
dH_2O	6320 µL		
Cresol Red/Suc	2000 µL	0.4/160 mg/mL	0.08/32 mg/mL
NH_4 Buffer	1000 µL	10X	1X
$MgCl_2$	300 µL	50 m*M*	1.5 m*M*
dNTPs	100 µL	20 m*M*	200 µ*M*
PRIMER B13F	160 µL	25 µ*M*	0.4 µ*M*
PRIMER B13RC	80 µL	25 µ*M*	0.2 µ*M*
Biotaq	40 µL	5U/ µL	1.5U/100 µL
HLA-B44 group			
dH_2O	6400 µL		
Cresol Red/Suc	2000 µL	0.4/160 mg/mL	0.08/32 mg/mL
NH_4 Buffer	1000 µL	10X	1X
$MgCl_2$	300 µL	50 m*M*	1.5 m*M*
dNTPs	100 µL	20 m*M* Each	200 µ*M* Each
Each Primer (X2)	80 µL	25pm/ µL	0.2 µ*M*
Biotaq	40 µL	5u/ µL	2u/100 µL
HLA-B35 group			
dH_2O	6400 µL		
Cresol Red/Suc	2000 µL	0.4/160 mg/mL	0.08/32 mg/mL
NH_4 Buffer	1000 µL	10X	1X
$MgCl_2$	300 µL	50 m*M*	1.5 m*M*
dNTPs	100 µL	20 m*M* each	200um each
Each Primer (X2)	80 µL	25 µ*M*	0.2 µ*M*
Biotaq	40 µL	5U/ µL	2U/100 µL

Table 3
PCR Amplification Conditions

Locus	Hold	Cycle	No. of Cycles	Hold	Hold
HLA-A Generic	96°C/5 min	96°C/1 min 60°C/30 s 72°C/1 min	35	72°C/5 min	15°C/forever
HLA-B Generic + HLA-B27 Testing	96°C/5 min	96°C/30 s 65°C/30 s 72°C/45 s	32	72°C/5 min	15°C/forever
HLA-DRB Generic	96°C/5 min	96°C/1 min 55°C/1 min 72°C/1 min	30	72°C/5 min	15°C/forever
HLA-DR3 -DR6 -DR11 Subtyping group	96°C/5 min	96°C/1 min 64°C/1 min 72°C/1 min then 96°C/1 min 56°C/1 min 72°C/1 min	10 20	72°C/5 min	15°C/forever
HLA-A2 Subtyping group	96°C/5 min	96°C/1 min 65°C/1 min 72°C/1 min then 96°C/1 min 62°C/30 s 72°C/1 min	15 27	72°C/5 min	15°C/forever
HLA-A24 -A29 -A26 -B13	96°C/5 min	96°C/1 min 64°C/1 min 72°C/1 min then	15 27	72°C/5 min	15°C/forever

(cont.)

Table 3 *(cont.)*
PCR Amplification Conditions

Locus	Hold	Cycle	No. of Cycles	Hold	Hold
HLA-A24 (cont.) Subtyping groups		96°C/1 min 56°C/1 min 72°C/1 min	27		
HLA-B7 -B15a -B15b Subtyping groups		96°C/1 min 96°C/5 min 58°C/1 min 72°C/1 min	38	72°C/5 min	15°C/forever
LA-B54 Subtyping group	96°C/5 min	96°C/1 min 65°C/1 min 72°C/1 min	35	72°C/5 min	15°C/forever
HLA-B44 Subtyping group	96°C/5 min	96°C/1 min 60°C/1 min 72°C/1 min	35	72°C/5 min	15°C/forever
HLA-B35 Subtyping group	96°C/5 min	96°C/20s 70°C/45s 72°C/25s	7	72°C/5 min	15°C/forever
		then 96°C/25s 65°C/50s 72°C/30s	25		
		then 96°C/30s 55°C/60s 72°C/90s	6		

3. **Tables 4–11** give the probes and probe patterns for the medium resolution system for HLA-A, -B, and -DR, and **Tables 12–32** gives the probes and probe patterns for allele resolution for the HLA-A and -B loci. (**Tables 4–32** may be found on pp. 87–112.)

3. Methods

*3.1. PCR Using 96-Well Plates (*see **Notes 1–5***)*

1. Heat DNA samples to be tested to 60°C for 5–10 min, vortex mix, and centrifuge for 5 s in a microcentrifuge.
2. Prepare 10 mL of mastermix for appropriate locus (**Table 2**). Use dH_2O of ultra-high purity quality. Dispense 100 µL slowly into tubes of the 96-well plate. Take care to avoid splashes and air bubbles at the bottom of the tubes. When all tubes have been filled, cover the 96-well plate with a sterile microtiter tray lid.
3. Add 1 µL DNA sample to each well from position 1A→1H, 2A→2H, etc. Only one row at a time should be uncovered by the lid. Leave two wells with master mix only, to act as negative controls and leave appropriate number of wells for control DNA (*see* **Note 9**). When a complete row of DNA samples have been added, place a strip of 8 caps over these samples, and press down gently. When DNA samples have been added to all tubes and caps are in place, use a cap sealing tool to ensure that all caps are pushed firmly into place.
4. Centrifuge the plate for 1 min at 500 *g*, place in PCR machine, and run appropriate cycle program (**Table 3**). After amplification, if the PCR samples are not to be processed immediately, store at –20°C.

3.2. Electrophoresis of PCR Samples

1. Add 4.5 g of agarose to 300 mL 1X TBE, boil and allow solution to cool to 65°C. It is important to stir agarose while cooling to prevent lumps forming.
2. While agarose is cooling, prepare 96-well-gel template by placing casting tray in the gel sealer. Take care to ensure that gel sealer is not over tightened otherwise casting tray may separate when agarose is added.
3. Place sealed casting tray on top of a leveling table and adjust the feet of the leveling table until the bubble in the spirit level is centered.
4. Once agarose has cooled to 65°C, add 15 µL of ethidium bromide (10 mg/mL) and mix gently (ethidium bromide is mutagenic).

5. Pour the molten agarose solution into the level casting tray. Immediately push any air bubbles to edges of the template using a pipet tip.
6. Insert four 24-slot combs into the gel, with equal spacing between combs. Allow gel to set for approx 1 h at room temperature.
7. Add 1000 mL of 1X TBE to an electrophoresis tank. Carefully remove combs from gel. Remove gel from the gel sealer. Place gel in tank containing 1X TBE buffer. Ensure gel is covered by buffer to a depth of 2 to 3 mm.
8. Add 4 μL of each PCR product to a 96-well Thermofast plate. Ensure product is in each well. Add 8 μL GLB to each well. Spin plate for 1 min to ensure mixing.
9. Load 10 μL of size marker into first well of each of the four rows.
10. Using an Octapipet, carefully load 10 μL of sample into each well of the gel. Care must be taken to ensure the octapipet is orientated properly when adding the samples to the gel.
11. Place the lid of the electrophoresis system on to the electrophoresis tank, connect the electrodes to the power pack and electrophorese the samples at 250 V, 250 mAmp for 20 min.
12. Once electrophoresis is complete, remove the gel from the tank, and photograph under UV light. Check size of PCR product against size marker to ensure correct product has been amplified (**Table 1**).

3.3. Dot Blotting of Membranes

We use the Robbins Hydra dot blotting machine, which enables us to make as many replicate membranes as required from the PCR product. Other laboratories use other equipment and some will dot blot by hand. If the preparation of the required number of membranes proves difficult, an alternative method is to dehybridize used membranes as follows.

1. Dehybridize a maximum of 3 membranes in 300 mL of each solution with shaking.
2. Rinse membranes in dH_2O for 5 min at room temperature.
3. Wash membranes in 0.4 *M* NaOH/0.1% SDS at 45°C for 30 min.
4. Wash membranes in 2X SSC for 30 min at room temperature.
5. Check dehybridization is complete by exposing membranes overnight to X-ray film and developing in usual manner.
6. Store membrane flat at 4°C in a sealed plastic bag if not using immediately.

3.4. Denaturation and Fixing of Blots

1. After PCR product has been dispensed onto the membranes, allow to air-dry for at least 20 min.
2. Carefully place membranes, DNA face up, onto 2 sheets thick (3 MM) Whatman paper soaked in 0.4 *M* NaOH. Leave for 10 min. When placing membranes onto Whatman paper, take care to ensure that membrane is not dragged over denaturation pad, that all of the membrane soaks up the 0.4 *M* NaOH, and that there are no air bubbles beneath the membrane.
3. Transfer each membrane onto Whatman paper (3 MM) soaked in 10X SSPE. Leave for 5 min.
4. Gently wash in 2X SSPE and allow to air-dry for at least 25 min.
5. Wrap membranes in Saran Wrap and place (DNA face down) on a UV transilluminator for 4 min. Ensure that all the UV lights are fully on during the procedure, and do not switch transilluminator off between each step. Place a glass plate on top of the membranes to hold them flat during this procedure. Store membranes wrapped in tin foil at 4°C if not using immediately.

3.5. 3' End Labeling of HLA Oligonucleotides

The labeling reagents are obtained in a kit from Boehringer (cat. no. 1362372).

1. Remove all reagents from freezer (except Terminal Transferase; this should be removed just before use) and allow to thaw. Vortex mix reagents briefly, and centrifuge in microcentrifuge for 5 s.
2. Combine the following: 4 µL reaction buffer (5X), 4 µL $CoCl_2$ (25 m*M*), 1 µL DIG-ddUTP (1 m*M*), 1 L Terminal Transferase (50 U), 100 pmol probe. Make up to 20 L with dH_2O. Vortex mix the samples briefly, centrifuge in microcentrifuge for 5 s, and incubate at 37°C for 30 min in a water bath.
3. Centrifuge for 5 s in microcentrifuge and place on ice for 5 min. Add 80 L dH_2O, vortex mix briefly, and centrifuge in microcentrifuge for 5 s. Aliquot in vol related to the amount of probe used (**Tables 4–7,** following p. 87) and store at –20°C.

3.6. Prehybridization, Hybridization, and SSPE Stringency Washes (see Notes 6–10)

Each probe is simultaneously hybridized to two different membranes, each containing 96 DNA samples.

1. Hand roll membranes lengthwise to form a cylinder. Place two membranes in a hybridization bottle. One membrane should have the DNA side of the membrane facing the glass, while the second membrane should have the DNA side facing inwards in the bottle.
2. Add 20 mL of freshly prepared hybridization buffer. Screw cap on tightly and clamp to the rotisserie of a Robbins incubator (preset at 45°C). Rotate the bottles for 1 h.
3. Just before the incubation is complete thaw appropriate aliquots of DIG-labeled oligonucleotide probe, vortex mix briefly, and centrifuge for 5 s in a microcentrifuge.
4. Add appropriate number of picomoles of probe (**Tables 4–7**) to 20 mL of prewarmed (45°C) hybridization buffer and mix by inversion.
5. Remove the hybridization bottle from the incubator and pour off the hybridization buffer into a disposable collection container. Add 20 mL of hybridization buffer containing DIG-labeled probe and incubate the bottle for 1 h at 45°C.
6. Remove the bottle from the incubator and pour off the fluid into a disposable collection container.
7. Add 100 mL of 2X SSPE/0.1% SDS. Recap the bottle and place inside a Robbins incubator (preset to 25°C) and incubate for 10 min. Make sure temperature does not rise above this.
8. Discard the fluid and repeat **step 7**.
9. Remove the bottle from the incubator. Uncap the bottle and, using forceps, carefully remove the membranes from the bottle, prior to discarding fluid, directly into a small plastic tray containing 200 mL 5X SSPE/0.1% SDS, which has been heated to the appropriate temperature (**Tables 4–7**). Place one membrane DNA face down and the other membrane DNA face up into the washing solution. Incubate with shaking for 40 min. Check temperature reading and record any variation on the hybridization record sheet. If the temperature varies more than 2°C above or below the required temperature, abandon this hybridization (*see* **Notes 6** and **7**).

10. Remove the membranes from the tray, blot dry, but do not allow the membrane to dry out. Wrap the membrane in Saran Wrap and store in tinfoil at 4°C, until ready to perform chemiluminescent detection.

3.7. Chemiluminescence (see *Note 12*)

All steps are performed at room temperature with shaking, using a platform shaker. Use separate enzyme storage boxes for different buffer solutions and keep light-tight. Use 1 enzyme box for a maximum of 3 membranes at the same time.

1. Add 240 mL anti-DIG-ALP conjugate in buffer 2 to the enzyme box. Place membranes into the boxes DNA side down. Incubate for 15 min on shaker.
2. Transfer membranes to 300 mL of washing buffer and incubate for 15 min on shaker. Discard washing buffer and replace with fresh washing buffer and incubate for a further 15 min.
3. Transfer membrane to 300 mL of buffer 3 and incubate for 5 min on shaker.
4. Remove from buffer 3, place 2 membranes back to back in a plastic bag. Add 20 mL of CSPD (1:100 dilution) and reseal the bag. Place the bag on a platform shaker, cover with tinfoil and shake for 5 min at room temperature.
5. Pour off CSPD fluid into 20-mL plastic tube for reuse (up to 5 times). Store at –20°C if using on more than 1 d, but note that CSPD should only be frozen once. Carefully remove the membrane from the bag, blot off excess liquid, and wrap in Saran Wrap.
6. Tape two membranes to the one X-ray film and place a second film on top. Expose the top film for 5 min and check the intensity of the dots. Depending on these results process the second film accordingly. It may be necessary to re-expose the membrane to a third or fourth film for a further period of time, depending on dot intensity.
7. Record the probe reaction for each sample and analyze according to the known patterns (**Tables 8–32**; pp. 86–87 and 91–111) using a computer programme (*see* **Note 13**).

4. Notes

1. When performing a PCR on 96 samples, there may be one or two samples that are not amplified. Therefore, we always run a gel to

ensure that we have product. This enables the SSOP method to be well controlled. If an amplification fails in the sequence-specific primer method, this would lead to an incorrect result. On some occasions, the product is deemed weak, and this sample will always be repeated. Good amplification always gives a clean and clear-cut SSOP hybridization, while almost all the problematic typing results we have encountered were due to poor amplification. Interpretation of weak hybridization signals can give an incorrect result.

2. When determining the conditions necessary for this technique, one important aspect is to ensure that the amplification works for both alleles. On some occasions, when we were determining the amplification conditions, we found that a product could be obtained whereby only one allele could be detected. Therefore, if a laboratory is setting up a technique from scratch, it should ensure that there is no differential amplification by testing various combinations of alleles.
3. We do not routinely determine the concentration of DNA in each isolation. When isolating DNA, the amount of TE buffer added to the pellet of DNA is judged by eye. However, we assess approx 10% of samples to ensure that the DNA is at an appropriate concentration. For our methods, we normally have the DNA concentration at approx 0.5 μg/μL.
4. When setting up a PCR, wear a separate laboratory coat, wear gloves and change them frequently, and perform all work in pre-PCR room using dedicated equipment. Pipets should not be removed from pre-PCR room. Pipets are labeled according to reagents and must only be used for these reagents. The use of tips with filters is advisable. When preparing the master mix thaw out the following reagents: $MgCl_2$, dNTPs, PCR buffer, and appropriate primers. Vortex mix each reagent briefly and centrifuge in a microcentrifuge for 5 s and place in an ice bucket (PCR buffer and $MgCl_2$ should be centrifuged for 2 min). *Taq* DNA polymerase should always be added last, after vortex mixing and centrifuging, and just prior to dispensing the master mix. The aliquoted master mixes should not be left on the bench too long (maximum time being 15 min, as *Taq* loses activity once diluted in buffer). Switch on PCR machine for at least 10 min prior to use to allow the machine to heat up. PCR machine should be situated in post-PCR room.
5. After setting up a PCR, wash work areas with 16% sodium hypochlorite. Soak all racks used to hold samples in 16% sodium hypochlorite for approx 30 min, and rinse thoroughly in water. Pipets should be

wiped with 2% sodium hypochlorite, followed by dH_2O. Wipe microcentrifuge, vortex, freezer handle, etc., with 16% sodium hypochlorite. Expose the working area, including pipets, etc., to UV light for 60 min.

6. In this laboratory, we do not use tetramethylammonium chloride (TMAC) owing to its toxic properties and the fact that in our experience it does not necessarily mean the use of one wash temperature. It would appear, that to follow our practice, a laboratory would require a large number of water baths. In this laboratory, one individual normally performs 15 hybridizations at the same time. Thus, if a laboratory is defining alleles at three loci (HLA-A, -B, -DR), probes can be selected for use at the same time according to their wash temperature (**Tables 4–7**). This eliminates the requirement for a large number of water baths.
7. If a laboratory is not performing tests on large numbers of samples, only one lot of membranes is hybridized. The picomoles given for each probe in the Tables is for the hybridization of one membrane. However, each probe can be hybridized to two different membranes in the same hybridization bottle, the amount of probe being doubled. The SSOP method is thus very suitable for typing large numbers of samples—we test 192 samples at the same time (96 on each membrane), which includes controls.
8. When we implement new probes to the system, we will initially use a wash temperature that is equivalent to the melting temperature of the probes. This is equal in °C to 2X (number of A + T bases) + 4X (number of G + C bases). Thereafter, we adjust the wash temperature by 1°C either up or down according to the probe reaction at the melting temperature. We also start with 20 pmol of probe and adjust accordingly.
9. When the probe conditions, i.e., number of picomoles and wash temperature, have been determined, it is worthwhile to keep a record on the performance of the probes, i.e., whether the probe is not giving an adequate signal with its positive control or whether it is cross-reacting with controls with which it should be negative. On occasions, the conditions for the probes need to be altered. This, in a way, is similar to HLA sera whereby after long-term storage, the specificities identified can change. If a probe appears to be giving strong false positive reactions, we will initially increase the wash temperature by 1°C, or if giving weak false positive reactions, we will decrease the probe concentration by approximately 20%. If a probe

appears to be giving false negative results, we will decrease the wash temperature by 1°C. If a probe is giving weak reactions, we will initially increase the probe concentration by approx 20%. One way to monitor the performance of the probes is to record the length of time needed for autoradiography exposure. If this varies too such an extent that it takes more than 30 min to achieve a good signal, the conditions of the probe should be altered.

10. Many of the probes used in this laboratory are DIG-labeled during their manufacture, adding the DIG moiety to 5' amino oligonucleotides by incubating with a DIG ester under mild alkali conditions.
11. Enough controls should be included so that each probe will have two positive reactions. In addition, control DNA should be included as negative controls. These contain alleles with sequences that are closely related to the sequence that the probe detects and with which the probe might cross-hybridize. This is especially important when initally determining the optimum conditions for the probe to work. To maintain consistency between membranes, we try to use the same controls. If a laboratory finds it difficult to have a large enough supply of the same control DNA, it may consider cloning control DNA by long range amplification *(**3**)*. Due to lack of sequence information this is only practical for HLA-A and -B at present. This gives material to use in as many tests as needed. This is especially important when the control DNA has been obtained from an outside source.
12. It is normal practice in this laboratory for chemiluminescent detection to be performed on 30 membranes at the same time. All membranes are processed up to the end of step two. Thereafter, membranes are processed in groups of six simultaneously, leaving the remaining membranes in the washing buffer.
13. In our laboratory, we always have two independent readings of the membrane. We do not believe in recording a result according to the strength of the reaction (e.g., 1, 2, 4, 6, 8, as in serology). The result should be positive or negative. If in doubt, it should be repeated. In the future, it would be beneficial to all laboratories if a scanning mechanism was available for reading the membranes, as mistakes are possible in the transmissions of results. We believe it is important that the probe patterns are not analyzed by eye. It would be far too easy to see the obvious allele(s) when examining the probe patterns rather that those that are obscure. To overcome this, laboratories should have a computer programme.

References

1. Opelz, G., Mytilineos, J., Scherer, S., et al. (1991) Survival of DNA HLA-DR typed and matched cadaver kidney transplants. *Lancet* **338**, 461–463.
2. Charron, D. and Fauchet, R. (eds.) (1997) *HLA. Genetic Diversity of HLA Functional and Medical Applications. Vol. 1, Workshop*. EDK, Paris.
3. Curran, M. D., Williams, F., Earle, J. A. P., et al. (1996) Long range PCR amplification as an alternative strategy for characterizing novel HLA-B alleles. *Eur. J. Immunogen.* **23**, 297–309.

Table 4
Probes Used for HLA-A Typing

Probe	Sequence 5' ------------------------------------3'	Wash Temp (°C)	Picomoles Used	Nucleotide Position
				Exon 2
W (A94)	TTCTTCACATCCGTGTC	50	25	22-38
2 (A89)	GGTATTTCTCCACATCCGT	56	20	17-35
A (56R)	GAGAGGCCTGAGTAT	46	20	163-177
B (62LQ)	TGGGACCTGCAGACA	48	25	178-192
C (62G)	GACGGGGAGACACGG	54	10	181-195
O (62RN)	GACCGGAACACACGG	52	10	181-195
D (62EG)	GAGGAGACAGGGAAA	46	20	184-198
Y (A276)	GGCCCACTCACAGACT	52	25	204-219
E (731)	TCACAGATTGACCGA	45	20	211-225
X (A290)	CTGACCGAGTGGACCT	51	20	218-233
R (A26)	TGACCGAGCGAACCTG	54	20	219-234
F (77S)	GAGAGCCTGCGGATC	50	10	226-240
				Exon 3
Z(A347)	CTCACACCATCCAGA	45	35	5-19
T (95V)	CACACCGTCCAGAGG	48	20	7-21
P (114EH)	TATGAACAGCACGCC	46	15	67-81
G (131R)	CGCTCTTGGACCGCG	52	20	121-136
H (142TK)	ACCACCAAGCACAAG	46	20	154-168
I (149T)	TGGGAGACGGCCCAT	50	20	169-183
J (150V)	GAGGCGGTCCATGCG	62	10	172-186
K (151R)	GCGGCCCGTGTGGCG	60	10	175-189
1 (A525)	TGAGGCGGAGCAGTTG	54	20	183-198
N (156Q)	GAGCAGCAGAGAGCC	52	10	190-204
Q (156W)	GAGCAGTGGAGAGCC	50	5	190-204
L (161D)	CTGGATGGCACGTGC	50	10	208-222
V (A551)	TGGAGGGCACGTGCGT	58	20	209-224
M (163R)	GAGGGCCGGTGCGTG	54	10	211-225
S (A355)	GGCGAGTGCGTGGAGTGGC	68	5	214-232
U (A357)	GGCGAGTGCGTGGACGGGC	68	5	214-232

Table 5
Probes Used for HLA-B Typing

Probe	Sequence 5' ---------------------------3'	Wash Temp (°C)	Picomoles Used	Nucleotide Position
				Exon 2
31 (B89)	GGTATTTCGACACCGCC	56	20	17-33
32 (B156)	GGACGGCACCCAGTT	52	20	84-98
33 (B168)	GTTCGTGCGGTTCGA	50	20	96-110
09 (BL09)	GAGTCCGAGAGAGGAGCC	57	3	123-140
01 (BL01)	GAGGAAGGAGCCGCGGGC	64	10	129-146
02 (BL02)	GAGGACGGAGCCCCGGGC	64	20	129-146
07 (BL07)	GAGGATGGCGCCCCGGGC	64	30	129-146
34 (B249)	TTGGGACGGGGAGAC	50	20	177-191
24 (BL24)	GGGAGACACAGATCTCCA	55	20	185-202
05 (BL05)	ACACAGATCTTCAAGACC	56	7	190-207
10 (BL10)	GATCTACAAGGCCCAGGC	58	5	195-212
12 (BL12)	ATCTGCAAGGCCAAGGCA	56	10	196-213
18 (BL18)	ACTGACCGAGTGAGCCTG	58	10	217-234
35 (B73)	ACTGACCGAGTGGGCCTG	63	20	217-234
20 (BL20)*	AGCGGAGCGCGGTGCGCA	64	20	233-250
21 (BL21)	CGGAACCTGCGCGGCTAC	62	20	235-252
22 (BL22)	CGGACCCTGCTCCGCTAC	61	20	235-252
23 (BL23)	CGGATCGCGCTCCGCTAC	62	20	235-252
				Exon 3
27 (BL27)	CTCACACTTGGCAGAGGA	56	10	5-22
36 (B348)	TCACACCATCCAGAGG	49	25	6-21
37 (B354)	CATCCAGGTGATGTAT	46	20	12-27
28 (BL28)	CCAGTGGATGTATGGCTG	56	20	15-32
38 (B361)	AGGATGTTTGGCTGC	48	20	19-33
26 (BL26)	CTGCGACCTGGGGCCCGA	65	20	30-47
30 (BL30)	GGCATAACCAGTTAGCCT	56	25	65-82
39 (B409)	TATGACCAGGACGCCT	55	20	67-82
40 (B427)	GACGGCAAAGATTACA	46	20	85-100
41 (B499)	ACCCAGCTCAAGTGG	47	20	157-171
42 (B505)	CGCAAGTTGGAGGC	46	20	163-176
43 (B532)	GAGCAGCTGAGAGCCT	52	20	190-205
44 (B539)	GAGAACCTACCTGGA	46	20	197-211
45 (B553a)	GAGGGCCTGTGCGT	48	20	211-224
46 (B553b)	GAGGGCACGTGCGT	48	20	211-224
47 (B566)	TGGAGTCGCTCCGC	48	20	224-237
48 (B597)	GAAGGACACGCTGGA	51	20	255-269
49 (B599)	AGGACAAGCTGGAGCG	52	20	257-272

*, Complimentary to coding sequence.

Table 6
Probes Used for HLA-DR Typing

Probe	Sequence 5' ---------------------------3'	Wash Temp (°C)	Picomoles Used	Nucleotide Position
				Exon 2
09 (1007)	GAAGCAGGATAAGTTTGA	50	5	24-41
03 (1008N)	GAGGAGGTTAAGTTTGAG	54	1	25-42
07 (1004)	GAGCAGGTTAAACATGAG	56	2	25-42
08 (1006)	TGGCAGGGTAAGTATAAG	50	5	25-42
06 (1003)	GTACTCTACGTCTGAGTG	56	2	27-44
02 (1002)	AGCCTAAGAGGGAGTGTC	56	15	29-46
18 (DR18)	CTACGGGTGAGTGTTAT	48	20	32-48
10 (2810)	GCGAGTGTGGAACCTGAT	56	5	66-83
01 (2801)	CGGTTGCTGGAAAGATGC	60	3	73-90
25 (DR25)	CGGTTCCTGGACAGATA	52	20	73-89
11 (DRB12)	CAGGAGGAGCTCCTGCGC	58	2	100-117
13 (DRB6)	CAGGAGGAGAACGTGCGC	62	3.5	100-117
22 (DR22)	CGGCCTAGCGCCGAGTA	58	25	163-179
05 (5703)	GCCTGATGAGGAGTACTG	54	10	165-182
15 DRB14/1	GGCCTGCTGCGGAGCACT	64	2	164-181
26 DRBALL	TGGAACAGCCAGAAGGAC	56	20	181-198
27 (DR27)	TGGAGCAGGCGCGG	50	20	203-216
14 (7031)	CTGGAAGACAAGCGGGCCG	60	15	202-220
16 (DRB13)	TGGAAGACGAGCGGGCCG	64	1.5	203-220
24 (DR24)	AGCGGAGGCGGGCCGAG	62	20	206-222
28 (DR28)	AGACAGGCGCGCCG	52	20	207-220
17 (7012) *	ACCGCGGCCCGCCTCTGC	66	15	207-224
23 (7005) *	ACCGCGGCCCGCTTCTGC	67	20	207-224
12 (DRB8)	GCGGGCCCTGGTGGACAC	64	10	213-230
04 (7004)	GGCCGGGTGGACAACTAC	62	0.5	217-234

*, Complimentary to coding sequence.

Table 7
Probes Used for HLA-DR 3/11/6 Group

Probe	Sequence 5' ---------------------------3'	Wash Temp (°C)	Picomoles Used	Nucleotide Position
				Exon 2
1 (DR19)	CGGTACCTGGACAGAT	50	20	73-88
2 (5703)	GCCTGATGAGGAGTACTG	54	10	165-182
3 (DRB14/1)	GGCCTGCTGCGGAGCACT	64	2	164-181
4 (7031)	CTGGAAGACAAGCGGGCCG	60	15	202-220
5 (DRB13)	TGGAAGACGAGCGGGCCG	64	1.5	203-220
6 (DR24)	AGCGGAGGCGGGCCGAG	62	20	206-222
7 (7012*)	ACCGCGGCCCGCCTCTGC	66	15	207-224
8 (7005*)	ACCGCGGCCCGCTTCTGC	65	20	207-224
9 (DRB8)	GCGGGCCCTGGTGGACAC	64	10	213-230
10 (7004)	GGCCGGGTGGACAACTAC	62	0.5	217-234
11 (5701)	GCCTGATGCCGAGTACTG	58	20	165-182

*, Complimentary to coding sequence.

Table 8
HLA-A SSOP Patterns

Probes	A	B	C	D	E	F	G	H	I	J	K	L	M	N	O	P	Q	R	S	T	U	V	W	X	Y	Z	1	2
HLA-A Alleles																												
0101/03/04N/05N/06							■			■			■					■					■		■	■		
0102							■			■			■					■							■	■		■
02011/012/07/09/13/15N/18/20/24/29/30/31/32N/33			■				■	■												■		■	■		■			
0202			■				■	■									■					■	■		■			
0203			■				■	■	■								■			■		■	■		■			
0204/171/172			■				■	■														■	■		■			
0205/08			■				■	■									■					■			■			
0206/10/21/28			■				■	■												■		■			■			
0211			■		■		■	■												■		■	■	IIIII				
0212/13/27			■				■	■						■						■		■	■		■			
0214			■				■	■														■			■			
0216			■				■	■											■	■			■		■			
0219			■				■							■						■		■	■		■			
0222			■				■	■									■			■		■	■		■			
0225			■				■	■	■											■		■	■		■			
0226			■				■	■												■		■	■		■		■	
0234			■				■	■												■		■	■					
0235			■				■	■												■		■	■	■				
03011/013/03N/04							■					■											■	■		■	■	
03012							■					■											■	■		■		
0302							■					■		■									■	■		■		
11011/012/02/03/05							■						■	■										■		■		
1104							■							■								■		■		■		
2301/03				■			■				■											■			■			■
2302				■			■				■						■					■			■			■
2402101/102L/031/032/05/09N/11N/15/17				■			■							■								■			■			■
2404				■			■							■				■				■			■			■
2406				■			■										■					■			■			■
2407				■			■							■								■						■
2408			■				■							■								■			■			■
2410				■			■						■	■											■			■
2413/22				■			■															■			■			■
2414				■			■							■						■		■			■			■
2416				■			■				■											■			■	■		
2418				■			■					■													■		■	■
2424							■				■				■							■						■
2501						■	■		■				■		■		■								■	■		
2502						■	■		■				■		■		■									■		
2503						■	■		■				■		■		■						■		■	■		
2601/02/10/12							■		■				■		■		■	■							■	■		
2603/05/06							■		■				■		■		■								■	■		
2604							■		■						■		■	■							■	■		
2607			■				■		■				■				■	■							■	■		
2608							■		■				■	■	■			■							■	■		
2609							■		■						■		■	■				■			■	■		
2611N							■						■		■		■	■							■	■		
2613							■		■				■		■		■	■								■		
2901/02/03/04		■					■				■							■				■				■		
3001	■						■									■						■		■		■		■
3002	■						■									■						■			■	■		■
3003							■									■						■			■	■		■
3004/06	■						■									■	■					■			■	■		■
3007	■			■			■									■						■			■	■		■
31012/02/03/04	■				■		■				■											■		IIIII		■		
3201						■	■				■											■	■		■	■		
3202						■	■							■								■	■		■	■		
3203							■				■											■	■		■	■		
3204						■	■					■											■		■	■	■	
3301/03/05					■		■				■				■							■		IIIII		■		
3304					■						■				■							■				■		

(cont.)

Table 8
HLA-A SSOP Patterns *(cont.)*

	A	B	C	D	E	F	G	H	I	J	K	L	M	N	O	P	Q	R	S	T	U	V	W	X	Y	Z	1	2
3401							■		■						■		■					■		■		■		
3402							■		■						■							■		■		■	■	
3601							■			■								■				■	■		■	■		
4301		■					■		■				■				■	■							■	■		
6601							■		■				■		■		■							■		■		
6602							■		■						■		■		■					■		■		
6603							■		■						■		■		■					■	■	■		
68011/012/02/07							■	■							■		■					■		■		■		
68031/032							■	■							■		■					■		■	■	■		
6804					■		■	■							■		■					■		IIIII		■		
6805							■	■							■		■					■			■	■		
6806							■	■							■	■	■					■		■		■		
6808							■	■							■							■		■		■		
6809							■	■						■	■							■		■		■		
6810/13/14							■	■									■					■		■		■		
6811N							■	■							■		■							■		■		
6812							■								■		■					■		■		■		
6901							■	■							■					■		■		■				
7401/02/03							■				■											■	■	■	■	■		
8001							■														■		■		■	■		

Vertical bars, Probe positive reactions, but unexpected from sequence.

Table 9
HLA-B SSOP Patterns

Probes	1	2	5	7	9	10	12	18	20	21	22	23	24	26	27	28	30	31	32	33	34	35	36	37	38	39	40	41	42	43	44	45	46	47	48	49
HLA-B Alleles																																				
07021/022/023/					■	■				■																										■
04/05/06/07/09/11																																				
0703/10/16					■					■																										■
0708			■		■					■																										■
0712					■	■				■				■																						■
0713										■																										■
0801/06/07/			■		■					■								■															■		■	
8/9/10																																				
0802			■		■				■									■															■		■	
0803			■		■							■						■															■		■	
0804/05					■					■								■															■		■	
1301/02				■					■				■				■											■		■						
1303				■					■				■				■													■		■				
1304				■					■				■																	■		■				
1536																																				
1401/02/04					■					■						■														■			■			
1403					■					■						■																	■			
1405/062					■					■																				■			■			
1501101/102N/04/				■						■			■																			■				
07/12/19/26N/30/																																				
32/33/34/35/38/																																				
45/57																																				
15012				■						■			■														■					■				
1502/21/31				■						■																				■		■				
1503					■					■			■																	■		■				
1505/25/39				■						■			■																	■		■				
1506				■						■			■												■					■		■				
1508			■	■						■																						■				
1509					■					■																	■			■		■				
1510/18/37					■					■																				■		■				
1511/15/28				■						■																						■				
1513/17				■								■																		■		■				
1514				■						■			■																			■		■		
1516				■								■			■															■		■				
1520				■						■			■	■																■		■				
1522		■	■							■																						■				
1523					■							■																		■		■				
1524				■								■	■																			■				
1527				■						■			■												■							■				
1529			■		■					■																				■		■				
1540				■						■			■																	■						
1542				■						■			■				■													■			■			
1543				■							■		■																			■				
1544				■						■																				■			■			
1546	■									■			■																			■				
1547/49					■					■			■																							
1548				■						■			■																		■	■				
1550				■						■			■																				■			
1551					■					■																						■				
1801/02/03/04/05/06		■								■									■											■			■			
1807		■	■							■									■														■			
2701					■		■		■																					■						
2702					■		■					■																		■						
2703/04/052/053/					■		■				■																			■						
06/07/09/10/																																				
11/13/14																																				
2708					■		■			■																				■						
2712					■					■																				■						
2716					■						■																			■						
3501/02/03/04/05/		■	■							■				■																■		■				
06/07/091/092/10																																				
11/13/21/24/27/																																				
29/30																																				
3508/14/18		■	■							■				■																		■				
	1	2	5	7	9	10	12	18	20	21	22	23	24	26	27	28	30	31	32	33	34	35	36	37	38	39	40	41	42	43	44	45	46	47	48	49
3512/16/17/22/32		■	■							■																				■		■				
3515/33		■	■							■				■																■						
3519	■		■							■				■																■		■				
3520		■								■				■																■		■				
3523		■	■							■				■											■					■		■				
3525			■		■					■				■																■		■				
3526					■					■				■																■		■				
3528		■								■			■	■																■		■				
3531		■	■							■																				■						
3701		■									■		■										■										■			
3702		■									■		■																	■						
3801					■							■																			■		■			
38021/022					■				■																						■		■			
3803					■				■				■																		■		■			
39011/013/03/04/					■					■																					■		■			
05/061/062/09/																																				
10/11/14/15/16/17																																				
39021/022/08/13					■					■			■																		■		■			

(cont.)

Table 9
HLA-B SSOP Patterns *(cont.)*

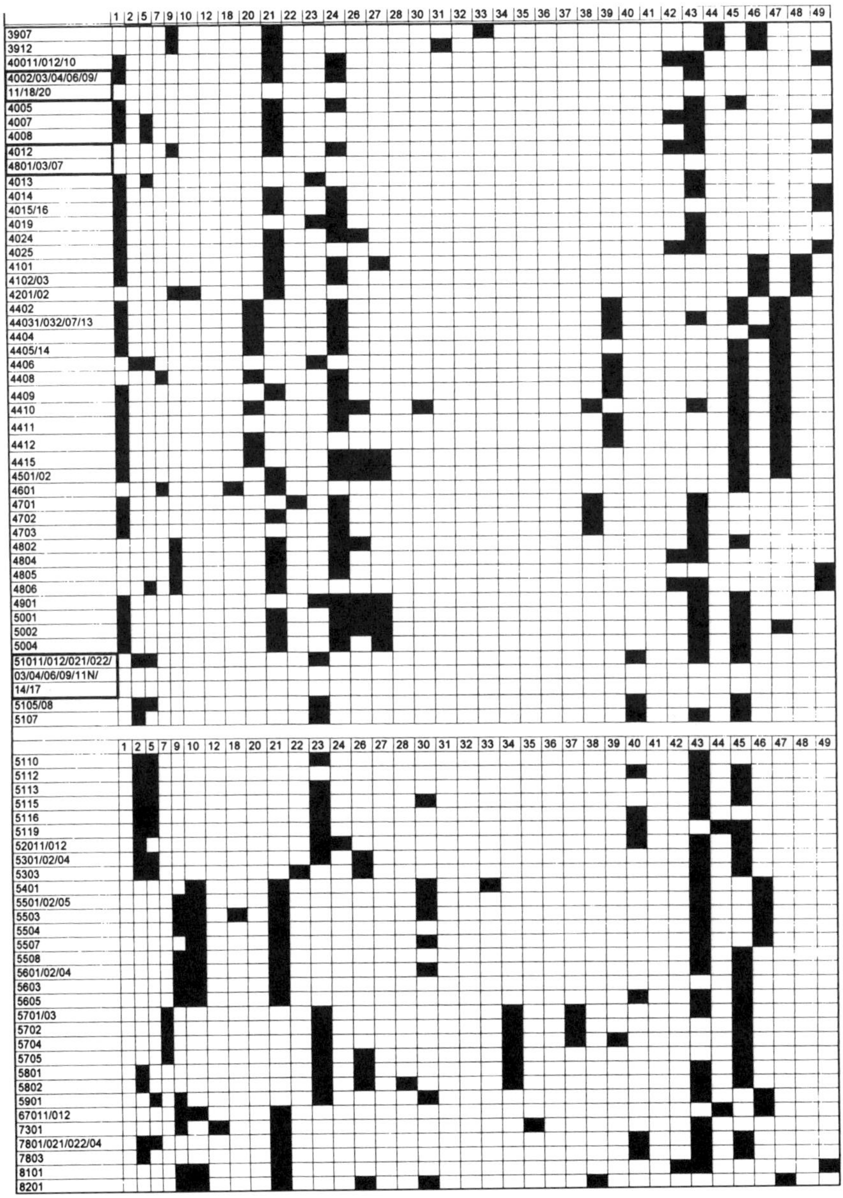

Table 10
HLA-DR SSOP Patterns

Probes	1	2	3	4	5	6	7	8	9	10	11	12	13	14	15	16	17	18	22	23	24	25	26	27	28
HLA-DR Alleles																									
0101/021/04/05	■																■						■		
01022	■																						■		
0103	■															■							■		
0106	■																						■	■	
03011/021/03/ 5/6/13				■		■							■										■		
03012/022/11/14/15						■							■										■		
0304/09				■		■																	■		
0307				■		■							■									■	■		
0308				■	■	■							■										■		
0310				■		■							■		■								■		
0312				■		■							■						■				■		
04011/012/13/ 16/21/26/33							■			■										■	■	■	■		
0402/14							■			■						■					■	■	■		
0404/08/19/23							■			■							■				■	■	■		
04051/052/10 28/29/30							■			■							■		■		■	■	■		
04031/032/06/07/ 20/27/32							■			■											■	■	■		
0409							■			■									■	■	■	■	■		
0411/17/24							■			■									■		■	■	■		
0412							■			■		■							■		■	■	■		
0415					■		■			■											■	■	■		
0418/25/31							■			■		■									■	■	■		
0422				■			■			■											■	■	■		
07011/03/04								■		■											■		■		
07012										■											■		■		
0801/032/ 06/10/12/16/17												■						■	■			■	■		
08021/022/041/ 042/043/07/09/11/ 13/15/19												■						■				■	■		
1415																									
0805/18																		■	■			■	■		
0808												■			■			■				■	■		
0814												■							■			■	■		
0820						■						■										■	■		
0821												■										■	■		
09012									■	■											■		■		
1001			■																				■		
11011/012/041/ 042/06/081/082/ 10/12/13/15/18/19/ 24/27/28/29/32					■	■																■	■		
11013					■	■																■	■		■
1102/03/11/14/21					■	■										■						■	■		
1105/30					■																	■	■		
1107				■	■	■																■	■		
1109					■	■							■									■	■		
1116/20					■	■							■			■						■	■		
1117					■	■															■	■	■		
1122					■		■															■	■		
1123/25					■	■						■										■	■		
1126/34					■	■											■					■	■		
1131/33/35 13071/072/11/14/25						■																■	■		
1201/021/022/ 032/06											■							■					■		■
1204					■						■							■					■		■
1205																		■					■		■
1301/02/16/20/ 28/29/31/35						■							■			■						■	■		
13031/032/33						■								■					■			■	■		

(cont.)

Table 10
HLA-DR SSOP Patterns *(cont.)*

	1	2	3	4	5	6	7	8	9	10	11	12	13	14	15	16	17	18	22	23	24	25	26	27	28
1304						■										■			■			■	■		
1305/06						■							■									■	■		
1433																									
1308/22/23/24/34						■										■						■	■		
1309						■							■									■	■	■	
1310						■							■	■								■	■		
1312/21/30						■													■			■	■		
1313						■						■							■			■	■		
1315/27						■							■			■							■		
1317																■		■				■	■		
1318						■						■	■									■	■		
1319						■										■							■		
1326						■							■										■		■
1332						■							■			■			■			■	■		
1401/07						■									■						■	■	■		
1402/06/29						■							■				■						■		
1403/12/27						■						■	■										■		
1404/28															■			■			■	■	■		
1405/08/14/23						■															■	■	■		
1409/17/30						■							■				■					■	■		
1410							■								■						■	■	■		
1411					■													■			■	■	■		
1413						■							■				■		■				■		
1416						■									■	■						■	■		
1418						■							■								■		■		
1419						■							■							■			■		
1420						■											■						■		
1421						■							■							■		■	■		
1422/25/32						■									■							■	■		
1424						■							■										■	■	
1426						■									■						■		■		
1431															■			■				■	■		
15011/012/021/		■																				■	■	■	■
022/04/05/06/																									
023/07/																									
16011/012/																									
021/022/03/05																									
1503/1607		■																					■	■	■
1508		■																				■		■	■
1604		■										■										■	■	■	■
1608		■											■									■	■	■	■

Table 11
HLA-DR3/11/6 Subtyping Group Patterns

Probes	1	2	3	4	5	6	7	8	9	10	11
HLA-DR Alleles											
03011/04/05/06/09	■									■	■
03012/11/14/15	■									■	■
03021/03/07											■
0308	■	■								■	■
0310	■		■							■	
0312/13	■									■	
11011/012/013/041/		■									
042/06/081/082/09/10/											
12/13/15/18/19/24/27/28/											
29/32											
1102/03/11/14/16/20/21		■			■						
1107		■								■	
1117		■				■					
1123/25		■							■		
1126/34		■					■				
1301/02/08/15/16/19/					■						■
20/22/23/24/28/29/34/35											
13031/032/33				■							
1304/31/32					■						
1305/06/071/09/11/14/25											■
1424/33											
03022											
13072/12/21/26/30											
1310				■							■
1313									■		
1318									■		■
1403/12/27											
0820											
1327	■				■						■
1401/07/26			■			■					
1402/06/09/17/20/29/30							■				■
1405/08/18						■					
1413							■				
1414/23						■					■
1416			■		■						
1419/21								■			■
1422/25/32			■								

Alleles not amplified: 1105, 1122, 1130, 1317, 1404, 1410, 1411, 1415, 1428, 1431. All 8's except 0820.

Table 12
Probes Used for HLA-A2 Subtyping

Probe	Sequence 5' ------------------------------3'	Wash Temp (°C)	Picomoles Used	Nucleotide Position
				Exon 2
A1 (1)	TCTTCACATCCGTGTC	50	20	23-38
A96 (8)	CTACACCTCCGTGTC	46	50	24-38
A152 (23)	ACGTGGACAACACGCAG	54	20	80-96
A2 (2)	GCGAGCCGGAGGAT	48	20	121-134
A262 (17)	ACACGGAAAGTGAAGG	48	20	190-205
A14 (14)	GACACGGAATGTGAAGGC	56	10	189-206
73I (3)	TCACAGATTGACCGA	45	20	211-225
A285 (12)	ACAGACTCACCGAGTG	50	20	213-228
				Exon 3
A5 (5)	TCACACCGTCCAGAGGAT	56	10	6-23
A4 (4)	CGTCCAGATGATGTATGG	54	10	12-29
A357 (22)	CCAGATGATGTTTGGC	48	20	15- 30
A11 (11)	CAGAGGATGTATGGCTG	52	10	16-32
A6 (6)	AGGATGTTTGGCTGCGA	60	5	19-35
A13 (13)	AGGATGTGTGGCTGCGA	58	5	19-35
A445 (26)	GCCCTGAACGAGGACCT	56	20	103-119
A478 (24)	CGGACAAGGCAGCT	52	20	137-150
A481 (25)	GACATGGCGGCTCAGAT	54	20	138-155
149T (7)	TGGGAGACGGCCCAT	50	20	169-183
A15 (15)	AT GAGGCGGAGCAGCA	52	20	182-197
A18 (18)	ATGTGGCGGAGCAGCA	56	10	182-197
A525 (27)	TGAGGCGGAGCAGTTG	54	20	183-198
156Q (9)	GAGCAGCAGAGAGCC	52	10	190-204
156W (16)	GAGCAGTGGAGAGCC	50	5	190-204
A10 (10)	GAGCAGTTGAGAGCCT	50	10	190-205
A355 (19)	GGCGAGTGCGTGGAGTGGC	68	5	214-232
				Exon 4
A236E (20)	AGGCCTGAAGGGGATGGA	60	20	154- 171
A835 (21)	CAGAGATAAACCTGCC	48	20	217- 232

Table 13
Probes Used for HLA-A 24[a] Group Subtyping

Probe	Sequence 5' ------------------------------3'	Wash Temp (°C)	Picomoles Used	Nucleotide Position
				Exon 2
A233 (11)	AGGAGGGGCCGGAGTA	54	10	161-176
56R (1)	GAGAGGCCTGAGTAT	46	20	163-177
62G (16)	GACGGGGAGACACGG	52	10	181-195
62EG (20)	GAGGAGACAGGGAAA	46	20	184-198
A275 (10)	AGGCCCAGTCACAGAC	56	20	203-218
A276 (5)	GGCCCACTCACAGACT	52	20	204-219
A290 (3)	CTGACCGAGTGGACCT	52	20	218-233
A26 (6)	TGACCGAGCGAACCTG	54	20	219-234
BL23 (12)	CGGATCGCGCTCCGCTAC	62	20	235-252
				Exon 3
95V (18)	CACACCGTCCAGAGG	48	20	7-21
A494 (13)	AGATCACCAAGCGCAAG	56	20	152-168
A495 (7)	GATCACCCAGCGCAA	48	10	153-167
A522 (14)	CCGTCGGGCGGAGCA	56	20	380-394
156Q (4)	GAGCAGCAGAGAGCC	52	10	190-204
156W (2)	GAGCAGTGGAGAGCC	50	5	190-204
A10 (19)	GAGCAGTTGAGAGCCT	50	10	190-205
A551 (21)	TGGAGGGCACGTGCGT	56	20	209-224
163R (17)	GAGGGCCGGTGCGTG	54	10	211-225
A564a (8)	CGTGGAGTGGCTCC	48	20	222-235
A564b (9)	CGTGGACGGGCTCC	50	25	222-235
				Intron 2
A24L(15)	GGGTCGAGGCCAG	46	30	229 – 241

[a]A24 group includes A*24, A*30, A*2301, and A*0102.

Table 14
Probes Used for HLA-A 29[a] Group Subtyping

Probe	Sequence 5' ---------------------------3'	Wash Temp (°C)	Picomoles Used	Nucleotide Position
				Exon 3
A369a (1)	TGGCTGCCACGTGG	48	20	27 - 40
A369b (3)	TGGCTGCGACGTGG	50	20	27 - 40
A564b (5)	CGTGGACGGGCTCC	50	25	222-235
A564a (6)	CGTGGAGTGGCTCC	50	25	222-235
A576 (2)	CCGCAGACACCTGGA	50	20	234-248
				Exon 4
A624 (4)	CCCCAAGACGCATAT	46	20	6 - 20

[a]A29 group includes A*24, A*33, and A*31012.

Table 15
Probes Used for HLA-A 26[a] Group Subtyping

Probe	Sequence 5' ------------------------3'	Wash Temp (°C)	Picomoles Used	Nucleotide Position
				Exon 2
A95 (1)	TCTACACTTCCGTGTC	46	50	23 - 38
A96 (2)	CTACACCTCCGTGTC	46	50	24 - 38
A97 (3)	TACACCTCCATGTCCC	50	30	25 - 40
A123a (4)	CGGGAAGCCCCGCTT	52	30	51 - 65
A123b (5)	CGGGGAGCCCCGCTT	54	20	51 - 65
62RN(24)	GACCGGAACACACGG	52	10	181-195
A262 (6)	ACACGGAAAGTGAAGG	48	20	190 - 205
A275 (17)	AGGCCCAGTCACAGAC	56	20	203-218
A276 (16)	GGCCCACTCACAGACT	52	20	204-219
73I(20)	TCACAGATTGACCGA	45	20	211-225
A285 (7)	ACAGACTCACCGAGTG	50	20	213 - 228
A290 (19)	CTGACCGAGTGGACCT	52	20	218-233
A292b (8)	GACCGAGCGAACCTG	50	20	220 - 234
A297 (9)	AGAGAACCTGGGGACC	52	20	225 - 241
				Exon 3
A405(23)	CGGGTACCGGCAGGACG	56	20	63-78
A408 (10)	GTACCAGCGGGACGCTT	56	20	66-82
A411 (11)	CCAGCAGAACGCTTA	46	20	69 - 83
A519 (14)	GGCCCGTGAGGCGGA	54	20	177-191
156W(21)	GAGCAGTGGAGAGCC	50	5	190-204
156Q (18)	GAGCAGCAGAGAGCC	52	10	190-204
A551 (15)	TGGAGGGCACGTGCGT	56	20	209-224
A553 (12)	GAGGGCCTGTGCGT	48	20	211 - 224
163R (13)	GAGGGCCGGTGCGTG	54	10	211-225
A355(22)	GGCGAGTGCGTGGAGTGGC	68	5	214-232

[a]A26 group includes A*11, A*25, A*26, A*34, A*4301, A*66, A*68, and A*6901.

Table 16
Probes Used for HLA-B7[a] Group Subtyping

Probe	Sequence 5' ----------------------------------3'	Wash Temp (°C)	Picomoles Used	Nucleotide Position
				Exon 2
11 (B96a)	CTACACCTCCGTGTC	48	50	23-37
12 (B96b)	CTACACCGCCGTGTC	50	20	23-37
28 (B98)	ACACCGCCGTGTCC	50	30	25-38
21 (B100)	ACCGCCATGTCCCG	48	20	27-40
23 (B252)	GGACCGGAACACACAG	52	20	179-194
31 (253)	GACCGGGAGACACAG	50	20	180-194
13 (BL24)	GGGAGACACAGATCTCCA	55	20	184-201
33 (B261)	CACACAGACCTTCAAGAC	54	20	187-204
1 (BL05)	ACACAGATCTTCAAGACC	55	7	189-206
3 (B263)	CACAGATCTACAAGACCA	56	20	190-207
2 (BL10)	GATCTACAAGGCCCAGGC	58	5	194-211
10 (B265a)	CAGATCTCCAAGACC	46	20	192-206
24 (B265b)	CAGATCTGCAAGACC	46	20	192-206
32 (B265c)	CAGATCTGCAAGGCC	50	20	192-206
22 (B287N)	AGACTTACCGAGAGAGCCTGC	66	10	214-234
16 (B288)	GACTGACCGAGAGAGCCT	58	20	215-232
8 (BL20)*	AGCGGAGCGCGGTGCGCA	64	20	232-249
9 (BL23)	CGGATCGCGCTCCGCTAC	62	20	234-251
				Exon 3
14 (BL27)	CTCACACTTGGCAGAGGA	56	10	4-21
19 (B353)	CTTGGCAGACGATGTA	48	20	10-25
4 (B355)	CTCCAGAGGATGTACGGC	58	20	12-29
7 (B358)	CAGAGCATGTACGGCT	50	20	15-30
20 (B360)	GAGGATGTCTGGCTGCG	58	20	17-33
15 (BL26)	CTGCGACCTGGGGCCCGA	65	20	29-46
17 (B405)	CGGGCATGACCAGT	46	20	62-75
5 (B408)	GCATAACCAGTACGCCTACG	62	20	65-84
25 (B412)	GACCAGTCCGCCTACG	54	20	69-84
30 (B413)	ACCAGTTCGCCTACG	48	20	70-84
18 (B530)	CGGAGCAGCGGAGA	48	20	187-200
6 (B531)	GGAGCAGGACAGAGCCT	56	20	188-204
29 (B532)	GAGCAGCTGAGAGCCT	52	20	189-204
26 (B535a)	CAGCGGAGAACCTAC	48	20	192-206
27 (B535b)	CAGCTGAGAACCTACC	50	20	192-207

[a]B7 group includes B*07, B*08, B*14, B*38, B*39, B*40 (4001, 407, 4016), B*41, B*42, B*45, B*48, B*50, B*4901, and B*6701.

*, Complimentary to coding sequence.

Table 17
Probes Used for HLA-B13[a] Group Subtyping

Probe	Sequence 5' ---------------------------------3'	Wash Temp (°C)	Picomoles Used	Nucleotide Position
				Exon 2
28 (B100)	ACCGCCATGTCCCG	48	20	27- 40
24 (BL01)	GAGGAAGGAGCCGCGGGC	64	10	128-145
11 (B240)	GCCGGAGCATTGGGA	50	20	167-181
18 (B252)	GGACCGGAACACACAG	52	20	179-194
27 (B265a)	CAGATCTCCAAGACC	46	20	192-206
26 (B265b)	CAGATCTGCAAGACC	46	20	192-206
25 (BL12)	ATCTGCAAGGCCAAGGCA	56	10	195-212
6 (B287N)	AGACTTACCGAGAGAGCCTGC	66	10	214-234
5 (B288)	GACTGACCGAGAGAGCCT	58	20	215-232
17 (B293)	ACCGAGAGGACCTGCG	54	20	220-235
8 (BL20)*	AGCGGAGCGCGGTGCGCA	64	20	232-249
9 (BL23)	CGGATCGCGCTCCGCTAC	62	20	234-251
10 (BL21)	CGGAACCTGCGCGGCTAC	62	20	234-251
				Exon 3
4 (B346)	TCTCACATCATCCAGAG	50	20	3-19
20 (B348)	TCACACCATCCAGAGG	49	25	5-20
3 (B353)	CTTGGCAGACGATGTA	48	20	10-25
7 (B355)	CTCCAGAGGATGTACGGC	58	20	12-29
30 (B357)	CCAGAATATGTATGGCTGC	56	20	14-19
13(B408)	GCATAACCAGTACGCCTACG	62	20	65-84
22 (B409)	TATGACCAGGACGCCT	55	20	66-81
12 (B411)	TGACCAGTACGCCTACG	56	20	68-84
15 (B411b)	CCACCAGCACGCCTA	52	20	68-82
19 (B412)	GACCAGTCCGCCTACG	54	20	69-84
29 (B445)	GCCCTGAAGGAGGACCT	56	20	102-118
2 (B499)	ACCCAGCTCAAGTGG	47	20	156-170
14 (B52Oa)	GCCCGTGAGGCGGA	50	20	177-190
16 (B520b)	GCCCGTGTGGCGGA	50	20	177-190
23 (B530)	CGGAGCAGCGGAGA	48	20	187-200
21 (B532)	GAGAACCTACCTGGA	52	20	189-204
1 (B553a)	GAGGGCCTGTGCGT	48	20	210-223

[a]B13 group includes B*13, B*18, B*27, B*37, B*40 (except 4001, 4007, 4012, 4016) B*47, B*57, B*1522, and B*8101.

Table 18
Probes Used for HLA-B15[a] Group Subtyping

Probe	Sequence 5' ------------------------------------3'	Wash Temp (°C)	Picomoles Used	Nucleotide Position
				Exon 2
19 (B252)	GGACCGGAACACACAG	52	20	180-195
11 (B255)	CCGGGAGATACAGATC	52	20	183-198
16 (BL24)	GGGAGACACAGATCTCCA	55	20	185-202
1 (BL05)	ACACAGATCTTCAAGACC	55	7	190-207
6 (B263)	CACAGATCTACAAGACCA	56	20	191-208
18 (B265a)	CAGATCTCCAAGACC	46	20	193-207
5 (BL18)	ACTGACCGAGTGAGCCTG	58	10	217-234
2 (BL23)	CGGATCGCGCTCCGCTAC	62	20	235-252
				Exon 3
17 (B353)	CTTGGCAGACGATGTA	48	20	11-26
14 (B355b)	CTCCAGACGATGTAC	46	20	13-27
15 (B355)	CTCCAGAGGATGTACGGC	58	20	13-30
8 (B358)	CAGAGCATGTACGGCT	50	20	16-31
20 (B360)	GAGGATGTCTGGCTGCG	58	20	18-34
3 (B361)	AGGATGTTTGGCTGC	48	20	19-33
10 (B362)	GGATGTAAGGCTGCG	50	20	20-34
13 (B373)	TGCGACCTGGGGC	46	20	31-43
7 (B408)	GCATAACCAGTACGCCTACG	62	20	66-85
12 (B455)	AGGACCTGCGCTCCTGGA	60	20	113-130
22 (B531b)	GGAGCAGTGGAGAGCCT	58	20	189-205
23 (B532)	GAGCAGCTGAGAGCCT	52	20	190-205
9 (B562)	TGCGTGGACGGGCTCC	56	20	220-235
47 (B566)	TGGAGTCGCTCCGC	48	20	224-237
21 (B575)	TCCGCAGACACCTGGAG	56	20	232-248

[a]B15 group includes certain B*15 alleles and B*4601.

Table 19
Probes Used for HLA-B15b[a] Group Subtyping

Probe	Sequence 5' -----------------------------------3'	Wash Temp (°C)	Picomoles Used	Nucleotide Position
				Exon 2
7 (BL09)	GAGTCCGAGAGAGGAGCC	57	3	123-140
15 (BL07)	GAGGATGGCGCCCCGGGC	64	30	129-146
11 (B252)	GGACCGGAACACACAG	52	20	180-195
16 (BL24)	GGGAGACACAGATCTCCA	55	20	185-202
2 (BL05)	ACACAGATCTTCAAGACC	55	7	190-207
17 (B265a)	CAGATCTCCAAGACC	46	20	193-207
13 (B265b)	CAGATCTGCAAGACC	46	20	193-207
8 (BL23)	CGGATCGCGCTCCGCTAC	62	20	235-252
19 (BL21)	CGGAACCTGCGCGGCTAC	62	20	235-252
				Exon 3
9 (B346)	TCTCACATCATCCAGAG	50	20	4-20
4 (BL27)	CTCACACTTGGCAGAGGA	56	10	5-22
18 (B355)	CTCCAGAGGATGTACGGC	58	20	13-30
5 (B361)	AGGATGTTTGGCTGC	48	20	19-33
3 (BL26)	CTGCGACCTGGGGCCCGA	65	20	30-47
21 (B405)	CGGGCATGACCAGT	46	20	63-76
22 (B411)	TGACCAGTACGCCTACG	56	20	69-85
12 (B412)	GACCAGTCCGCCTACG	54	20	70-85
6 (B427)	GACGGCAAAGATTACA	46	20	85-100
20 (B520a)	GCCCGTGAGGCGGA	50	20	178-191
10 (B520b)	GCCCGTGTGGCGGA	50	20	178-191
23 (B531b)	GGAGCAGTGGAGAGCCT	58	20	189-205
1 (B532)	GAGCAGCTGAGAGCCT	52	20	190-205
14 (B575)	TCCGCAGACACCTGGAG	56	20	233-249

[a]B15b group includes certain B*15 alleles and B*4012.

Table 20
Probes Used for HLA-B35[a] Group Subtyping

Probe	Sequence 5' ----------------------------------3'	Wash Temp (°C)	Picomoles Used	Nucleotide Position
				Exon 2
26 (B113)	GGCCCGTCCGCGG	50	20	40-52
13 (BL01)	GAGGAAGGAGCCGCGGGC	64	10	128-145
14 (B252)	GGACCGGAACACACAG	52	20	179-194
24(B253)	GACCGGGAGACACAG	50	20	180-194
7 (BL05)	ACACAGATCTTCAAGACC	55	7	189-206
9 (B265a)	CAGATCTCCAAGACC	46	20	192-206
2 (B287N)	AGACTTACCGAGAGAGCCTGC	66	10	214-234
1 (B288)	GACTGACCGAGAGAGCCT	58	20	215-232
				Exon 3
4 (B346)	TCTCACATCACTCCAGAG	50	20	3-19
31 (B349)	CACATCATCCAGAGCATGT	56	20	6-24
11 (B353)	CTTGGCAGACGATGTA	48	20	10-25
8 (B355)	CTCCAGAGGATGTACGGC	58	20	12-29
3 (BL28)	CCAGTGGATGTATGGCTG	56	20	14-31
15 (B358)	CAGAGCATGTACGGCT	50	20	15-30
29 (B358b)	CAGAGGATGTACGGCT	50	20	15-30
22 (B373)	TGCGACCTGGGGC	46	20	30-42
28 (B389)	ACCGGGCGCTTCCTCCG	54	20	46-62
27 (B403)	CGCGGGCATAACCAG	50	20	60-74
19 (B405)	CGGGCATGACCAGT	46	20	62-75
23 (B408)	GCATAACCAGTACGCCTACG	62	20	65-84
20 (B412)	GACCAGTCCGCCTACG	54	20	69-84
30 (B413)	ACCAGTTCGCCTACG	48	20	70-84
21 (B455)	AGGACCTGCGCTCCTGGA	60	20	112-129
16 (B520a)	GCCCGTGAGGCGGA	50	20	177-190
17 (B520b)	GCCCGTGTGGCGGA	50	20	177-190
18 (B530)	CGGAGCAGCGGAGA	48	20	187-200
10 (B531)	GGAGCAGGACAGAGCCT	56	20	188-204
32 (B532)	GAGCAGCTGAGAGCCT	52	20	189-204
25 (B551)	TGGAGGGCGAGTGCGTG	58	20	208-224
12 (B564)	CGTGGAGGGGCTCC	52	20	221-234
5 (B575)	TCCGCAGACACCTGGAG	56	20	232-248
6 (B575b)	TCCGCAGATACCTGGAG	56	20	232-248

[a]B35 group includes B*35, B*51, B*52, B*53, B*58, and B*78.

Table 21
Probes Used for HLA-B44[a] Subtyping

Probe	Sequence 5' -----3'	Wash Temp (°C)	Picomoles Used	Nucleotide Position
				Exon 2
10 (B186b)	CGACGCCGCGAGTCC	56	20	113-127
6 (B186a)	CGACGCCACGAGTCC	52	20	114-127
4 (BL07)	GAGGATGGCGCCCCGGGC	64	30	128-145
12 (BL20)*	AGCGGAGCGCGGTGCGCA	64	20	232-249
13 (B305)	TGCGCACCCCGCTC	50	20	232-245
5 (BL21)	CGGAACCTGCGCGGCTAC	62	20	234-251
3 (BL23)	CGGATCGCGCTCCGCTAC	62	20	234-251
				Exon 3
7 (BL30)	GGCATAACCAGTTAGCCT	54	25	64-81
11 (B409)	TATGACCAGGACGCCT	55	20	66-81
2 (B411)	TGACCAGTACGCCTACG	56	20	68-84
9 (B530)	CGGAGCAGCGGAGA	48	20	187-200
1 (B531)	GGAGCAGGACAGAGCCT	56	20	188-204
8 (B532)	GAGCAGCTGAGAGCCT	52	20	189-204

*, Complimentary to coding sequence.

Table 22
Probes Used for HLA-B54[a] Subtyping

Probe	Sequence 5' -----3'	Wash Temp (°C)	Picomoles Used	Nucleotide Position
				Exon 2
9 (B157)	GACGACACCCAGTTCGT	54	20	85-101
10 (B239)	GGCCGGCGTATTGG	50	20	167-180
1 (BL10)	GATCTACAAGGCCCAGGC	58	5	195-212
14 (B288)	GACTGACCGAGAGAGCCT	58	20	216-233
2 (BL18)	ACTGACCGAGTGAGCCTG	58	10	217-234
				Exon 3
4 (B353)	CTTGGCAGACGATGTA	48	20	11-26
5 (B355)	CTCCAGAGGATGTACGGC	58	20	13-30
15 (B372)	CTGCGACGTGGGGCC	54	20	30-44
11 (B373)	TGCGACCTGGGGC	46	20	31-43
3 (BL30)	GGCATAACCAGTTAGCCT	54	25	65-82
6 (B408)	GCATAACCAGTACGCCTACG	62	20	66-85
12 (B520a)	GCCCGTGAGGCGGA	50	20	178-191
13 (B520b)	GCCCGTGTGGCGGA	50	20	178-191
7 (B553a)	GAGGGCCTGTGCGT	48	20	211-224
8 (B553b)	GAGGGCACGTGCGT	48	20	211-224

[a]B54 group includes B*5401, B*55, B*56, and B*5901.

Table 23
HLA-A*02 SSOP Subtyping Group Patterns

Probes	1	2	3	4	5	6	7	8	9	10	11	12	13	14	15	16	17	18	19	20	21	22	23	24	25	26	27
HLA-A*02 Alleles																											
02011/012/	■				■					■	■	■					■										
30/31/32N/34																											
0202	■	■									■	■				■	■										
0203	■				■		■				■	■				■	■										
0204	■			■						■		■					■										
0205		■						■			■	■				■	■										
0206/28					■			■		■	■	■					■										
0207	■				■					■		■	■				■										
0208		■						■			■	■		■		■											
0209	■				■					■	■	■					■			■							
0210					■	■		■		■		■					■										
0211	■		■		■					■	■						■										
0212	■				■				■		■	■					■	■									
0213	■				■				■		■	■			■		■										
0214		■						■		■	■	■					■										
0215N	■				■					■		■	■				■				■						
0216	■				■					■	■	■					■		■								
02171/172	■									■		■					■					■					
0218	■				■					■		■	■				■							■			
0219	■				■				■		■	■					■	■							■		
0220	■				■					■	■	■		■													
0221					■			■		■	■	■					■						■				
0222	■				■						■	■				■	■										
0224	■				■					■	■	■					■									■	
0225	■				■		■			■	■	■					■										
0226	■				■					■	■	■					■										■
0227	■				■				■		■	■					■										
0229	■				■					■	■	■															
0233	■				■					■		■					■										
0235/013	■				■					■	■						■										

Table 24
HLA-A24 Subtyping Group Patterns

Probes	1	2	3	4	5	6	7	8	9	10	11	12	13	14	15	16	17	18	19	20	21
HLA-A* Alleles																					
0102					■	■			■		■		■				■				
2301/03					■		■		■		■	■							■	■	■
2302		■			■		■		■		■	■								■	■
2402101/09N/11N/				■	■				■		■	■	■							■	■
15/17																					
24031/032				■	■			■			■	■	■							■	■
2404				■	■	■			■		■		■							■	■
2405				■	■		■		■		■	■								■	■
2406		■			■				■		■	■	■							■	■
2407				■					■	■	■	■	■							■	■
2402102L				■	■				■		■	■	■		■					■	■
2408				■	■				■		■	■	■			■					■
2410				■	■			■			■	■	■				■			■	
2413					■				■		■	■	■						■	■	■
2414				■	■				■		■	■	■					■		■	■
2418					■			■			■	■	■						■	■	
2422					■			■			■	■	■							■	■
2424							■		■	■	■	■							■		■
3001	■		■				■	■		■									■		■
3002	■				■		■	■						■					■		■
3003					■		■	■			■			■					■		■
3004	■	■			■		■	■													■
3006	■	■			■		■	■													■
3007	■				■		■	■						■					■	■	■

Table 25
HLA-A29 Subtyping Group Patterns

Probe	1	2	3	4	5	6
HLA-A Alleles						
2416			■			■
2901	■			■		■
2902/04/			■	■		■
31012/02/03/04						
2903			■	■	■	
3301/04/05		■	■			■
3303			■	■		■

Table 26
HLA-A26 Subtyping Group Patterns

Probes	1	2	3	4	5	6	7	8	9	10	11	12	13	14	15	16	17	18	19	20	21	22	23	24
HLA-A* Alleles																								
11011/012/05		■			■								■				■	■	■				■	
1102		■		■									■				■	■	■				■	
1103		■			■								■	■			■	■	■				■	
1104		■			■										■		■	■	■				■	
2501		■			■								■			■					■			■
2502		■			■								■				■				■			■
2601/10/11N/12		■			■			■					■			■					■			■
2602		■			■			■			■		■			■					■			■
2603		■			■		■						■			■					■			■
2604		■			■			■				■				■					■			■
2605		■			■				■				■			■					■			■
2606		■			■		■			■			■			■					■			■
2607		■			■	■		■					■			■					■			
2608		■			■			■					■			■		■						■
2609		■			■			■							■	■					■			■
2613		■			■			■					■				■				■			■
3401		■			■	■									■		■		■		■			■
3402/6808		■			■										■		■		■				■	■
4301		■			■			■					■			■					■			
6601		■			■								■				■		■		■			■
6602		■			■												■		■		■	■		■
6603		■			■											■			■		■	■		■
68011		■			■										■		■		■		■		■	■
68012/11N/12	■				■										■		■		■		■		■	■
6802			■		■										■		■		■		■			■
68031/032	■				■										■	■			■		■		■	■
6804		■			■										■				▥	■	■		■	■
6805	■				■		■								■	■					■		■	■
6806/07	■				■										■		■		■		■			■
6809	■				■										■		■	■	■				■	■
6810/13/14		■			■										■		■		■		■		■	
6901		■			■										■		■		■					■
0206/10/14/21/28		■			■	■	■								■	■								
0205		■			■	■	■								■	■					■			
0208		■			■		■								■	■					■			

Table 27
HLA-B7 Subtyping Group Patterns

Probes / HLA-B Alleles	1	2	3	4	5	6	7	8	9	10	11	12	13	14	15	16	17	18	19	20	21	22	23	24	25	26	27	28	29	30	31	32	33
07021/022/023		■					■				■					■	■	■					■										
0710							■				■					■	■	■					■									■	
0703			■				■				■					■	■	■					■										
0704		■				■	■				■					■	■						■										
0705/06		■			■		■				■					■		■					■										
0707		■		■							■					■	■	■					■										
0708	■						■				■					■	■	■					■										
0709		■					■				■					■	■	■					■		■								
0711		■					■				■						■	■					■		■								
0712		■									■				■	■	■	■					■										
0713							■				■						■	■													■		
0716			■				■				■						■	■				■	■										
0801/08N	■				■	■	■									■					■		■										
0802	■				■	■	■	■													■		■										
0803	■				■	■	■		■												■		■										
0804					■	■	■			■						■					■		■										
0805					■	■	■									■					■		■										■
0806	■				■		■											■			■		■										
0807	■					■	■									■	■				■		■										
0809	■				■	■										■			■		■		■										
0810	■				■	■	■									■					■												
1401											■					■							■	■					■	■			
1402/062												■				■							■	■				■	■	■			
1403												■				■		■					■	■				■		■			
1404												■				■							■					■	■	■			
1405							■					■				■							■	■				■	■	■			
3801				■					■		■												■	■			■			■			
38021/022				■				■			■												■	■			■			■			
3803				■				■		■	■		■			■											■			■	■		
39011/013				■							■					■							■	■			■			■			
39021/022				■						■	■		■			■											■			■	■		
3903							■				■					■							■	■			■			■			
3904				■												■					■		■	■			■			■			
3905				■							■											■	■	■			■			■			
39061/062											■					■			■				■	■			■			■			
3907				■							■						■					■	■	■	■		■						
3908				■						■	■		■					■				■				■				■	■		
3909											■					■				■			■	■			■			■			
3910			■	■							■					■							■				■			■			
3911				■							■							■				■	■	■		■				■			
3912				■												■							■	■			■	■		■			
3913				■						■	■		■									■					■			■	■		
3914					■		■				■					■							■	■			■						
3915				■							■					■	■						■	■			■			■			
3916			■								■					■							■				■			■			
3917			■	■							■					■							■				■						
40011/012				■	■					■			■								■	■							■		■		
4007	■			■	■																■	■							■		■		
4016					■		■			■			■					■			■	■									■		
4101					■	■				■			■	■							■	■									■		
4102					■	■	■			■			■								■	■									■		
4103				■	■	■				■			■								■	■									■		
4801/04					■		■			■	■		■									■							■		■		
4802										■	■		■		■		■					■			■				■		■		
4803				■	■					■	■		■									■							■		■		
4805					■		■			■	■		■					■				■									■		
4806	■				■		■				■											■	■						■				
4807							■			■	■		■									■							■	■	■		
4201		■			■	■	■				■					■							■										
4202		■			■	■	■									■							■										
4501						■				■			■	■	■						■	■									■		
4502						■				■			■	■	■						■	■								■	■		
4901									■	■			■	■	■						■								■		■		
5001/02										■			■	■	■						■	■							■		■		
5004										■			■	■							■	■							■		■		
67011/012		■		■							■					■							■				■			■			

Table 28
HLA-B13 Subtyping Group Patterns

Probes	1	2	3	4	5	6	7	8	9	10	11	12	13	14	15	16	17	18	19	20	21	22	23	24	25	26	27	28	29	30
HLA-B Alleles																														
1301		■		■				■								■					■						■	■		
1302		■	■					■								■					■						■	■		
1303	■		■					■								■					■						■	■		
1304	■		■					■								■			■		■						■	■		
1801						■	■			■						■		■	■		■						■			
1802						■				■						■		■	■		■						■			■
1803					■		■			■						■		■	■		■						■			
1804						■	■			■						■		■	■		■						■	■		
1805						■	■			■						■		■	■		■						■		■	
1806							■			■						■		■	■		■						■			
1807						■	■			■						■		■	■		■									
2701								■								■					■				■					■
2702									■							■					■				■					■
2703											■					■	■				■				■					■
2704					■									■							■				■					■
27052/053/13																■	■				■				■					■
2706					■							■		■							■				■					■
2707													■			■	■				■				■					
2708					■					■						■					■				■					■
2709															■	■	■				■				■					
2710														■			■				■				■					■
2711					■								■			■					■				■					
2712					■					■						■					■					■				■
2714			■													■	■				■				■					
2716																■	■				■					■				■
3701																■	■			■							■			
3702																■	■				■						■			■
4002						■				■			■			■					■			■			■			
4003						■				■						■			■		■			■			■			
4004				■		■				■			■			■					■			■			■			
4005	■					■				■			■	■							■			■			■			
4006			■			■				■			■			■					■			■			■			
4008						■				■			■			■		■			■			■						
4009/24						■				■		■				■					■			■			■			
4010/14						■	■			■			■			■					■			■			■	■		
4011						■	■			■			■			■					■			■			■			
4013									■				■			■					■			■						
4015						■				■			■	■									■	■			■	■		
4018						■				■		■				■					■			■			■			
4019									■				■			■					■			■			■			
4020						■	■			■						■			■		■			■			■			
4025						■	■			■			■			■		■			■			■			■	■		
4701																■	■				■			■			■	■		
4702						■				■						■					■			■			■	■		
4703						■										■					■			■			■	■		
5701	■								■							■			■		■							■		
5702	■								■				■			■							■					■		
5703	■								■				■			■					■							■		
5704	■								■							■						■	■					■		
5705	■			■					■							■							■					■		
1522	■					■	■			■				■				■	■											
8101					■					■			■			■		■			■									
4402	■			■				■								■						■		■			■	■		
44031/032/07/13	■			■				■								■					■	■		■			■	■		
4404				■				■								■						■	■	■			■	■		
4405	■			■				■				■				■								■			■	■		
4406	■			■					■							■		■				■						■		
4408	■			■				■								■						■					■	■		
4409	■			■						■						■						■		■			■	■		
4410	■			■				■								■					■			■			■	■		
4411	■			■												■						■		■			■	■		
4412	■			■				■								■		■				■		■			■	■		
4414	■			■				■								■											■	■		
4415	■					■		■								■								■			■	■		

Table 29
HLA-B15a Subtyping Group Patterns

Probes HLA-B Alleles	1	2	3	4	5	6	7	8	9	10	11	12	13	14	15	16	17	18	19	20	21	22
1501101/01102N/012/43/ 46/50/57															■	■		■				■
1504																■	■	■				■
1507/45								■								■		■				■
1508	■														■				■			■
1511						■									■				■			■
1512/19									■						■	■		■				■
1514				■											■	■		■				■
1515															■			■	■			■
1524		■													■	■		■				■
1526N										■						■		■				■
1527			■													■		■				■
1528											■				■			■				■
1530							■								■	■		■				■
1532																■		■		■		■
1533												■			■	■		■				■
1534													■		■	■		■				■
1535														■		■		■				■
1538															■	■		■			■	■
1547/49															■	■		■				
1548							■								■	■		■				
1551															■				■			
4601					■										■							■

HLA-B15b Subtyping Group Patterns

Probes HLA-B Alleles	1	2	3	4	5	6	7	8	9	10	11	12	13	14	15	16	17	18	19	20	21	22
1502	■								■		■	■			■		■		■	■		
1503	■						■					■				■	■	■	■	■	■	
1505	■									■		■			■	■	■	■	■		■	
1506	■				■							■			■	■	■		■	■		
1509	■					■	■				■		■					■	■	■		
1510	■						■				■		■					■	■	■	■	■
1513	■							■	■		■	■			■		■			■		
1516	■			■				■				■			■					■	■	
1517	■							■							■			■		■		
1518	■						■				■	■	■					■	■	■	■	
1520	■		■						■	■		■			■	■	■		■		■	
1521/44	■								■		■	■	■		■				■	■		
1523	■						■	■			■	■	■					■		■	■	
1525	■								■			■			■	■	■		■	■		
1529	■	■					■				■	■						■	■	■	■	
1531	■									■	■	■			■		■	■	■			
1536	■								■			■			■	■	■			■		
1537	■						■				■		■	■				■	■	■	■	■
1539/40	■											■			■	■	■	■	■	■	■	
1542	■														■	■	■		■	■		
1547/49							■					■				■	■	■	■	■	■	
1548										■					■	■	■	■	■			
1551							■			■	■	■	■					■	■		■	
4012	■						■			■						■	■	■	■			

Table 30
HLA-B35 Subtyping Group Patterns

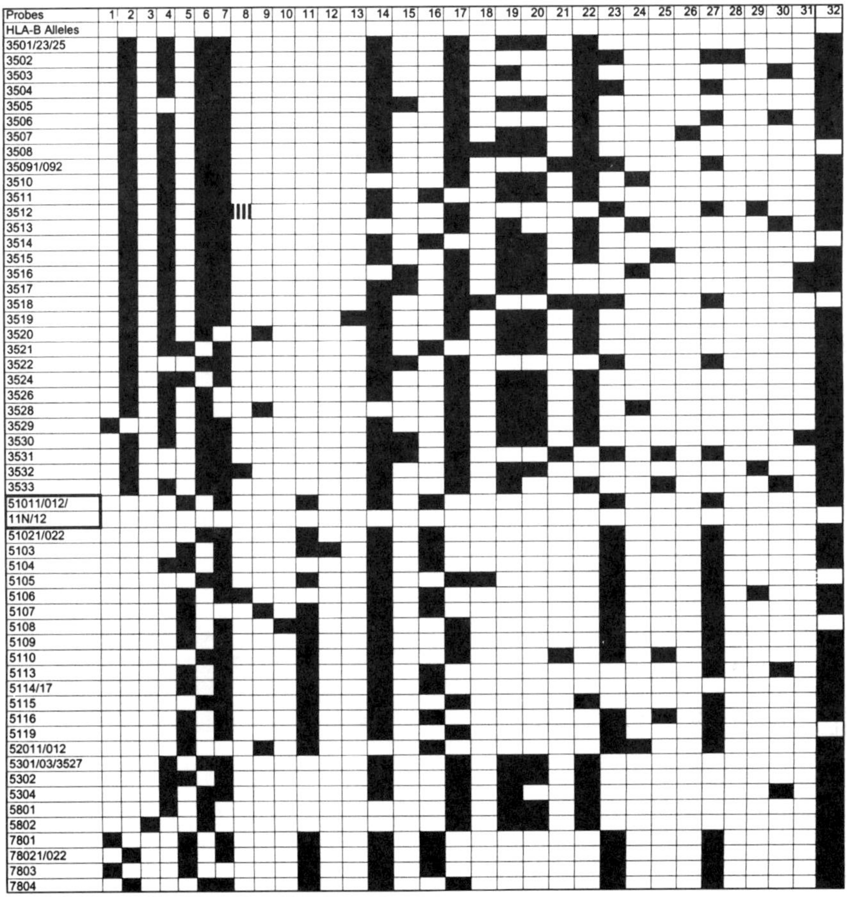

Vertical bars, probe positive reactions, but unexpected from sequence.

Table 31
HLA-B44 Subtyping Group Patterns

Probes	1	2	3	4	5	6	7	8	9	10	11	12	13
HLA-B Alleles													
4402/12	■					■					■	■	
44031/032/13						■		■			■	■	
4404						■			■		■	■	
4405	■	■				■						■	
4406	■		■							■	■		
4407								■		■	■	■	
4408	■			■						■	■	■	
4409	■				■	■					■		
4410						■	■	■				■	
4411	■					■					■		■
4414	■					■						■	

Table 32
HLA-B54 Subtyping Group Patterns

Probes	1	2	3	4	5	6	7	8	9	10	11	12	13	14	15
HLA-B Alleles															
5401	■		■	■				■			■		■	■	
5507															
5501	■		■	■				■			■	■		■	
5502	■		■	■				■			■		■	■	
5503	■	■	■	■				■			■	■			
5504	■					■		■					■	■	■
5505	■		■	■				■		■	■	■		■	
5508	■				■	■	■						■	■	■
5601	■		■	■			■				■		■	■	
5602	■		■		■		■				■		■	■	
5603	■				■		■		■			■		■	■
5604	■		■		■		■						■	■	■
5605	■			■		■	■					■		■	■
5901			■	■				■			■		■		

6

HLA-DPA1 and -DPB1 Typing Using the PCR and Nonradioactive Sequence-Specific Oligonucleotide Probes

Lori L. Steiner, Priscilla V. Moonsamy, Teodorica L. Bugawan, and Ann B. Begovich

1. Introduction

The class II molecules of the human major histocompatibility complex are cell surface glycoproteins encoded by a series of genes located on the short arm of chromosome 6 in the HLA-D region. These molecules consist of an α and β chain that associate as heterodimers on the surface of antigen presenting cells. They serve as receptors for processed peptides derived primarily from membrane and extracellular proteins that they present to $CD4^+$ lymphocytes.

The HLA-D region contains many genes that encode polypeptides that play a central role in the regulation of the immune process. Within the HLA-D region, there are three subregions, HLA-DR, -DQ, and -DP, which contain the classical class II genes. A great deal is known about the structure and function of the HLA-DR and -DQ molecules, but until recently, relatively little was known about the DP molecule and the genes encoding it. This is most likely due to the low levels of cell surface expression of the DP molecule, making it difficult to generate DP-specific serologic

From: *Methods in Molecular Biology, vol. 210: MHC Protocols*
Edited by: S. H. Powis and Robert W. Vaughan © Humana Press Inc., Totowa, NJ

reagents. In addition, because of this low level of expression, the DP molecule also appears to elicit a weak response in the primary mixed lymphocyte reaction (MLR) *(1)*. Consequently, the two techniques that were invaluable in the initial characterization of the variability in the DR and DQ molecules proved ineffective for DP.

The development of the polymerase chain reaction (PCR) *(2)* has revolutionized the field of molecular biology and, when used with other techniques, has permitted a detailed characterization of multiple genes and gene families in a relatively short period of time. One such area in which this technique has proved extremely useful is the characterization of the genes encoding the HLA molecules, including DPA1 and DPB1. Prior to the development of this technology, the standard method for DP typing was the primed lymphocyte typing (PLT) assay *(3–5)*. This cellular assay, which was time-consuming, difficult to perform, and relied on specifically primed T-cells, detected only six different DP specificities, DPw1–DPw6. Other cellular *(6)*, biochemical *(7)*, and restriction fragment length polymorphism (RFLP) *(8)* analyses suggested that these six specificities were an underestimate of the actual degree of polymorphism within DP.

Using the PCR to amplify genomic DNA and cDNA and a variety of different techniques to characterize the resulting PCR product, we now know that both the DPA1 and DPB1 molecules are highly variable. To date, the nucleotide sequences of 77 DPB1 and 11 DPA1 alleles have been reported *(9,10)*. At the DPB1 locus, 72 of these 77 alleles encode unique amino acid sequences, while the remaining five encode silent nucleotide changes. Comparison of these DPB1 sequences reveals an unusual pattern of polymorphism; DPB1 variation is almost exclusively localized to 18 amino acid residues within six regions of variability in the first extracellular domain of the protein (which is encoded by the second exon of the DPB1 gene). In addition, the majority of the nucleotide substitutions observed in the second exon are nonsynonymous amino acid replacement changes. Within each region of variability, which can range from one to five amino acids in length, there are: (i) a limited number of polymorphic residues (n = 2–4) at each amino acid position; and (ii) between three and six common polymorphic sequence motifs, few of which are allele-specific. Instead, the shuffling of these limited numbers of sequence motifs in the six regions of variability results in the formation of the various alleles. This shuffling of sequence motifs, which leads to a patchwork pattern of polymorphism, is characteristic of the DPB1 locus.

Of the 11 DPA1 alleles, eight encode unique amino acid sequences, while the remaining three contain silent nucleotide substitutions. In addition to being less diverse, the second exon of the DPA1 locus also contains more silent (synonymous) nucleotide changes than found in DPB1; only nine amino acid positions are variable in the first domain of DPA1 compared to 18 in the first domain of DPB1. At each of these nine residues only two amino acids have been observed.

A variety of different populations have now been typed for the DPA1 and DPB1 loci ***(11–13)***. The results show that in most populations, there is one predominant DPA1 and DPB1 allele; the identity of this common allele is dependent on the ethnic origin of the population. The DP molecule has also been shown to play a role in susceptibility to certain autoimmune disorders including pauciarticular juvenile rheumatoid arthritis ***(14–17)***, type I diabetes ***(18,19)***, and chronic beryllium disease, an environmentally-induced lung disorder ***(20,21)***. Together, these observations suggest that the DP molecule may be more important functionally than originally thought. Consequently, methods for DPA1 and DPB1 typing using the PCR and nonradioactive sequence-specific oligonucleotide probes (SSOPs) have been developed and are described in this chapter.

2. Materials

2.1. Preparation of Genomic DNA

In order to obtain the best PCR amplification results, one should start with a pure sample of genomic DNA. There are many kits and methods available for purifying DNA, and we recommend any one of the following: PureGene DNA Isolation Kit (Gentra Systems, Minneapolis, MN, USA), QIAamp® Blood Kit (Qiagen, Valencia, CA, USA), or standard phenol-chloroform extraction of genomic DNA *(22)*.

2.2. Control DNAs

To ensure that each probe has the correct specificity and that the assay is performed and interpreted correctly, a no-DNA control as well as a positive control DNA for each probe must be included in every assay. (A positive control for one probe in a region serves as a negative control for another probe within the same region.) Either genomic DNA or DNA from cloned PCR product can be used. In our laboratory, we use DNA isolated from the following B lymphoblastoid cell lines.

1. DPA1-Typing:

Cell Line	DPA1 Type
LBUF	DPA1*02011
CB6B	DPA1*02021
AMAI	DPA1*0301
T7526	DPA1*0401
SK*	DPA1*0104

*We do not have an available cell line with the DPA1*0104 allele; consequently we have cloned and purified the second exon of DPA1*0104 from a DNA sample, SK, carrying this allele to use as a source of control DNA.

2. DPB1-Typing:

Cell Line	DPB1 Type
LKT3	DPB1*0501
LBUF	DPB1*1701
TER81	DPB1*0101,*1301
JY	DPB1*0201,*0401
BIN40	DPB1*0301,*0601
PLH	DPB1*1501

CRK	DPB1*01011,*11011
AH696*	DPB1*11012
NG78*	DPB1*3201
SE53*	DPB1*3801
T93*	DPB1*4101
C23*	DPB1*6001
C53*	DPB1*6101N

*We do not have available cell lines with the DPB1*11012, *3201, *3801, *4101, *6001, and *6101N alleles. Consequently, we have cloned and purified the DPB1 second exons from DNA samples carrying these alleles to use as sources of control DNAs.

2.3. DP Amplifications

1. Micropipets (designated clean) and aerosol-resistant micropipet tips for PCR setup (Rainin, Emeryville, CA, USA). These pipets should be used exclusively for PCR setup and should not come into contact with PCR product or any reagents exposed to PCR product.
2. 2X PCR premixture: 100 m*M* KCl, 20 m*M* Tris-HCl, pH 8.3, 30% glycerol, 0.375 m*M* dTTP, 0.375 m*M* dGTP, 0.375 m*M* dCTP, 0.375 m*M* dATP, 3 m*M* $MgCl_2$, and 0.1 U/µL *Taq* DNA polymerase. Fifty microliters of this 2X premixture will be used for each 100-µL amplification reaction. This premixture should be prepared in a designated clean area. It can be made in bulk and stored at 2°–8°C until use.
3. DPA1 locus-specific primers DPA1-F: 5'-ACATTTTGTCG TGTTTTTCTCTA-3' and DPA1-R: 5'-GAAGGTCAACC CGATGTC-3' (Rozemuller et al., in preparation), each at 50 µ*M*. These primers are intronic and flank exon 2.
4. DPB1 locus-specific primers UG19: 5'-GCTGCAGGAGA GTGGCGCCTCCGCTCAT-3' and UG21: 5'-CGGATCCGG CCCAAAGCCCTCACTC-3' *(23)*, each at 50 µ*M*. These primers are intronic and flank exon 2.
5. DPA1 group-specific primer: *see* **Table 1**.
6. DPB1 group-specific primers: *see* **Table 1**.
7. Thermocycler: GeneAmp® PCR System 9600 or GeneAmp PCR System 2400 (Perkin Elmer, Norwalk, CT, USA).

Table 1
DP Group-Specific Primers

Primer name	Locus	Sequence variant[a]	Primer sequence	Alleles amplified[b]	Amplification profile
AB139[c]	DPA1	A@11(F)	GGATCCATGTGTCAACTTATGCCGC	0103, 02013	95°C-20 sec/60°C -1 min/72°C -20 sec
PM012[d]	DPB1	L@11(F)	AGAGAATTACGTGTACCAGTT	2101, 3601	95°C-15 sec/65°C -1 min/72°C -15 sec
PM023[e]	DPB1	A@55(R)	GTCCTTCTGGCTGTTCCAGTACTCCGCAG	01011, 0401, 26012, 5501	95°C-30 sec/65°C -1 min/72°C -30 sec
PM024[e]	DPB1	D@55(R)	GTCCTTCTGGCTGTTCCAGTACTCCTCAT	02012, 0402, 4801, 4901, 5101	95°C-15 sec/70°C -1 min/72°C -15 sec
AB1002[d]	DPB1	E@56(F)	GAGCTGGGGCGGCCTGATGA	02012, 0402	95°C-30 sec/65°C -1 min/72°C -30 sec
PM026[e]	DPB1	L@65(R)	TGCCCGCTTCTCCTCCAGGAG	0301, 1401, 2501, 4501,	95°C-15 sec/70°C -1 min/72°C -15 sec
PM016[e]	DPB1	V@76(R)	CTCGTAGTTGTGTCTGCATAC	01011, 0301, 0801, 0901, 1001, 1401, 2501, 2901, 3501, 4501, 5001	95°C-15 sec/65°C -1 min/72°C -15 sec

[a]Indicates amino acid distinguished at the 3' end of the primer and the location of the amino acid in the protein alignment. (F) indicates forward primer; (R) indicates reverse primer.

[b]Alleles listed are those present in common ambiguous genotypes that need to be resolved; it is not a complete list of all alleles that will be amplified by each primer.

[c]Used in combination with the 3' primer DPA1-R.

[d]Used in combination with the 3' primer UG21.

[e]Used in combination with the 5' primer UG19.

2.4. Gel Electrophoresis

1. 10X TB: 0.89 *M* Tris, 0.89 *M* boric acid, 0.025 *M* Na_2EDTA. To make the working 1X TB buffer, dilute 10X TB 1:10 in sterile distilled water.
2. 3% NuSieve® (FMC, Rockland, ME, USA), 1% agarose gel in 1X TB.
3. 100 mg/mL Ethidium bromide (EtBr) (Sigma, St. Louis, MO, USA).
4. Microwave oven.
5. Gel loading dye: 0.25% bromophenol blue, 0.25% xylene cyanol, 30% glycerol.
6. Molecular weight marker: φX174 DNA-*Hae*III digest (New England Biolabs, Beverly, MA, USA).
7. Electrophoresis gel box and power supply, such as the Minisub™ DNA Cell or Wide Minisub™ Cell and PowerPac 300 (Bio-Rad, Hercules, CA, USA).
8. Designated post-PCR micropipets and aerosol-resistant micropipet tips for all post-PCR work (Rainin). These pipets should be used for any reagent that contains PCR product or comes into contact with PCR product; they should never be used in the designated clean area or for PCR setup.

2.5. Dot Blotting

1. Denaturation solution: 0.4 N NaOH, 25 mM EDTA, 0.01% Orange II dye (Fluka, St. Louis, MO, USA).
2. Biodyne® B nylon membrane (Pall BioSupport Division, Port Washington, NY, USA) cut to fit the dot blot apparatus. Laboratories may wish to invest in silk-screening dot position numbers on the membrane to aid in interpretation. We recommend Palmer Display (San Leandro, CA, USA).
3. Dot blotting apparatus: for automated dot blotting, use the Hydra-96 Microdispenser (Robbins Scientific, Sunnyvale, CA, USA). For manual dot blotting, use either the Convertible™ Filtration Manifold System (BRL Life Technologies, Rockville, MD, USA) or the Bio-Dot™ apparatus (Bio-Rad).
4. Vacuum source.
5. UV Stratalinker® (Stratagene, La Jolla, CA, USA).
6. A pair of forceps for handling membranes.
7. A multichannel pipettor designated for post-PCR work.

2.6. Hybridization

1. 20X SSPE: 0.11 *N* NaOH, 3.6 *M* NaCl, 0.2 *M* NaH_2PO_4, 0.02 *M* EDTA.
2. 20% Sodium dodecyl sulfate (SDS).
3. 17 Horseradish peroxidase (HRP)-labeled DPA1 SSOPs (**Table 2**) and 35 HRP-labeled DPB1 SSOPs (**Table 3**). Oligonucleotide probes can be ordered commercially with HRP covalently linked to the 5' end. Suggested vendors are Tri-Link (La Jolla, CA, USA), and CyberSyn (Lenni, PA, USA). Probes should be diluted in a solution of 0.5 *M* NaCl, 50 m*M* Na_3PO_4, pH 7.5, to a final concentration of 2 μ*M* and stored at 4°C (do not freeze).
4. Seal-A-Meal® bags, 8 x 6 in. (Dazey, New Century, KS, USA).
5. Impulse sealer (American International Electric, Santa Fe Springs, CA, USA).
6. Glass bowls, such as Pyrex® or Kimax® crystallizing dishes, size 150 x 75 mm or 170 x 90 mm, watch glasses to cover glass bowls, and vinyl-coated lead weights (VWR, USA).
7. Shaking water baths with plastic bubble covers and temperature control, such as the Hot Shaker (Bellco, Vineland, NJ, USA).
8. A submersible thermometer for each water bath, used to monitor the water bath temperature independently of the machine's own internal temperature controls.
9. Dulbecco's phosphate buffer saline (PBS) (2.68 m*M* KCl, 137 m*M* NaCl, 1.47 m*M* KH_2PO_4, 8 m*M* Na_2HPO_4, pH 7.4.)
10. Hybridization and stringent wash solutions are listed in **Tables 2** and **3**.

    Example of how to make 500 mL of 1X SSPE/0.5% SDS hybridization solution: Add 25 mL of 20X SSPE to 462.5 mL water. Mix well, add 12.5 mL of 20% SDS, and mix again. Do not add SDS solution directly to SSPE without adding water first or SDS will precipitate out. Once the solution is made, SDS can precipitate out if the room temperature drops below 20°C. To resuspend, heat solution in a 50°C water bath and mix well.

Table 2
DPA1 SSOPs: Sequences and Hybridization Conditions

Probe name	Sequence variant[a]	HRP-SSOP sequence 5' to 3'[b]	Hybridization[c]	Wash[d]
PVM1	8-YAAF	CTTATGCCGCGTTTGTAC	5x	1x
PVM3	8-YAMF	CTTATGCCATGTTTGTAC	1x	1x
PVM27	13-VQTH1	GTACAGACGCATAGA	1x	0.4x
PVM28	12-VQTH2	TTTGTACAGACCCATAGA	1x	0.4x (15min)
PVM5	28-EDEQ	AGATGAGCAGTTCTATGT	1x	1x (15min)
PVM22	25-DDEM1	ATTTGATGACGATGAGAT	5x	0.4x
PVM30	25-DDEM2	TCTCATCTTCATCAAAT	3x	1x
AB135	36-DKK1	CTGGATAAAAAGGAGAC	5x	1x
AB130	36-DKK2	TCCTTCTTGTCCAGAT	3x	1x
PVM31	36-DKK3	GTCTCCTTCTTATCCAG	1x	1x
PVM24	41-TVWHLE	ACCGTCTGGCATCTGAG	1x	0.1x (15 min)
PVM6	47-FGQA	AGTTTGGCCAAGCCTTTT	2x	0.2x
PVM9	63-AISN	TTGCTATATCGAACAACA	5x	1x (15 min)
AB138	63-AILN	TGTTGTTCAATATAGCAA	5x	1x
PVM11	70-IAIQ	TTGAATATCGCTATCCAG	5x	1x
LS004	81-QATN	CAGGCCACCAACGGTACG	5x	0.4x (15 min)
LS005	81-QAAN	CGTACCATTGGCGGCCTG	5x	0.4x (15 min)

[a]Indicates the polymorphic sequence motif detected by the probe and the 5' residue at which the probe starts.

[b]Probes are labeled at the 5' end with HRP.

[c]All hybridization solutions contain 0.5% SDS in addition to SSPE. Table indicates the SSPE concentration used for each probe (20X SSPE: 0.11 NaOH, 3.6 M NaCl, 0.2 M NaH_2PO_4, 0.02 M EDTA). All hybridizations are done in a 42°C water bath for 30 min.

[d]All wash solutions contain 0.1% SDS in addition to SSPE. Table indicates the SSPE concentration used for each probe. All washes are done in a 42°C water bath for 12 min unless otherwise indicated.

Table 3
DPB1 SSOPs: Sequences and Hybridization Conditions

Probe name	Sequence variant[a]	HRP-SSOP sequence 5' to 3' [b]	Hybridization (°C)[c]	Wash (°C)[d]
DPB96	5-LFQG	GAATTACCTTTTCCGGGGACG	4X/50°	1X/50°
AB127	6-VYQL	CGTAACTGGTACACGTAA	4X/50°	1X/50°
DPB4	6-VHQL	TTACGTGCACCAGTTACG	4X/50°	1X/50°
DPB7	7-VYQG	CGTCCCTGGTACACGTA	4X/50°	1X/50°
AB117	33-EEFARF	AGGAGTTCGCGCGCTT	5X/55°	0.1X/50°
DPB11	33-EEFVRF	AGCGCACGAACTCCT	4X/50°	1X/50°
DPB12	33-EELVRF	AGGAGCTCGTGCGCTT	4X/50°	1X/50°
DPB18	31-QEYARF	AACCGGCAGGAGTACG	4X/50°	1X/50°
DPB8	32-EEYARF	GGGAGGAGTACGCG	4X/50°	1X/50°
DPB19	54-AAE	CCTGCTGCGGAGTACT	4X/50°	1X/50°
DB101	53-DEE	CCAGTACTCCTCATCAGGC	1X/50°	0.1X/42° (12 min)
DPB27	54-EAE	CCTGAGGCGGAGTACT	4X/50°	1X/50°
DPB92	55-DED	GTTCCAGTAGTCCTCATC	4X/50°	1X/50°
AB112	54-DEV	CCTGATGAGGTGTACTG	3X/50°	0.2X/42° (12 min)
DB34	64-ILEEK	GACATCCTGGAGGAGAAGC	2X/55°	0.1X/42°
DPB127	64-ILEEE	CTCCTCCTCCAGGATGTC	4X/50°	1X/50°
DPB104	64-LLEEK	GACCTCCTGGGGGAGAAGC	4X/50°	1X/50°
DB62	64-LLEEE	GACCTCCTGGAGGAGGAG	3X/50°	0.1X/42°
DB63	64-LLEER	GACCTCCTGGAGGAGAGG	2X/55°	0.1X/42° (12 min)
PVM32	62-FLEEE	CCTCCAGGAAGTCCTTCT	1X/42°	0.2X/42° (12 min)
DPB109	64-LL*EK	ACCTCCTGTAGGAGAAG	4X/42°	1X/42°
DPB110	63-NLEEK	AAGGACAACCTGGAG	4X/42°	1X/42°

(cont.)

Table 3 *(cont.)*
DPB1 SSOPs: Sequences and Hybridization Conditions

Probe name	Sequence variant[a]	HRP-SSOP sequence 5' to 3'[b]	Hybridization (°C)[c]	Wash (°C)[d]
AB96	73-M	TGTCTGCACATCCTGTCCG	1X/42°	0.2X/50°
AB97	73-V	TGTCTGCATACCCTGTCCG	2X/42°	0.2X/50°
AB98	73-I	CGGACAGGATATGCAGACA	2X/42°	0.4X/50°
DPB65	82-GGPM	GGGCCCGCCCAGCTC	4X/50°	1X/50°
AB123	81-VGPM	CGAGCTGGTCGGGCCCA	3X/55°	0.1X/55° (10 min)
DB77	83-DEAV	CTGGACGAGGCCGTG	3X/50°	0.1X/42° (12 min)
PVM16	14-ECYPFNG	CCATTAAACGGGTAGCAT	1X/42°	0.4X/42° (12 min)
PVM17	14-ECYAFNG	ATGCTACGCGTTTAATGG	5X/42°	0.1X/42° (12 min)
PVM12	41-DVGEFR2	GACGTGGGAGAGTTCCGG	2X/50°	0.1X/50° (12 min)
PVM13	41-DVGEFR1	CCGGAACTCCCCCACGTC	2X/50°	0.1X/50° (12 min)
PVM20	29-IYNREE2	CTCCTGCCTGTTGTAGA	5X/42°	0.4X/42° (12 min)
PVM21	29-IYNREE1	TCTACAACCGGCAGGAG	5X/42°	0.1X/42° (12 min)
DPB131	37-RFDSDV	GCTTCGACAGCGACGT	4X/50°	1X/50°

[a]Indicates the polymorphic sequence motif detected by the probe and the 5' residue at which the probe starts.

[b]Probes are labeled at the 5' end with HRP.

[c]All hybridization solutions contain 0.5% SDS in addition to SSPE. Table indicates the SSPE concentration used for each probe (20X SSPE: 0.11 NaOH, 3.6 M NaCl, 0.2 M NaH_2PO_4, 0.02 M EDTA). All hybridizations are done for 30 min.

[d]All wash solutions contain 0.1% SDS in addition to SSPE. Table indicates the SSPE concentration used for each probe. All washes are done for 15 min unless otherwise indicated.

2.7. Detection

1. 0.1 *M* sodium citrate, pH 5.0.
2. 2 mg/mL 3,3',5,5'-Tetramethylbenzidine (TMB) (Fluka) in 100% ethanol.
3. 30% Hydrogen peroxide (J.T. Baker®, Phillipsburg, NJ, USA).
4. Glass bowl.
5. Rotating platform, such as the Gyrotory® Shaker (New Brunswick Scientific, Edison, NJ, USA).

2.8. Stripping of Probes for Reuse of Membranes

1. 18% Sodium sulfite (Na_2SO_3) stored in an amber bottle.
2. 0.1% SDS solution.
3. Glass bowl.
4. Microwave oven.

2.9. Interpretation

1. While the results of the DPA1 probe patterns can be interpreted manually, it is strongly recommended that laboratories interested in doing DPB1 typing consult a software engineer about designing a pattern matching program for interpreting the DPB1 probe hit patterns.
2. Polaroid or charge-coupled device (CCD) camera.
3. Probe hit patterns for DPA1 (**Table 4**) and DPB1 (**Table 5**).
4. Probe hit scoresheets for DPA1 (**Table 6**) and DPB1 (**Table 7**).

3. Methods

3.1. Preparation of Genomic DNA

Choose any of the recommended kits/methods to obtain genomic DNA for amplification. DNA prepared without the aid of a commercial kit should be resuspended and stored in 1X TE (0.01 *M* Tris, 0.1 m*M* EDTA , pH 8.0). DNA should be prepared in a designated clean area and not come into contact with PCR product or any reagents, equipment or materials exposed to PCR product.

Table 4
DPA1 Probe Hit Patterns

Probe	DPA1 Allele										
	0103	0104	0105	02011	02012	02013	02021	02022	0301	0302	0401
8-YAAF	+	+	+	+	+	+					+
8-YAMF							+	+	+	+	
13-VQTH1	+	+	+		+	+					+
12-VQTH2				+			+	+	+	+	
28-EDEQ				+	+	+	+	+			
25-DDEM1		+									
25-DDEM2	+		+	+	+	+	+	+	+	+	
36-DKK1				+	+						
36-DKK2	+	+	+			+		+	+	+	+
36-DKK3							+				
47-FGQA	+	+	+						+	+	
63-AISN									+		
63-AILN	+	+	+	+	+	+	+	+		+	+
70-IAIQ											+
81-QATN	+	+							+	+	
81-QAAN			+	+	+	+	+	+			+
40-TVWHLE	+	+	+	+	+	+	+	+	+	+	+

Table 5
DPA1 Probe Hit Patterns

Probe	DPB1 Allele																																							
	01011	01012	02012	02013	0202	0301*a*	0401	0402	0501	0601	0801	0901	1001	11011	11012	1301	1401	1501	1601	1701	1801	1901	20011*b*	20012*b*	2101	2201	2301	2401	2501	26011	26012	2701	2801	2901	3001	3101	3201	3301	3401	3501	3601
5-LFQG			+	+	+		+	+	+		+								+			+				+	+	+					+			+	+	+	+		
6-VYQL						+				+				+	+	+							+	+	+				+	+	+	+		+							+
6-VHQL												+	+				+			+															+					+	
7-VYQG	+	+																+			+																				
33-EEFARF							+																					+					+			+		+			
33-EEFVRF			+	+		+		+		+	+	+	+				+		+	+	+	+	+	+			+		+					+	+		+			+	
33-EELVRF					+				+																+	+													+		+
31-QEYARF														+	+			+																							
32-EEYARF	+	+														+														+	+	+									
54-AAE	+	+					+							+	+	+		+									+			+	+	+				+		+	+		
53-DEE			+	?				+			+		+						+		+								+				+								
54-EAE					+				+													+			+	+		+							+						+
55-DED						+				+		+					+			+			+	+										+						+	
54-DEV																																					+				
64-ILEEK	+	+					+	+	+												+						+	+		+	+	+								+	+
64-ILEEE			+	+	+						+	+	+			+			+	+		+			+	+									+		+	+			
64-LLEEK						+											+						+	+					+				+			+			+		

(cont.)

Table 5
DPA1 Probe Hit Patterns *(cont.)*

Probe	DPB1 Allele																																							
	01011	01012	02012	02013	0202	0301[a]	0401	0402	0501	0601	0801	0901	1001	11011	11012	1301	1401	1501	1601	1701	1801	1901	20011[b]	20012[b]	2101	2201	2301	2401	2501	26011	26012	2701	2801	2901	3001	3101	3201	3301	3401	3501	3601
64-LLEEE										+																								+							
64-LLEER														+	+			+																							
62-FLEEE																																									
63-NLEEK																																									
64-LL*EK																																									
73-M			+	+	+		+	+	+	+				+	+			+	+	+	+		+	+	+	+	+	+				+	+		+	+	+	+	+		+
73-V	+	+				+					+	+	+				+												+	+	+			+						+	
73-I																+						+																			
82-GGPM			+	+	+		+	+																			+	+									+	+			
81-VGPM																		+			+												+						+		
83-DEAV	+	+				+			+	+	+	+	+	+	+	+	+		+	+		+	+	+	+	+			+	+	+	+		+	+	+				+	+
41-DVGEFR1		+												+	+			+												+											
41-DVGEFR2	+		+	+	+	+	+	+	+	+	+	+	+			+	+		+	+	+	+	+	+	+	+	+	+	+		+	+	+	+	+	+	+	+	+	+	+
14-ECYPFNG																																									
14-ECYAFNG	+	+	+	+	+	+	+	+	+	+	+	+	+	+	+	+	+	+	+	+	+	+	+	+	+	+	+	+	+	+	+	+	+	+	+	+	+	+	+	+	+
29-IYNREE1															+																										
29-IYNREE2														+				+																							
37-RFDSDV	+	+	+	+	+	+	+	+	+	+	+	+	+	+	+	+	+	+	+	+	+	+	+	+	+	+	+	+	+	+	+	+	+	+	+	+	+	+	+	+	+

(cont.)

Table 5 *(cont.)*
DPA1 Probe Hit Patterns

Probe	3701	3801	3901	4001	4101	4401	4501	4601	4701	4801	4901	5001	5101	5201	5301	5401	5501	5601	5701	5801	5901	6001	6101 N	6201	6301	6401 N	6501	6601	6701	6801	6901	7001 *a*	7101	7201	7301	7401
5-LFQG		+	+	+	+			+	+	+	+		+		+				+		+	+		+	+		+			+			+	+	+	
6-VYQL	+					+								+				+					+								+	+				+
6-VHQL							+									+	+			+								+	+							
7-VYQG												+																								
33-EEFARF													+					+										+						+		
33-EEFVRF	+				+		+	+	+			+		+		+	+		+		+	+	+			+			+	+	+	+	+		+	
33-EELVRF		+				+				+										+				+	+											
31-QEYARF																																				+
32-EEYARF			+	+							+				+												+									
54-AAE			+	+										+			+	+		+				+	+		+	+	+				+	+		+
53-DEE	+				+		+			+	+		+		+						+	+								+					+	
54-EAE		+							+							+																				
55-DED						+		+				+							+				+			+					+	+				
54-DEV																																				
64-ILEEK		+	+	+							+		+		+									+	+		+	+		+						
64-ILEEE	+							+	+	+						+	+			+													+			
64-LLEEK							+					+		+				+	+		+		f						+			+		+	+	

(cont.)

Table 5 *(cont.)*
DPA1 Probe Hit Patterns

Probe	3701	3801	3901	4001	4101	4401	4501	4601	4701	4801	4901	5001	5101	5201	5301	5401	5501	5601	5701	5801	5901	6001	6101N	6201	6301	6401N	6501	6601	6701	6801	6901	7001[a]	7101	7201	7301	7401
64-LLEEE						+																				+										
64-LLEER																															+					+
62-FLEEE					+																															
63-NLEEK																						+														
64-LL*EK																							+													
73-M		+	+	+	+			+	+	+	+		+		+		+			+	+	+		+	+	+		+			+		+	+		+
73-V	+					+	+					+		+		+		+	+				+				+		+	+		+			+	
73-I																																				
82-GGPM			+		+			+	+	+	+		+								+	+						+					+	+	+	
81-VGPM				+											+									+												+
83-DEAV	+	+				+	+					+		+		+	+	+	+	+			+		+	+	+		?	+	+	+				
41-DVGEFR1																																				+
41-DVGEFR2	+	+	+	+	+	+	+	+	+	+	+	+	+	+	+	+	+	+	+	+	+	+	+	+	+	+	+	+	+	+	+	+	+	+	+	+
14-ECYPFNG		+																																		
14-ECYAFNG	+		+	+	+	+	+	+	+	+	+	+	+	+	+	+	+	+	+	+	+	+	+	+	+	+	+	+	+	+	+	+	+	+	+	+
29-IYNREE1																																				
29-IYNREE2																																				+
37-RFDSDV	+	+	+	+	+	+	+	+	+	+	+	+	+	+	+	+	+	+	+	+	+	+	+	+	+	+	+	+	+	+	+	+	+	+	+	+

[a]DPB1*0301 and the newly discovered *7001 allele have identical probe hybridization patterns. They differ by a single nucleotide in codon 9 that results in a predicted amino acid (Y to D) change; however, this single nucleotide does not destabilize the hybridization of the probe for the VYQL motif (AB127) with the VDQL sequence motif. To resolve these two alleles a new probe capable of detecting the nucleotide difference between these two alleles is under development.

[b]The probe hybridization patterns for alleles *20011 and *20012 are identical; they differ by a single nucleotide at position 3 in codon 91.

?, There is a single nucleotide difference between the probe and the target sequence. DNA was unavailable, so specificity of the probe is unknown.

f (faint), A single nucleotide difference between the probe and target sequence destabilizes the probe binding and decreases the intensity of the signal.

Table 6
DPA1 Probe Hit Scoresheet

		0103, 0104, 0105, 02011/2/3, 0401	02021/2, 0301, 0302	0103, 0104, 0105, 02012/3, 0401	02011, 02021/2, 0301, 0302	02011/2/3, 02021/2	0104	All but 0104 & 0401	02011, 02012	All but 02011/2 & 02021	02021	All	0103, 0104, 0105, 0301, 0302	0301	All but 0301	0401	0103, 0104, 0301, 0302,	0105, 02011/2/3, 02021/2 0401
Sample[a]	**DPA1 Type**	**YAAF**	**YAMF**	**VQTH1**	**VQTH2**	**EDEQ**	**DDEM1**	**DDEM2**	**DKK1**	**DKK2**	**DKK3**	**ALL**	**FGQA**	**AISN**	**AILN**	**IAIQ**	**QATN**	**QAAN**
1 LBUF	02011	+			+	+		+	+			+			+			+
2 CB6B	02021		+		+	+		+			+	+			+			+
3 AMAI	0301		+		+			+		+		+	+	+			+	
4 T7526	0401	+		+						+		+			+	+		+
5 SK	0104	+		+			+			+		+	+		+		+	
6																		
7																		
8																		
9																		
10																		
11																		
12																		

[a]Samples LBUF, CB6B, AMAI, T7526, and SK are the control DNA panel.

Table 7
DPB1 Probe Hit Scoresheet

Sample[a]	DPB1 Type	LFQG	VYQL	VHQL	VYQG	EEFARF	EEFVRF	EELVRF	QEYARF	EEYARF	AAE	DEE	EAE	DED	DEV	ILEEK	ILEEE	LLEEK	LLEEE	LLEER	FLEEE	NLEEK	LL*EK	M	V	I	GGPM	VGPM	DEAV	DVGEFR1	DVGEFR2	ECYPFNG	ECYAFNG	IYNREE1	IYNRRE2	RFDSDV
1 LKT3	0501	+						+					+			+								+					+		+		+			+
2 LBUF	1701			+			+							+			+							+					+		+		+			+
3 TER81	0101, 1301		+		+					+	+					+	+								+	+			+		+		+			+
4 JY	0201, 0401	+				+	+				+	+				+	+							+			+				+		+			+
5 BIN40	0301, 0601		+				+							+				+	+					+	+				+		+		+			+
6 PLH	1501				+				+		+									+				+				+		+			+		+	+
7 CRK	01011, 11011		+		+				+	+	+					+				+				+	+				+	+	+		+		+	+
8 AH696	11012		+						+		+									+				+					+	+			+	+		+
9 NG78	3201	+					+								+		+							+			+				+		+			+
10 SE53	3801	+						+					+			+								+					+		+	+				+
11 T93	4101	+					+					+									+			+			+				+		+			+
12 C23	6001	+					+					+										+		+			+				+		+			+
1 C53	6101N		+				+							+				f					+		+				+		+		+			+
2																																				
3																																				
4																																				
5																																				
6																																				
7																																				
8																																				
9																																				
10																																				
11																																				
12																																				

[a]Samples of LKT3, LBUF, TER81, JY, BIN40, PLH, CRK, AH696, NG78, SE53, T93, C23, and C53 are the control DNA panel. f (faint), A single nucleotide difference between the probe and target sequence destabilizes the probe binding and decreases the intensity of the signal.

3.2. DP Amplifications

Set up amplifications in a designated clean area.

1. DPA1: Add 50 µL of the DP amplification premixture, 1 µL each of primers DPA1-F and DPA1-R (each at 50 µM), 200 ng of control DNA or sample, and sterile water to a final vol of 100 µL. Include at least one no-DNA control (PCR mixture with water instead of DNA) in each tray and/or for each amplification premixture. Cap the reaction tubes, place in the thermal cycler, and start the following amplification program: Soak at 95°C for 5 min, then 35 cycles of 95°C for 25 s, 55°C for 45 s, 72°C for 45 s, then soak at 72°C for 5 min, and hold at 4–10°C forever.
2. DPB1: Add 50 µL of DP amplification premixture, 1 µL each of primers UG19 and UG21 (each at 50 µM), 200 ng of control DNA or sample, and sterile distilled water to a final vol of 100 µL. Include at least one no-DNA control in each tray and/or for each amplification premixture. Cap the reaction tubes, place in the thermal cycler, and start the following amplification program: Soak at 95°C for 5 min, then 35 cycles of 95°C for 15 s, 65°C for 1 min, 72°C for 15 s, then soak at 72°C for 5 min, and hold at 4–10°C forever.

   Once the amplification is complete, all materials and procedures from this point forward are considered post-PCR.

3.3. Gel Electrophoresis

1. To determine the efficiency of amplification, examine 3–5 µL of amplicon combined with 1 to 2 µL of gel loading dye on a 3% NuSieve/1% agarose gel stained with EtBr. Use approx 2.5 µL of 100 mg/mL EtBr per 100 mL of agarose, and carefully mix EtBr with melted agarose prior to pouring. Run at 100 V until the faster of the two running dyes is at the bottom of the gel. Amplicons of both DPA1 and DPB1 should appear as single, intense bands of just over 300 bp and will run approx the same distance as the fifth fragment of the ϕX174 DNA-*Hae*III molecular weight marker. Amplifications resulting in weak bands on the gel should be repeated. There should be no amplicon in the no-DNA control lane; if a band is present, discard all amplifications and repeat PCR setup with entirely new reagents (*see* **Notes 1–3**).

3.4. Dot Blotting

1. For the DPA1-SSOP assay, it is most convenient to blot the amplicon onto nine membranes, thus enabling two sets of hybridization reactions.
   a. Automated dot blotting (Hydra microdispenser): Denature remaining amplicon (approx 90–95 µL) in 100 µL of denaturation solution. Using the Hydra microdispenser, ensure mixing of amplicon with denaturation solution by dispensing denaturation solution into amplicon, then filling, emptying, and refilling the glass capillary tubes with denatured amplicon. Program the Hydra to dispense 20 µL per dot onto a dry membrane applied to a manifold equipped with 96 holes and attached to a vacuum source. (Be sure to use forceps and wear gloves when handling membranes; do not touch membranes with bare hands.) Repeat blotting 20 µL per dot per membrane until nine membranes are made.
   b. Manual dot blotting: For nine membranes, add approx 550 µL of denaturation solution to amplicon using a multichannel pipettor. Pipet up and down to ensure mixing. Attach the dot blotter to a vacuum and, following the dot blotter manufacturer's protocol, blot 70 µL of denatured amplicon onto membranes prewet in distilled water.
2. For the DPB1-SSOP assay, it is most convenient to blot the amplicon onto 14 membranes, thus enabling three sets of hybridizations.
   a. Automated dot blotting: Using the Hydra microdispenser, denature remaining amplicon (approx 90–95 µL) in 200 µL of denaturation solution. Dispense 20 µL onto a dry membrane applied to a manifold equipped with 96 holes and attached to a vacuum source. Repeat blotting 20 µL per dot per membrane until 14 membranes are made.
   b. Manual dot blotting: Using a multichannel pipettor, add approx 900 µL of denaturation solution to amplicon, mix, and blot 70 µL per dot per prewet membrane until 14 membranes are made.
3. After DNA has been blotted, immobilize the DNA onto the membrane by UV crosslinking with a Stratalinker at 50 mJ/cm^2.
4. Rinse unbound DNA by boiling the membranes in a glass bowl filled with distilled water or 0.1% SDS for approx 10 min in a microwave oven.

3.5. Hybridization

Each individual probe has its own optimal hybridization and wash conditions. Users should pay close attention to the concentration of SSPE used in the hybridization and wash solutions for each probe (**Tables 2** and **3**). Attention should also be given to the temperatures at which these hybridizations and washes are carried out. A submersible thermometer should be placed in all water baths so that the temperature can be monitored independently of the display on the machine.

1. Place each membrane in a separate plastic Seal-A-Meal bag and add 10 mL of hybridization solution per 96-sample membrane. (If hybridizing half a membrane, less than 10 mL is adequate; simply use enough solution to cover the membrane.) Add 1 μL of probe (at 2 μM) per mL of hybridization solution and seal each bag with a heat sealer. Ensure that no air is left in the bags when sealing and make the seal as close to the membrane as possible. Submerge the bags containing the membranes in a water bath preheated to the desired temperature (**Tables 2** and **3**) and place lead weights on the corners of the bags to keep them submerged. Do not put the weights directly on top of the membranes, as this can interfere with the hybridization. Set the shaker to approximately 60 rpm and incubate for at least 30 min. To ensure that the water bath remains at the correct temperature, do not remove the lid during the hybridization step.
2. Follow the hybridization step with the indicated stringent wash step (**Tables 2** and **3**). Remove the membranes from the bags and place them in glass bowls containing stringent wash solution prewarmed in water baths to the desired wash solution temperatures. Make sure there is enough solution to cover the membranes and allow the membranes to move freely within the bowls when the water bath is shaking. Cover each bowl with a watch glass held in place with a lead weight, close the water bath lid, and set the shaker speed to approximately 60 rpm. Wash for lengths of time and at temperatures indicated in **Tables 2** and **3**.
3. After the stringent wash step, immediately remove the membranes and place them in a bowl of PBS solution at room temperature with enough liquid to cover all of them. Membranes can be stored in this manner until the detection step.

3.6. Detection

1. Hybridization of the HRP-labeled probe to the immobilized PCR product is detected by using the colorless soluble substrate TMB, which is converted to a blue precipitate by HRP in the presence of hydrogen peroxide. In a glass bowl shaking moderately, combine sodium citrate and TMB in a ratio of 20:1. (A total vol of approx 200 mL will be enough to develop about 10 membranes.) Add hydrogen peroxide to a final concentration of 0.0015%, then add two or three membranes at a time. As soon as a blue precipitate appears, transfer membranes to a glass bowl shaking moderately and containing enough water to cover all membranes. This will stop the color development. Shake for 5 min and replace water. It is best to develop only a few membranes at a time in order to prevent overdevelopment. The best indicator of the proper time to remove the membranes from the development solution is the hybridization patterns of the control samples. As soon as the controls turn blue for the correct probe, remove the membranes. They are fully developed when the dots with the appropriate positive control DNAs are blue and the dots with the negative control DNAs are still white.
2. Immediately record results by photographing each membrane with a Polaroid or CCD camera. Do not allow the membranes to sit for very long after development, as the background signal from nonspecific hybridization may increase, making it difficult to interpret the results (*see* **Notes 4–7**).
3. Record probe hit patterns for each sample using the probe hit scoresheets in **Tables 6** (DPA1) and **7** (DPB1).

3.7. Stripping of Probes for Reuse of Membranes

1. The DNA immobilized on the membranes is stable, so the membranes can be repeatedly reused for subsequent hybridizations after the bound probe is removed. It is best to strip the probes from the membranes as soon as the results have been recorded. If the membranes are left for longer than a few hours, the blue precipitate becomes difficult to remove, and the membranes may become discolored. First, to remove the blue precipitate, submerge the membranes in a bowl of warm distilled water (approx 750 mL for 10 membranes) mixed with 5–10 mL of 18% Na_2SO_3. Shake until the

blue color disappears. Rinse the membranes thoroughly in distilled water to remove the Na_2SO_3. Second, to remove the probe, submerge membranes completely in a 0.1% SDS solution (approx 750 mL per 10 membranes) and heat to boiling (10–15 min) in a microwave oven. Rinse the membranes in distilled water. They are now ready to be used with the next set of probes. When the assay is complete, the membranes can be air-dried and stored in Seal-a-Meal bags.

3.8. Interpretation

1. Interpretation of DPA1 hybridization results can be performed by recording the probe hybridization patterns on a scoresheet like the one shown in **Table 6**. Interpretation can be done manually using the probe reactivity patterns provided at the top of the scoresheet. With the present number of 11 DPA1 alleles, one ambiguous genotype may arise when interpreting probe hybridization results; the DPA1*0103,*02022 genotype cannot be distinguished from DPA1*02013,*0302. The group-specific primer AB139 (**Table 1**) should be used to selectively amplify either the *0103 or *02013 allele. The resulting amplicon can then be typed with the original 17 SSOPs, and the phase of the sequence motifs detected by the SSOPs can be determined, resolving the ambiguity.
2. Because there are over 75 DPB1 alleles, and because a large majority of the alleles result from shuffling of the same sequence motifs in the six regions of variability, manual interpretation of DPB1 hybridization results becomes quite tedious and difficult. Consequently, interpretation software is highly recommended. Depending on the population being typed, one may obtain a high percentage of ambiguous types in which the phase of the sequence motifs (as indicated by positive probes) cannot be determined. These ambiguities can be resolved by doing a second amplification using a group-specific primer to selectively amplify one of the two alleles in a heterozygous sample (**Table 1**). The resulting amplicon should then be typed with a subset of the original probes, establishing the phase of the sequence motifs. **Table 8** outlines the most common ambiguities uncovered in our analyses of over 3500 samples and the sequence-specific primers and probes used to resolve them. As the ambiguous genotypes will vary between populations, the user may have to design additional group-specific primers (*see* **Note 8**).

Table 8
Common Ambiguous DPB1 Genotypes

Ambiguous genotypes	Sequence-specific primer	Alleles amplified	Probes to resolve genotype
01011, 0301 ↔ 26012, 5001	A@55(R)	01011, 26012	VYQL, VYQG, EEFVRF, EEYARF, AAE, DED
01011, 2001 ↔ 2701, 5001	V@76(R)	01011, 5001	VYQL, VYQG, EEFVRF, EEYARF, AAE, DED, IK, LK, M, V
0201, 0202 ↔ 4701, 4801	D@55(R)	0201, 4801	LFQG, EEFVRF, EELVRF, DEE, EAE
0201, 0301 ↔ 2501, 4601	V@76(R)	0301, 2501	LFQG, VYQL, EEFVRF, DEE, DED, IE, LK, M, V
0201, 0401 ↔ 0402, 3301 ↔ 5101, 7101[a]	E@56(F)	0201, 0402, 5101	AAE, DEE, IK, IE, M, GGPM
0201, 0501 ↔ 0402, 2201	E@56(F)	0201, 0402	DEE, EAE, IK, IE, M, GGPM, DEAV
0201, 0901 ↔ 1001, 4601	V@76(R)	0901, 1001	LFQG, VHQL, EEFVRF, DEE, DED, IE, M, V
0201, 1401 ↔ 4501, 4601 ↔ 1701, 7301	V@76(R)	1401, 4501, 7301	LFQG, VHQL, EEFVRF, DEE, DED, IE, LK, M, V
0201, 3501 ↔ 0402, 0901	V@76(R)	0901, 3501	LFQG, VHQL, EEFVRF, DEE, DED, IK, IE, M, V
0301, 0601 ↔ 2001, 2901 ↔ 0301, 6401N[b]	V@76(R)	0301, 2901	VYQL, EEFVRF, DED, LK, LE, M, V
0301, 1001 ↔ 0901, 2501 ↔ 1401,3701	L@65(R)	0301, 1401, 2501	VYQL, VHQL, EEFVRF, DEE, DED, LK, IE
0301, 1601 ↔ 0801, 2001	V@76(R)	0301, 0801	LFQG, VYQL, EEFVRF, DEE, DED, IE, LK, M, V
0301, 1701 ↔ 0901, 2001	V@76(R)	0301, 0901	VYQL, VHQL, EEFVRF, DED, IE, LK, M, V
0401, 0402 ↔ 2301, 5101	D@55(R)	0402, 5101	LFQG, EEFARF, EEFVRF, AAE, DEE
0401, 0901 ↔ 3301, 3501	V@76(R)	0901, 3501	LFQG, VHQL, EEFARF, EEFVRF, AAE, DED, IK, IE M, V,
0401, 3001 ↔ 2401, 5501	A@55(R)	0401, 5501	LFQG, VHQL, EEFARF, EEFVRF, AAE, EAE
0402, 3901 ↔ 2301, 4901	D@55(R)	0402, 4901	LFQG, EEFVRF, EEYARF, AAE, DEE
0501, 0901 ↔ 2201, 3501	V@76(R)	0901, 3501	LFQG, VHQL, EEFVRF, EELVRF, EAE, DED, IK, IE, M, V
0501, 2101 ↔ 2201, 3601	L@11(F)	2101, 3601	LFQG, VYQL, EELVRF, EAE, IK, IE, M, DEAV
0901, 3601 ↔ 2101, 3501	V@76(R)	0901, 3501	VYQL, VHQL, EEFVRF, EELVRF, EAE, DED, IK, IE, M, V
0901, 4501 ↔ 1001, 1401	L@65(R)	1401, 4501	VHQL, EEFVRF, DEE, DED, IE, LK

[a]With the recent discovery of the DPB1*7101 allele, the *5101,*7101 genotype was introduced into this ambiguous combination. Using the sequence-specific primer E@56, the *0402 and *5101 alleles have the same hybridization pattern; consequently the *0402,*3301 and *5101,*7101 genotypes cannot be distinguished. However, the DPB1*3301, *5101, and *7101 alleles are so rare that all samples with this ambiguous probe hybridization pattern have been shown to be *0201,*0401.

[b]With the recent report of the DPB1*6401N allele, the *0301,*6401N genotype was introduced into this ambiguous combination. Using the sequence-specific primer V@76 the *0301,*0601 and *0301*6401N genotypes cannot be resolved; however, this ambiguity can be resolved by introducing a probe for the stop codon found at position 7 in the rare *6410N allele.

4. Notes

Below are possible problems one may encounter in performing the DPA1 and DPB1 typing assays. For each possible problem, one or more solutions are presented.

1. Complete PCR dropout (no PCR amplification of any sample). The PCR mixture may have been prepared incorrectly, or the wrong amplification program may have been used. Repeat the amplification with new reagents and check that the correct amplification profile is used.
2. Sporadic PCR dropouts or weak amplifications. (i) The sample may have contained an insufficient amount of DNA; add more sample to amplification. (ii) An inhibitor may have been present in the DNA sample; the two most common inhibitory problems are heme carried over from the sample preparation or too much EDTA in the solution used to resuspend the genomic DNA. Repeat sample extraction, removing all heme. Check the concentration of EDTA in the buffer used to resuspend DNA; it should be less than 1 m*M*. (iii) DNA may not have been added; repeat amplification.
3. Positive band in the no-DNA control lane after gel electrophoresis. Reaction mixture may be contaminated with PCR product or genomic DNA. Discard amplification and repeat with entirely new reagents.
4. Probe signal is positive on negative controls. (i) Cross-hybridization may have occurred because the stringency of the wash step was too low (the temperature in the water bath was too low or the salt concentration in the hybridization or wash solutions was too high). Check the water bath temperature using the submersible thermometer. If the temperature was accurate, prepare new hybridization and wash solutions. (ii) The membrane may have been left too long in development solution; strip probe from the membrane, then repeat hybridization, wash, and development steps, removing the membrane as soon as the positive control dots begin to turn blue.
5. No probe signal present. (i) Probe may not have been added; repeat hybridization, wash, and development steps. (ii) Stringency may have been too high. Check water bath temperature to see if it was too high. If the temperature was accurate, prepare new hybridization and wash solutions. (iii) Probe may have stopped working (HRP inactivated); strip probe from the membrane, repeat hybridization, wash, and development steps, and if no results are obtained, re-order the probe.

6. Weak probe signals on certain samples. (i) An insufficient amount of DNA may have been blotted on the membrane; compare the control probe (DPA1: 40-TVWHLE; DPB1: 37-RFDSDV) intensity on the weak sample with that of the other positive samples. If the questionable sample has a weak control probe signal as well, it is probably positive for the faint probe; however, it is recommended that the sample be re-amplified and typed again. (ii) The probe may not have hybridized equally well to all of the samples on the membrane. Repeat hybridization, making sure that the membrane is completely covered by the hybridization solution and that the bag is completely submerged in the water bath. (iii) The sample may contain a mutant sequence in the region complementary to the probe, preventing efficient hybridization. If the same results are obtained a second time, consider cloning and sequencing the sample to confirm the sequence.
7. Sample has a unique probe hybridization pattern. Make sure probes display the hybridization patterns consistent with the expected patterns for the positive controls. If they are correct, the sample may contain a new allele. If the same results are obtained for the sample a second time, clone and sequence it to confirm the sequence.
8. Finally, as the sequences of new alleles are reported in the literature, additional ambiguities in the interpretation of the results might be introduced. Additional probes and group-specific primers may have to be designed to resolve these ambiguities. However, the decision to add additional reagents to an assay should depend on the frequency of the new allele in the population the user is studying, as well as on the level of resolution the user wishes to achieve. Although the sequences of new DP alleles are constantly being reported, most appear to be extremely infrequent. For example, the probe hybridization pattern of the DPA1*0203 allele, which was found in a single Brazilian individual ***(24)***, introduced a new ambiguity in the DPA1-typing system described here. The heterozygous genotypes DPA1*02011,02013 and DPA1*02013,0203 cannot be distinguished. This ambiguity could be resolved by introducing a new probe for the methionine residue at position 31 in the DPA1 molecule; however, the frequency of both the DPA1*02013 and 0203 alleles is so low (DPA1*02013 was found in a single individual in the Cameroon *[**10**]*), that it is very unlikely that this genotype will appear in any population study.

Acknowledgments

This work was supported in part by National Institute of Health (NIH) Grant No. AI29042. We are grateful to S. Mack and H. Erlich for valuable comments on this manuscript.

References

1. Termijtelen, A., Naipal-van den Berge, S., Suwandi-Thung, L., and van Rood, J. J. (1984) The influence of matching for SB on MLC typing is significant but marginal. *Scand. J. Immunol.* **19**, 265–268.
2. Saiki, R. K., Scharf, S., Faloona, F., et al. (1985) Enzymatic amplification of β-globin genome sequences and restriction site analysis for diagnosis of sickle cell anemia. *Science* **230**, 1350–1354.
3. Wank, R., Schendel, D. J., Hansen, J. A., and Dupont, B. (1978) The lymphocyte restimulation system: evaluation by intra-HLA-D group priming. *Immunogenetics* **6**, 107–115.
4. Mawas, C., Charmot, D., and Mercier, P. (1980) Split of HLA-D into two regions alpha and beta by a recombination between HLA-D and GLO. I. Study in a family and primed lymphocyte typing for determinants coded by the beta region. *Tissue Antigens* **15**, 458–466.
5. Shaw, S., Johnson, A. H., and Shearer, G. M. (1980) Evidence for a new segregant series of B cell antigens that are encoded in the HLA-D region and that stimulate secondary allogeneic proliferative and cytotoxic responses. *J. Exp. Med.* **152**, 565–580.
6. Odum, N., Hofmann B., Hyldig-Nielsen, J. J., et al. (1987) A new supertypic HLA-DP related determinant detected by primed lymphocyte typing. *Tissue Antigens* **29**, 101–109.
7. Lotteau, V., Teyton, L., Tongio, M.-M., et al. (1987). Biochemical polymorphism of the HLA-DP heavy chain. *Immunogenetics* **25**, 403–407.
8. Hyldig-Neilsen, J. J., Morling, N., Odum, N. H., et al. (1987) Restriction fragment length polymorphism of the HLA-DP subregion and correlations to HLA-DP phenotypes. *Proc. Natl. Acad. Sci. USA* **84**, 1644–1648.
9. Bodmer, J. G., Marsh, S. G. E., Albert, E. D., et al. (1997) Nomenclature for factors of the HLA system, 1996. *Tissue Antigens* **49**, 297–321.

10. Steiner, L. L., Cavalli, A., Zimmerman, P. A., et al. (1998) Three new DP alleles identified in sub-Saharan Africa: DPB1*7401, DPA1*02013, and DPA1*0302. *Tiss. Antig.* **51**, 653–657.
11. Tsuji, K., Aizawa, M., and Sasazuki, T. (eds.) (1992) *HLA 1991: Proceedings of the Eleventh International Workshop and Conference.* Oxford Scientific Publications, Oxford, England.
12. Terasaki, P. I. and Gjertson, D. W. (eds.) (1997) *HLA 1997.* UCLA Tissue Typing Laboratory, Los Angeles.
13. Charron, D. (ed.) (1997) *Genetic Diversity of HLA: Functional and Medical Implications.* EDK, Paris.
14. Hoffman, R. W., Shaw, S., Francis, L. C., et al. (1986) HLA-DP antigens in patients with pauciarticular juvenile rheumatoid arthritis. *Arthritis Rheum.* **29**, 1057–1062.
15. Odum, N., Morling, N., Friis, J., et al. (1986) Increased frequency of HLA-DPw2 in pauciarticular onset juvenile chronic arthritis. *Tissue Antigens* **28**, 245–250.
16. Begovich, A. B., Bugawan, T. L., Nepom, B. S., Klitz, W., Nepom, G. T., and Erlich, H. A. (1989) A specific HLA-DPB allele is associated with pauciarticular juvenile rheumatoid arthritis but not adult rheumatoid arthritis. *Proc. Natl. Acad. Sci. USA* **86**, 9489–9493.
17. Fernandez-Vina, M. A., Fink, C. W., and Stasney, P. (1990) HLA antigens in juvenile arthritis. Pauciarticular and polyarticular juvenile arthritis are immunogenetically distinct. *Arthritis Rheum.* **33**, 1787–1797.
18. Erlich, H. A., Rotter, J. I., Chang, J. D., et al. (1996) Association of HLA-DPB1*0301 with IDDM in Mexican-Americans. *Diabetes* **45**, 610–614.
19. Noble, J. A., Valdes, A. M., Cook, M., Klitz, W., Thomson, G., and Erlich, H. A. (1996) The role of HLA class II genes in insulin-dependent diabetes mellitus: molecular analysis of 180 Caucasian, multiplex families. *Am. J. Hum. Genet.* **59**, 1134–1148.
20. Richeldi, L., Sorrentino, R., and Saltini, C. (1993) HLA-DPB1 glutamate 69: a genetic marker of beryllium disease. *Science* **262**, 242–244.
21. Newman, L. S. (1993) To Be^{2+} or not to Be^{2+}: immunogenetics and occupational exposure. *Science* **262**, 197–198.
22. Maniatis, T., Fritsch, E. F., and Sambrook, J., (eds.) (1982) *Molecular Cloning: A Laboratory Manual.* CSH Laboratory Press, Cold Spring Harbor, NY.

23. Bugawan, T. L., Begovich, A. B., and Erlich, H. A. (1990) Rapid HLA-DPB typing using enzymatically amplified DNA and nonradioactive sequence-specific oligonucleotide probes. *Immunogenetics* **32**, 231–241.

24. Muntau, B., Thye, T., Pirmez, C., and Horstmann, R. D. (1997) A novel DPA1 allele (DPA1*0203) composed of known epitopes. *Tissue Antigens* **49**, 668–669.

7

PCR–Sequence-Specific Primer Typing of HLA Class I and Class II Alleles

Mike Bunce

1. Introduction

1.1. Background

The use of molecular typing methods for defining human leukocyte antigen (HLA) Class I and Class II alleles is now commonplace. Molecular methods offer a greater accuracy than traditional serological methods *(1–5)* which is reflected in many disease and transplantation studies *(6–9)*. This chapter describes a polymerase chain reaction sequence-specific primer (PCR-SSP) typing system that is applicable to the identification of all HLA class I and class II alleles, as well as alleles from non-HLA loci *(10,11)* . PCR-SSP is also variably known as allele-specific PCR or amplification refractory mutation system (ARMS).

Most PCR-SSP systems feature multiple small volume PCRs, in which each reaction is specific for an allele or, more commonly, a group of alleles that correspond to a serologically defined antigen. PCR-SSP specificity is derived from matching the 3' end of one or both primers with the target DNA sequence (*see* **Fig. 1**), thus allowing the identification of any single point mutation to be identified within one or two PCRs *(12,13)*. To type an individual

From: *Methods in Molecular Biology, vol. 210: MHC Protocols*
Edited by: S. H. Powis and Robert W. Vaughan © Humana Press Inc., Totowa, NJ

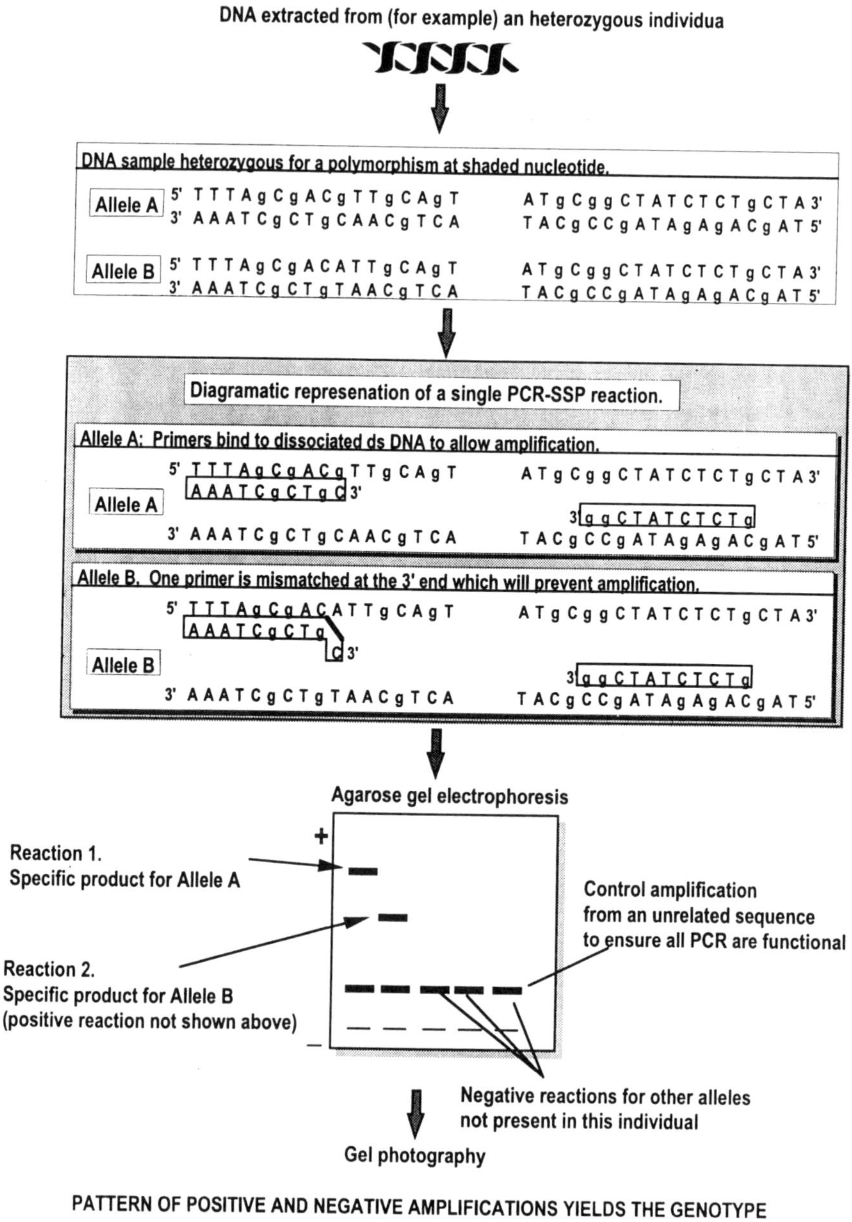

Fig. 1. Principles of PCR-SSP

completely, at any given locus multiple PCR-SSP reactions are set up and subjected to PCR under identical conditions. The presence or absence of PCR amplification is detected in the gel electrophoresis step with visualization by ethidium bromide incorporation. An important feature of SSP is that each individual reaction contains primers to amplify a so called housekeeping gene ***(14)***, which detects possible PCR inhibition and thus acts as a positive control. Without this positive control, it would be difficult to discriminate between a failed PCR and a negative PCR, and hence all homozygous results would be questionable.

PCR-SSP works because *Taq* DNA polymerase lacks 3' to 5'-exonucleolytic proofreading activity ***(15,16)***. Such an activity would correct the mismatched terminal base of an SSP in a mismatched primer-template complex and, subsequently, permit efficient priming with the repaired primer. For efficient SSP amplification without false priming, the conditions need to be highly stringent, as it is theoretically possible for 3' mismatch extension ***(12,13,17)***. SSP stringency is multifactorial, relying on the concentration of all the PCR constituents such as target DNA, *Taq*, dNTPs, Tris, and free magnesium. PCR stringency kinetics also relies on individual primer factors, such as primer sequence, length, and type of primer-template mismatches. Published PCR-SSP methods generally define a single locus ***(3,14,18–29)*** or they may be, as described here, a combination of loci known as Phototyping ***(30)***. The methods in this chapter are designed for medium resolution typing of *HLA-A, B, C, DRB1, DRB3, DRB4, DRB5,* and *DQB1*, which is updated from our previous publications ***(30)***. The reactions described can be used as a whole set, or they can be broken up into various locus-specific units.

1.2. Design of PCR Primers and PCR Primer Mixes

Consistent design of PCR primers along with use of the most up-to-date sequence alignments, are key features of successful and accurate PCR-SSP HLA typing. All primers are initially designed to have a primer-template annealing temperature of 60° or 62°C, based on the popular formula 2X (number of A and T bases) + 4X (number of G and C bases) = annealing temperature in °C. Ideally, primers should have an even ratio of GC to AT bases, but this is not always possible to achieve, and in fact, some primers work well in PCR-SSP with 100% GC content. Where possible, primers are designed with the specificity-dependent (the mismatched) nucleotide on the terminal 3'- nucleotide but internal mismatches in a primer may also significantly contribute to a primer's specificity, as shown in **Fig. 2**.

Purchase or synthesize primers (*see* **Table 1**) as desalted oligonucleotides (Cruachem, Glasgow) on a 25 OD (approx 0.2 μ*M*) scale that are resuspended in double-distilled water (ddH_2O) at a concentration of 2000 μg/mL and stored frozen until required. Generally, primers can also be left at 4°C for long periods.

The theoretical specificity of a PCR-SSP primer mixture is derived from the intersection of both primers' specificities. Thus, if sense primer matches *HLA-A*0101* and *A*0102*, and the antisense primer matches *HLA-A*0101* and *A*0103*, then if PCR stringency is maintained, the primer mixture will be specific for *HLA-A*0101*. One of the key factors in maintaining PCR stringency is the concentration of the primers used: the concentrations given in **Table 2** are to be used as a guide only as the optimal concentrations should be determined empirically within individual laboratories.

Fig. 2. *(opposite page)* PCR-SSP reactions may use internal mismatches in primers for specificity as well as 3'-mismatches. There are seven different *HLA-B* sequences found between positions 259 and 272. All four bases can be found at position 272 coupled with a dimorphic motif at nucleotides 259 and 261. Primer mixture 73 uses a combination of primer 192 and 214 to identify many *B*15* alleles, but it is mainly used for discriminating between the *B*1501*-like group of alleles and the *B*1502* and *B*1513* alleles. The mismatches at position 12 and 14 of primer 192 are sufficient to destabilize primer-template annealing in *B*1502* and *B*1503*-positive individuals and, thus, allow discrimination between *B*1501/3* and *B*1502/13* groups. Primers that utilize internal mismatches for their specificity must be carefully titrated and properly tested before use to prevent false-positive amplification of closely related alleles.

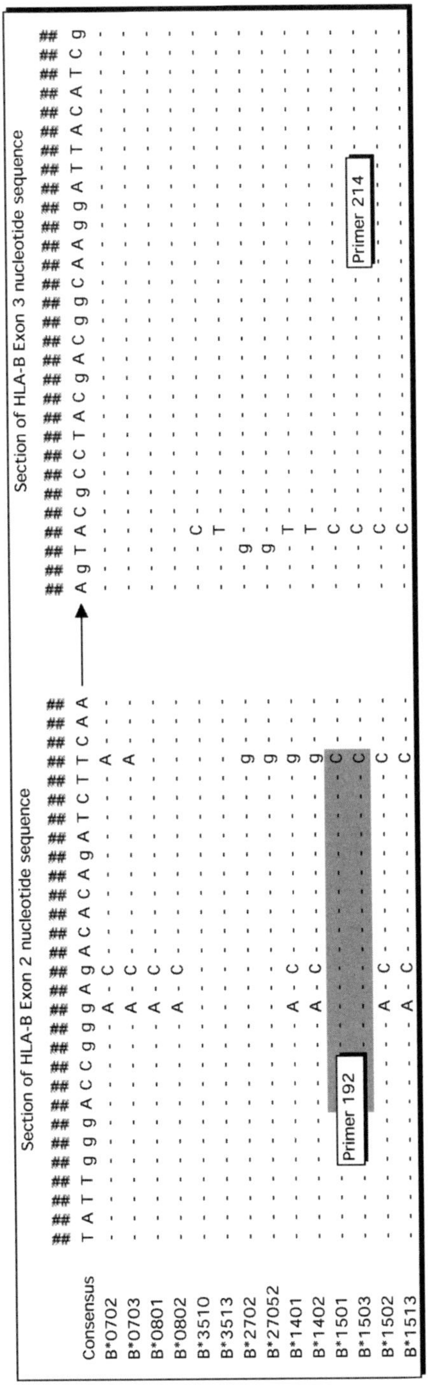
Section of HLA-B Exon 2 nucleotide sequence
Section of HLA-B Exon 3 nucleotide sequence
Consensus
TATTgggACCgggAgACACAgATCTTCAA
AgTACgCCTACgACggCAAggATTACATCg
B*0702
B*0703
B*0801
B*0802
B*3510
B*3513
B*2702
B*27052
B*1401
B*1402
B*1501
B*1503
B*1502
B*1513
Primer 192
Primer 214

Table 1a
Class I Primers

Class I sense primers

Primer Identity	Annealing Position	Primer sequences
130	402-419	5-CCgCgggTATgACCAgTC
159	325-343	5-TACAACCAgAgCgAggCCA
160	326-343	5-ACAACCAgAgCgAggCCg
165	158-176	5-ACgACACgCAgTTCgTgCA
173	78-98	5-CCACTCCATgAggTATTTCTT
174	241-259	5-CCggAgTATTgggACCTgC
187	217-234	5-gCgCCgTggATAgAgCAA
188	190-206	5-gCCgCgAgTCCgAggAC
189	254-272	5-ACCggAACACACAgATCTg
192	254-272	5-ACCgggAgACACAgATCTC
193	243-261	5-ggAgTATTgggACCggAAC
194	268-285	5-AACATgAAggCCTCCgCg
195	253-272	5-gACCggAACACACAgATCTT
197	230-246	5-AgCAggAggggCCggAA
202	124-141	5-ggggAgCCCCgCTTCATT
203	265-283	5-CAgATCTACAAggCCCAgg
205	83-103	5-CCATgAggTATTTCTACACCg
206	253-272	5-gACCggAACACACAgATCTA
207	294-311	5-CCgAgAgAgCCTgCggAA
208	293-311	5-ACCgAgAgAACCTgCggAT
209	189-206	5-CgCCgCgAgTCCgAgAgA
239	406-423	5-gggTACCAgCAggACgCT
240	259-278	5-gAgACACAgAAgTACAAgCg
242	295-312	5-CgAgAgAgCCTgCggAAC
243	192-209	5-CgCgAgTCCgAggATggC
246	78-97	5-CCACTCCATgAggTATTTCC
251	300-317	5-ggACCTgCggACCCTgCT
271	125-142	5-gggAgCCCCgCTTCATCT
272	189-206	5-CgCCACgAgTCCgAggAA
278	300-317	5-gAgCCTgCggACCCTgCT
280	149-167	5-gCTACgTggACgACACgCT
284	247-265	5-TATTgggACgAggAgACAg
286	186-203	5-CgACgCCgCgAgCCAgAA
288	283-302	5-TCACAgACTgACCgAgCgAA
290	264-282	5-ACggAATgTgAAggCCCAg
291	184-200	5-AgCgACgCCgCgAgCCA
292	239-257	5-ggCCggAgTATTgggACgA
294	283-302	5-TCACAgACTgACCgAgAgAg
295	110-126	5-CCCggCCCggCAgTggA
296	222-240	5-gTggATAgAgCAggAgggT
312	262-282	5-ACACAgATCTACAAgACCAAC
313	252-270	5-ggACCgggAgACACAgAAC
366	294-312	5-CCgAgTgAACCTgCggAAA
367	322-341	5-TACTACAACCAgAgCgAggA
368	284-302	5-CACAgACTgACCgAgTgAg
369	196-213	5-AgTCCAAgAggggAgCCg
371	78-98	5-CCACTCCATgAggTATTTCTC
395	190-206	5-gCCgCgAgTTCgAgAgg
402	275-292	5-AggCCCACTCACAgACTC
434	78-98	5-CCACTCCATgAggTATTTCAC
435	303-319	5-CCTgCgCACCgCgCTCC
451	406-423	5-gggTACCggCAggACgCT
475	264-282	5-ACggAAAgTgAAggCCCAg
1605	342-361	5-CAggTCTCACACCCTCCAgT

Class I antisense primers

Primer Identity	Annealing Position	Primer sequences
126	477-494	5-TgAgCCgCCgTgTCCgCA
127	361-379	5-ggTCgCAgCCATACATCCA
143	368-385	5-gCCCCAggTCgCAgCCAA
145	559-576	5-gAgCCACTCCACgCACTC
146	539-557	5-CCCTCCAggTAggCTCTCT
157	474-490	5-CCgCCgTgTCCgCggCA
166	302-318	5-gCgCAggTTCCgCAggC
167	559-576	5-gAgCCACTCCACgCACCg
168	559-576	5-gAgCCACTCCACgCACgT
170	538-556	5-CCTCCAggTAggCTCTCTg
171	391-407	5-CCgCggAggAAgCgCCA
183	368-384	5-CCCCAggTCgCAgCCAg
184	512-528	5-CgCACgggCCgCCTCCA
212	538-556	5-CCTCCAggTAggCTCTgTC
213	387-402	5-gAggAggCgCCCgTCg
214	419-435	5-CTTgCCgTCgTAggCgg
215	420-438	5-ATCCTTgCCgTCgTAggCT
216	435-454	5-CgTTCAgggCgATgTAATCT
217	544-561	5-CgTgCCCTCCAggTAggT
218	559-576	5-gAgCCACTCCACgCACTC
219	572-589	5-CCAggTATCTgCggAgCg
220	603-619	5-CCgCgCgCTCCAgCgTg
221	605-622	5-TACCAgCgCgCTCCAgCT
223	353-372	5-gCCATACATCCTCTggATgA
224	361-379	5-CgTCgCAgCCATACATCAC
225	527-544	5-CTCTCAgCTgCTCCgCCT
228	411-428	5-TCgTAggCgTCCTggTgg
229	499-516	5-CTCCAACTTgCgCTgggA
232	246-265	5-gTgTgTTCCggTCCCAATAT
234	319-337	5-CgCTCTggTTgTAgTAgCg
235	787-806	5-gCCCACTTCTggAAggTTCT
236	354-371	5-CCATACATCgTCTgCCAA
237	302-318	5-gCgCAggTTCCgCAggC
238	559-576	5-gAgCCACTCCACgCACAg
241	463-479	5-gCCgCggTCCAggAgCT
244	571-588	5-CAggTATCTgCggAgCCC
247	463-479	5-gCggCggTCCAggAgCg
249	538-556	5-CCTCCAggTAggCTCTCAA
250	299-316	5-gCAggTTCCgCAggCTCT
276	499-515	5-TCCCACTTgCgCTgggT
277	387-403	5-ggAggAAgCgCCCgTCg
281	280-298	5-CTCggTCAgTCTgTgCCTT
282	280-299	5-TCTCggTAAgTCTgTgCCTT
285	412-430	5-CgTCgTAggCgTACTggTC
287	559-576	5-gAgCCCgTCCACgCACTC
298	423-443	5-ATgTAATCCTTgCCgTCgTAA
299	555-572	5-CACTCCACgCACgTgCCA
300	448-466	5-AgCgCAggTCCTCgTTCAA
301	414-431	5-CCgTCgTAggCgTgCTgT
302	453-471	5-CCAAgAgCgCAggTCCTCT
315	368-384	5-CCCCAggTCgCAgCCAC
317	526-543	5-TCTCAgCTgCTCCgCCgT
377	538-556	5-CCTCCAggTAggCTCTCCA
378	853-870	5-CAgCCCCTCgTgCTgCAT
379	601-618	5-CgCgCgCTgCAgCgTCTT
382	538-556	5-CCTCCAggTAggCTCTCAg
388	361-379	5-ggTCgCAgCCAAACATCCA
389	589-608	5-AgCgTCTCCTTCCCATTCTT
392	419-436	5-CCTTgCCgTCgTAggCgA
393	412-429	5-gTCgTAggCgTCCTggTC
394	363-382	5-CCACgTCgCAgCCATACATT
429	259-278	5-gCCTTCACATTCCgTgTgTT
431	559-575	5-AgCCCgTCCACgCACCg
438	317-335	5-CTCTggTTgTAgTAgCggA
486	257-276	5-CTTCACATTCCgTgTCTCCT

1.3. PCR Using Phototyping Methods

The methods described here were initially described for Phototyping *(30)*. The basic tenet is that multiple primer mixes consisting of water, cresol red, allele-specific and control-specific prim-

Table 1b
Class II Primers

Class II sense primers

Primer identity	Annealing Position	Primer sequences
36	22-43	5- T T g T g g C A g C T T A A g T T T g A A T
41	20-38	5- T C C T g T g g C A g C C T A A g A g
44	88-109	5- T A C T T C C A T A A C C A g g A g g A g A
46	60-77	5- g A C g g A g C g g g T g C g g T A
47	17-38	5- g T T T C T T g g A g C A g g T T A A A C A
48	21-40	5- C C T g T g g C A g g g T A A g T A T A
50	26-46	5- A g T A C T C T A C g g g T g A g T g T T
52	17-39	5- g T T T C T T g A A g C A g g A T A A g T T T
53	73-91	5- C g g T T g C T g g A A A g A C g C g
61	17-38	5- g T T T C T T g C A g C A g g A T A A g T A
68	17-38	5- g T T T C T T g g A g T A C T C T A C g T C
69	18-38	5- T T T C T T g g A g C T g C g T A A g T C
70	17-38	5- g T T T C T T g g A g C T g C T T A A g T C
76	135-152	5- g g A g T A C C g g g C g g T g A g
77	44-63	5- g C T A C T T C A C C A A C g g g A C C
79	61-77	5- A C g g A g C g C g T g C g g g g
82	70-89	5- g T g C g T C T T g T g A g C A g A A g
181	60-77	5- g A C g g A g C g C g T g C g T T A
263	75-94	5- g T T C C T g g A C A g A T A C T T C C
264	75-94	5- g T T C C T g g A g A g A T A C T T C C
270	-20-1	5- g A T C g T T C g T g T C C C C A C A A
273	19-38	5- T T C T T g g A g T A C T C T A C g g g
283	-20-1	5- g A T C g T T C g T g T C C C C A C A g
347	18-38	5- T T T C g T g C T C C A g T T T A A g g C
348	127-144	5- g A C g T g g g g g T g T A C C g C
349	63-81	5- g g A g C g C g T g C g T C T T g T A
353	70-89	5- g T g C g T C T T g T g A C C A g A T A
1465	59-77	5- g g A C g g A g C g C g T g C g T C T

Class II antisense primers

Primer Identity	Annealing Position	Primer sequences
37	257-276	5- C T g C A C T g T g A A g C T C T C A C
38	257-276	5- C T g C A C T g T g A A g C T C T C C A
39	199-216	5- C C g C C T C T g C T C C A g g A g
40	199-217	5- C C C g C T C g T C T T C C A g g A T
49	232-252	5- C C C g T A g T T g T g T C T g C A C A C
51	220-239	5- C T g C A g T A g g T g T C C A C C A g
54	173-192	5- C T g g C T g T T C C A g T A C T C C T
58	221-240	5- T C T g C A A T A g g T g T C C A C C T
78	231-251	5- T g g T A g T T g T g T C T g C A T A C g
102	169-187	5- T g T T C C A g T A C T C g g C g C T
104	199-217	5- C C C g C C T g T C T T C C A g g A A
107	170-188	5- C T g T T C C A g T g C T C C g C A g
111	250-266	5- C C g C g g A A C g C C A C C T C
112	250-267	5- C g T g C g g A g C T C C A A C T g
151	231-251	5- C C g T A g T T g T g T C T g C A g T A A
152	250-267	5- C C C g C g g T A C g C C A C C T C
252	211-226	5- C C A C C g C g g C C C g C g C
255	212-228	5- g T C C A C C C g g C C C C g C T
256	170-188	5- C T g T T C C A g g A C T C g g C g A
258	208-224	5- A C C g C g g C C C g C C T g T C
259	211-227	5- T C C A C C g C g g C C C g C T C
261	169-188	5- C T g T T C C A g T A C T C g g C A T C
268	220-239	5- C T g C A g T A A T T g T C C A C C C g
314	126-144	5- C T g g T A C T C C C C C A g g T C A
350	220-238	5- T g C A C A C C g T g T C C A A C T C
351	221-238	5- T g C A C A C C C T g T C C A C C g
354	170-188	5- C T g T T C C A g T A C T C g g C g g
485	211-227	5- T C C A C C g C g g C C C g C T T
491	140-156	5- C T C C g T C A C C g C C C g g T
492	171-189	5- g C T g T T C C A g T A C T C A g C g
866	130-147	5' C g C C T g g T A C T C C C C C A g
1466	170-189	5- g C T g T T C C A g T A C T C g g C g T

Table 1c
Control Primers

796bp control pair	
63	5- T g C C A A g T g g A g C A C C C A A
64	5- g C A T C T T g C T C T g T g C A g A T

256bp control pair	
210	5- A T g A T g T T g A C C T T T C C A g g g
211	5- T T C T g T A A C T T T T C A T C A g T T g C

ers, are synthesized, tested, and stored in 1-mL primer mixture vol. A typing set collected from these stored primer mixes is dispensed in 5-μL vol under mineral oil in 96-well PCR plates. Separate from the primer mixes, a PCR buffer containing all the other ingredients of PCR is made up and stored frozen in aliquots awaiting the addition of DNA and *Taq* DNA polymerase. DNA is then added to a predetermined vol of the PCR buffer, and 8 μL of this mixture is added to each well of the PCR plate prior to PCR amplification and agarose gel electrophoresis. This method allows extreme flexibility in the design and incorporation of any new primer mixes.

Table 2a
Primer Mix Information for Reactions 1–72

Lane	PM Number	Sense primer	Sense concentration	antisensense primer	Antisense concentration	Amplicon size	Locus	Alleles amplified
1	151	286		431		629	A	*0101, *0102, *0104N.
2	3	296		302		489	A	*0201, *0202, *0203, *0204, *0205, *0206, *0207, *0208, *0209, *0210, *0211, *0212, *0213, *0214, *0215N, *0216, *0217, *02172, *0218, *0219, *0220, *0221, *0222, *0225, *0226.
3	4	291		299		628	A	*0301, *0302, *0303N, *0304.
4	2	286		168		630	A	*3601.
5	5	284		302		464	A	*2301, *2402, *2402102L, *2403, *2404, *2405, *2406, *2407, *2409N, *2410, *2411N, *2413, *2414.
6	6	292		249		557	A	*2301, *2413.
7	7	292		170		557	A	*2402, *2402102L, *2403, *2404, *2405, *2407, *2409N, *2410, *2411N, *2414.
8	8	239		167		170	A	*2501, *2502, *2601, *2603, *2605, *2607, *2608, *4301, *6601, *6602, *6603.
9	9	193		298		440	A	*2501, *2502, *2601, *2602, *2603, *2604, *2605, *2606, *2608, *2609, *3401, *3402, *6601, *6602, *6603.
10	10	294		298		400	A	*2501, *2502.
11	11	288		298		400	A	*2601, *2602, *2604, *2607, *2608, *2609, *4301.
12	12	174		298		442	A	*4301.
13	13	239		168		170	A	*3401, *3402, *2609.
14	14	290/475		298		419	A	*2502, *3401, *3402, *6601, *6602.
15	15	290		145/167		552	A	*1101, *1102, *1103, *2502, *6601.
16	152	174		300		465	A	*2901, *2902, *2903.
17	18	295		301		561	A	*3001, *3002, *3003, *3004, *3006.
18	184	434		486		198	A	*31012.
19	153	173		234		259	A	*3201, *3202.
20	154	434		429		200	A	*3301, *3303.
21	22	173		300		628	A	*3201, *3202, *7401, *7402, *7403.
22	23	290		302		447	A	*68011, *68012, *6802, *6806, *6901.
23	24	290		171		383	A	*6901.
24	25	367		287		494	A	*8001.
25	112	367		394		300	A	*0101, *0102, *0104N, *1101, *1102, *1103, *1104, *3402, *3601, *8001.
26	165	435/208/251/278		235		1331-1341	B	Bw4
27	31	242		235		1339	B	Bw6 (not B*7301)
28	34	193		221		1207	B	*07021, *07022, *07023, *0703, *0704, *0705, *0706, *0707, *0708, *8101.
29	35	312		221		1188	B	*0703.
30	36	195		212		543	B	*0801, *0802, *0803, *4406, *5108.
31	155	195		220		606	B	*0801, *0802, *0803.

32	37	280	225	635	B	*2704, *2706, *2710, *4005, *4901, *5001, *5002,
33	38	208	215	385	B	*4901, *5901,
34	39	246	215	600	B	*4501, *4901, *5001, *5002,
35	40	207	219	535	B	*1514, *4409, *4501, *5002, *8201,
36	160	202/272	285/393	480-545	B	*3519, *4003, *4009, *4018, *4402, *44031, *44032, *4404, *4405, *4406, *4407, *4409,
37	42	192	220	605	B	*4101, *4102, *4103,
38	43	272	276	566	B	*3519, *4002, *4003, *4004, *4005, *4006, *4008, *4009, *4011, *4014, *4015, *4016, *4017, *4018, *4101, *4102, *4103, *4402, *44031, *44032, *4404, *4405, *4407, *4409, *4410, *4501, *4701, *4702, *4703, *4901, *5001, *5002
39	44	192	247	465	B	*1533, *40011, *40012, *4002, *4003, *4004, *4005, *4006, *4009, *4010, *4011, *4012, *4014, *4015, *4016, *4017, *4018, *4101, *4102, *4103, *4801, *4803,
40	45	272	218	627	B	*40011, *40012, *4002, *4003, *4004, *4006, *4007, *4008, *4009, *4010, *4011, *4014, *4016, *4017, *4018, *4701, *4702, *4703,
41	46	280	229	607	B	*40011, *40012, *4007,
42	47	243	215	486	B	*1301, *1302, *1303,
43	48	197	127	389	B	*1401, *1402, *1403, *1404,
44	49	285	232	182	B	*1402, *1403, *1405, *3904,
45	50	207	217	507	B	*39011, *39013, *39021, *39022, *3903, *3904, *3905, *39061, *39062, *3907, *3908, *3909, *3910, *3911, *3912, *67011, *67012,
46	51	206	217	548	B	*3910, *67011, *67012,
47	172	208/435	217	498/508	B	*3801, *38021, *38022,
48	53	209	217	612	B	*3801, *38021, *38022, *39011, *39013, *39021, *39022, *3903, *3904, *3905, *39061, *39062, *3907, *3908, *3909, *3910, *3911, *3912, *67011, *67012,
49	54	208	223	319	B	*1513, *4406, *5104, *5301, *5302, *5801,
50	56	194	213	374	B	*5801, *5802,
51	55	194	224	351	B	*5701, *5702, *5703, *5704,
52	166	243	438	143	B	*1301, *1302, *1303, *1304, *1513, *1516, *1517, *1524, *4408, *5701, *5702, *5703, *5704,
53	57	187	214	458	B	*1801, *1802, *1803, *1805,
54	58	209	236	422	B	*39061, *39062, *5501, *5502, *5503, *5505, *5601, *5901, *7301,
55	59	242	215	383	B	*4501, *5001, *5002, *5401, *5501, *5502, *5505, *5601, *5602, *5604, *8201,
56	60	203	238	551	B	*5601, *5602, *5603, *5604,
57	61	395	236	421	B	*5401,
58	62	280	281/282	149/150	B	*2701, *2702, *2703, *2704, *27052, *27053, *2706, *2707, *2708, *2709, *2710, *2711,
59	63	188	212	606	B	*3701, *4406, *5108,
60	64	192	228/392	422	B	*3701, *39021, *39022, *3908,
61	65	192	228	414	B	*3702, *4701, *4702, *4703,
62	66	367	236	289	B	*7301,
63	67	203	220	594	B	*4201, *4202,
64	68	209	229	567	B	*4012, *4801, *4803, *8101,
65	69	194	225	516	B	*1516, *1517,
66	70	240	241	460	B	*4601,
67	72	243	250	124	B	*1501, *1502, *1504, *1505, *1506, *1507, *1508, *1511, *1512, *1514, *1515, *1519, *1520, *1521, *1525, *1526N, *1527, *1528, *1530, *1531, *1532, *1533, *1534, *1535, *1538, *1539,
68	73	192	214	421	B	*1304, *1501, *1503, *1504, *1505, *1506, *1507, *1512, *1514, *1519, *1520, *1524, *1525, *1526N, *1527, *1532, *1533, *1534, *1535, *1538, *1539, *4003, *4802,
69	156	271	223	487	B	*4802,
70	74	209	214	486	B	*1503, *1518, *1523, *1529, *3907, *4802, *5603,
71	75	271	238	691	B&C	*1503, *1509, *1510, *1518, *1523, *1529, *1537, *4802, Cw*0703,
72	76	189	238	562	B	*1509, *1510, *1518, *1521, *1523, *1537,

Table 2b
Primer Mix Information for Reactions 73–144

Lane	PM Number	Sense primer	Sense concentration	antisense primer	Antisense concentration	Amplicon size	Locus	Alleles amplified
73	77	243		219/244		636-637	B	*4408, *1512, *1514, *1519.
74	78	193		377		553	A&B	A*2501, A*2502, A*2601, A*2602, A*2603, A*2604, A*2605, A*2606, A*2609, A*3401, A*6601, A*6602, A*6603, A*68011, A*68012, A*6802, A*6803, A*68032, A*6804, A*6805, A*6806, *1508, *1511, *1515, *1522, *5603.
75	79	193		223		369	B	*1502, *1513, *1521, *3501, *3502, *3503, *3504, *3506, *3507, *3508, *35091, *35092, *3511, *3512, *3515, *3518, *3519, *3520, *3521, *4406, *5104, *5301, *5302.
76	80	188		237		128	B	*1522, *1801, *1802, *1803, *1805, *3501, *3502, *3503, *3504, *3505, *3506, *3507, *3508, *35091, *35092, *3510, *3511, *3512, *3513, *3515, *3516, *3517, *3518, *3520, *3521, *7801, *78021, *78022.
77	81	195		213/277		389-390	B	*3501, *3502, *3503, *3504, *3505, *3506, *3507, *3508, *35091, *35092, *3511, *3515, *3517, *3518, *3519, *3521, *5301, *5302.
78	82	207		216		400	B	*1509, *7801, *78021, *78022.
79	83	208		216		401	B	*51011, *51012, *51021, *51022, *5103, *5104, *5105, *5106, *5107, *5108, *5109, *5111N, *52011, *52012.
80	84	193		216		451	B	*1509, *51011, *51012, *51021, *51022, *5103, *5104, *5105, *5106, *5107, *5108, *5109, *5111N, *7801, *78021, *78022.
81	85	192		216		440	B	*52011, *52012.
82	86	368		315		340	Cw	*0102, *0103.
83	87	366		145		522	Cw	*02021, *02022, *02023, *02024, *1701, *1702.
84	88	368		389		564	Cw	*0302, *0303, *0304.
85	89	366		143		331	Cw	*04011, *04012, *0402, *0403, *0404, *1801, *1802.
86	90	366		379		564	Cw	*0501.
87	91	367		127		297	Cw	*0602.
88	92	130		378		1056	Cw	*0701, *0702, *0703, *0705, *0706, *0707, *0708.
89	93	313		184		516	Cw	*0701, *0706, *0707.
90	94	367		183		302	Cw	*0702, *0703.
91	95	367		238		494	Cw	A*2604, *0703.
92	96	367		379		536	Cw	*0704.
93	182	165		166/317		160/625	Cw	Multiplex: 160bp only = *0802. 160bp + 625bp = *0801, *0803
94	99	159		389		523	Cw	*0303.
95	100	160		389		522	Cw	*0302, *0304.
96	101	368		126		450	Cw	*12021, *12022, *1301.
97	102	369		126		538	Cw	*12021, *12022.
98	103	368		157		446	Cw	*1203.
99	104	371		388		541	Cw	*1402, *1403.
100	106	366		223		318	Cw	*1502, *1503, *1504, *15051, *15052, *1506.
101	107	366		382		502	Cw	*0404, *0707, *1502, *1503, *1504, *15051, *15052, *1506, *1701, *1702.
102	109	368		146		513	Cw	*1601.
103	110	366		146		503	Cw	*1602.
104	111	366		377		502	Cw	*02021, *02022, *02023, *02024, *0602, *12041, *12042, *1205.
105	113	36		39		194	DRB1	*0101, *01021, *01022, *0104.
106	114	36		40		195	DRB1	*0103.
107	115	41		252		206	DRB1	*15011, *15012, *15021, *15022, *15023, *1503, *1504, *1505, *1506.
108	185	41		491		136	DRB1	*16011, *16012, *16021, *16022, *1603, *1604, *1605, *1607, *1608.
109	162	68		255		211	DRB1	*03011, *03012, *03021, *03022, *0303, *0304, *0305, *0306, *0307, *0308, *0309, *0310, *0311, *1107.
110	118	46		38		216	DRB1	*03011, *03012, *0304, *0306, *0308, *0310, *0311, *1327.
111	119	44		37		188	DRB1&3	*03021, *03022, *0305, *1109, *1120, *1302, *1305, *1326, *1329, *1331, *1402, *1403, *1409, *1413, *1419, *1424, *1427, *1430, *1608, DRB3*0206.
112	120	47		37/38		259	DRB1	*04011, *04012, *0402, *0403, *0404, *04051, *04052, *0406, *0407, *0408, *0409, *0410, *0411, *0412, *0413, *0414, *0415, *0416, *0417, *0418, *0419, *0420, *0421, *0422, *0424, *0425, *0426, *1122, *1410

113	121	48		49		231	DRB1	*0701, *0703,
114	122	50		51/102		161/213	DRB1	*0801, *08021, *08022, *08032, *08041, *08042, *08043, *0805, *0806, *0807, *0808, *0809, *0810, *0811, *0812, *0813, *0815, *0816, *0817, *0818, *0819, *1415,
115	123	52		49		235	DRB1	*09012,
116	124	53		37		203	DRB1	*1001,
117	125	68/273		54		173-175	DRB1	*0308, *11011 to *1121, *1123 to *1129, *1204, *1411
118	126	50		256		162	DRB1	*1201, *12021, *12022, *12032, *1205,
119	183	68/273		259		208/210	DRB1	*1102, *1103, *1111, *1114, *1116, *1120, *1121, *1301, *1302, *1304, *1308, *1315, *1316, *1317, *1319, *1320, *1322, *1323, *1324, *1327, *1328, *1329, *1331, *1332, *1416,
120	127	263		259		151/152	DRB1	*1116, *1120, *1301, *1302, *1308, *1309, *1316, *1320, *1327, *1328, *1329, *1331, *1332, *1416,
121	168	44		40/485		129/139	DRB1	*1116, *1120, *1301, *1302, *1306, *1310, *1315, *1316, *1327, *1328, *1331, *1332, *1419, *1421
122	169	44		258		136	DRB1	*1109, *1305, *1306, *1318, *1326, *1403, *1412, *1427, *1608,
123	167	68		485		210	DRB1	*13031, *13032, *1310, *1333, *1419, *1421,
124	170	68		102		170	DRB1	*13031, *13032, *1304, *1312, *1313, *1321, *1330, *1332, *1333, *1413,
125	131	68		104		200	DRB1	*11011, *11012, *11013, *1103, *11041, *11042, *1106, *1109, *1110, *1111, *1112, *1115, *1123, *1124, *1125, *1127, *1128, *1129, *1305, *1307, *1311, *1314, *1318, *1321, *1324, *1326, *1422, *1425, *1427,
126	132	68		261		171	DRB1	*03011 to *0307, *0309,*0311,*1301,*1302,*1305,*1306,*1307,*1308,*1309,*1310,*1311,*1314,*1315,*1316,*1318,*1319, *1320,*1322,*1323,*1324,*1325,*1327,*1328,*1329,*1402,*1403,*1406,
127	171	263/264		51/58		164/165	DRB1	*1401, *1403, *1404, *1405, *1407, *1408, *1410, *1411, *1412, *1414, *1415, *1418, *1423, *1426, *1427, *1428, *0809, *1318, *1117,
128	134	68		107		171	DRB1	*0310, *1401, *1407, *1416, *1422, *1425, *1426,
129	135	44		39/51		128/151	DRB1&3	*03011, *03012, *03021, *03022, *0303, *0305, *0306, *0307, *0308, *0310, *0311, *1402, *1406, *1409, *1413, *1417, *1418, *1419, *1421, *1429, *1430, *1318, *1403, *1412, *1427, DRB3*0206,
130	186	61		492		172	DRB5	*01011, *01012, *0102, *0103, *0104, *0105, *0106, *0107, *0108N, *0109, *0201, *0202, *0203, *0204,
131	164	69		268		221	DRB3	*01011, *01012, *01013, *01014, *0103,
132	138	76		151		116	DRB3	*0201, *0202, *0203, *0205, *0206, *0207, *0208,
133	139	70		38		259	DRB3	*0201, *0204, *0301, *0302,
134	140	283		314		151	DRB4	*0101, *0102, *0103101, *0104, *0105, *0201N
135	1054	270		866		167	DRB4	*0103102N
136	142	77		350		194	DQB1	*0305,
137	143	82		112		197	DQB1	*0201, *0202, *0203,
138	144	77		78		207	DQB1	*0401, *0402,
139	145	79		152		206	DQB1	*0501, *0502, *05031, *05032, *0504,
140	146	347/348		111		139/248	DQB1	*06011, *0602, *0603, *0610, *0611, *0613, *0614,
141	147	349		350/351		175	DQB1	*0603, *0604, *06051, *06052, *0606, *0607, *0608, *0609, *0612, *0614,
142	148	181		112		207	DQB1	*03011, *03012, *0304,
143	149	353		354		118	DQB1	*0302, *0307,
144	268	1465		1466		130	DQB1	*0203, *03032, *0306,

Table 2c
Supplementary Primer Mix Information for Reactions

Lane	PM Number	Sense primer	Sense concentration	antisensense primer	Antisense concentration	Amplicon size	Locus	Alleles amplified
	1048	402		249		521	A	*0201, *0204, *0206, *0207, *0209, *0210, *0214, *0215N, *0216, *0217, *0218, *0220, *0221, *0224, *0225
	945	208		146		504	A	*2402, *2403, *2405, *2407, *2408, *2409N, *2410, *2411N, *2414
	1055	202		236		487	B	*51011, *51012, *5103, *5107, *5108, *5109, *5110, *52012, *7801, *78022,
	1017	368		127		335	Cw	*0102, *0103, *1203, *1402, *1403, *1601, *16041
	1076	1605		377		454	Cw	*0602, *1203, *12041, *12042, *1205, *16041

1.4. Other Important Factors

Good quality DNA is paramount for successful PCR-SSP. Sodium citrate or EDTA anticoagulated blood is preferred to heparinized blood, as heparin is a severe inhibitor of PCR and especially PCR-SSP *(31)*. If heparinized blood is the only source, then the DNA extraction protocol described below should allow for satisfactory typing.

The basic set of 144 PCR-SSP primer mixes described in this publication are the same as the Phototyping set previously described *(30)*. This basic set has set has been updated and corrected to include alleles sequenced since that publication. Some of the recently sequenced alleles (*A*0224, A*2408, Cw*16041*) are not amplified in any of the original Phototyping PCR-SSP reactions and so, for completeness, a supplementary set of mixes is listed (**Table 2c**), which covers these alleles. A primer mixture which amplifies *HLA-B*5110*, is also given, as this allele would only be amplified in the Bw4 mixture of the 144-reaction set. **Table 3** lists the alleles considered in this publication.

2. Materials

2.1. DNA Preparation

1. Red cell lysis buffer (RCLB): 0.144 *M* NH_4Cl, 1 m*M* $NaHCO_3$. Dissolve 15.4 g of NH_4Cl and 1.68 g of $NaHCO_3$ in 2 L of ddH_2O.
2. Nuclear lysis buffer (NLB): 10 m*M* Tris-HCl, pH 8.2, 0.4 *M* NaCl, 2 m*M* disodium EDTA, pH 8.0. Dissolve 23.37 g of NaCl in 900 mL of distilled water. Add 10 mL 1 *M* Tris-HCl, pH 8.2, and 10 mL Na_2EDTA, pH 8.0, and make up to 1 L with dH_2O.
3. 10% w/v sodium dodecyl sulfate (SDS): in a fume hood, dissolve 100 g of SDS in 1 L of ddH_2O. Store at approx 20°C to prevent precipitate forming.
4. NLB+SDS buffer: Combine 300 mL of NLB with 20 mL of 10% w/v SDS. Store at approx 20°C to prevent precipitate forming.
5. 95% Ethanol.
6. 70% Ethanol.
7. 6 *M* NaCl.

2.2. Ingredients for Primer Mixes

1. Synthesize or purchase primers (*see* **Table 1**) resuspended in ddH_2O at a concentration of 2000 µg/mL and stored frozen until required.
2. Cresol red 6 mg/mL stock: dissolve 3 g of cresol red sodium salt (Sigma, St. Louis, MO, USA; C9877) in 500 mL ddH_2O. Pass through a 0.22-µm filter and store frozen in 1-mL vol.
3. Stock solutions of control primers: for the 796 bp control amplification used in all primer mixes except the Bw4 and Bw6 mixes, autoclave 2 L of ddH_2O. When cold, add 1.5 µL/mL each of primer 63 and primer 64 along with 10 µL/mL of cresol red and filter through a 0.22-µm filter. For the 256 bp control amplification used in the Bw4 and Bw6 mixes, autoclave 20 mL of ddH_2O. When cold, add 3 µL/mL each of primers 210 and 211 along with 10 µL/mL of cresol red and filter through a 0.22-µm filter. The control stocks must be tested (*see* **Notes 1** and **2**) prior to freezing in aliquots.

2.3. PCR Ingredients

1. 25 m*M* magnesium chloride: add 1 mL 1*M* stock solution (Sigma; M1028) to 39mL ddH_2O. Note that stock solutions go off over a period of 6 mo, so when a new bottle is opened, dispense in 1-mL vol and freeze.
2. dNTP mixture: combine 0.4 mL of each dNTP (dCTP, dATP, dTTP, dGTP) from 100 m*M* stock (Promega, Madison, WI, USA; U1240).
3. 10X Base buffer: 670 m*M* Tris-HCl, pH 8.9, 166 m*M* ammonium sulfate, 1% (v/v) Tween 20 (polyoxyethylenesorbitan monolaurate) (Sigma; P9416). Dissolve 40.568g Tris-base in 400 mL dH_2O and adjust to pH 8.9 with concentrated HCl. Dissolve 10.96 g of ammonium sulfate in the Tris solution. Filter through a 0.22-µm filter into an autoclaved bottle. Add 5 mL of Tween and make up to 500 mL with ddH_2O. Store at –70°C.
4. PCR buffer (200 mL recipe). This when combined with all other PCR ingredients, gives a final MgCl concentration of 1.9 m*M*. 33.3 mL x 10 base buffer, 25.1 mL 25 m*M* fresh MgCl, 140.4 mL freshly autoclaved water, 1.215 mL dNTP mix (i.e., all 4 mixed together). Test the PCR buffer (*see* **Note 3**) before freezing in 13.3-mL aliquots.

Table 3
HLA-A, B, Cw, DRB1, DRB3, DRB4, DRB5, and DQB1 Alleles Considered for This Publication

HLA-A	HLA-B		HLA-Cw	HLA-DRB1		HLA-DRB3/4/5	HLA-DQB1
A*0101	B*07021	B*39011	Cw*0102	DRB1*0101	DRB1*1119	DRB3*01011	DQB1*0201
A*0102	B*07022	B*39013	Cw*0103	DRB1*01021	DRB1*1120	DRB3*01012	DQB1*0202
A*0104N	B*07023	B*39021	Cw*02021	DRB1*01022	DRB1*1121	DRB3*01013	DQB1*0203
A*0201	B*0703	B*39022	Cw*02022	DRB1*0103	DRB1*1122	DRB3*01014	DQB1*03011
A*0202	B*0704	B*3903	Cw*02023	DRB1*0104	DRB1*1123	DRB3*0102	DQB1*03012
A*0203	B*0705	B*3904	Cw*02024	DRB1*03011	DRB1*1124	DRB3*0103	DQB1*0302
A*0204	B*0706	B*3905	Cw*0302	DRB1*03012	DRB1*1125	DRB3*0201	DQB1*03032
A*0205	B*0707	B*39061	Cw*0303	DRB1*03021	DRB1*1126	DRB3*0202	DQB1*0304
A*0206	B*0708	B*39062	Cw*0304	DRB1*03022	DRB1*1127	DRB3*0203	DQB1*0305
A*0207	B*0801	B*3907	Cw*04011	DRB1*0303	DRB1*1128	DRB3*0204	DQB1*0306
A*0208	B*0802	B*3908	Cw*04012	DRB1*0304	DRB1*1129	DRB3*0205	DQB1*0307
A*0209	B*0803	B*3909	Cw*0402	DRB1*0305	DRB1*1130	DRB3*0206	DQB1*0401
A*0210	B*0804	B*3910	Cw*0403	DRB1*0306	DRB1*1131	DRB3*0207	DQB1*0402
A*0211	B*0805	B*3911	Cw*0404	DRB1*0307	DRB1*1201	DRB3*0208	DQB1*0501
A*0212	B*1301	B*3912	Cw*0501	DRB1*0308	DRB1*12021	DRB3*0301	DQB1*0502
A*0213	B*1302	B*40011	Cw*0602	DRB1*0309	DRB1*12022	DRB3*0302	DQB1*05031
A*0214	B*1303	B*40012	Cw*0701	DRB1*0310	DRB1*12032	DRB3*0303	DQB1*05032
A*0215N	B*1304	B*4002	Cw*0702	DRB1*0311	DRB1*1204	DRB4*01011	DQB1*0504
A*0216	B*1401	B*4003	Cw*0703	DRB1*04011	DRB1*1205	DRB4*0102	DQB1*06011
A*02171	B*1402	B*4004	Cw*0704	DRB1*04012	DRB1*1301	DRB4*0103	DQB1*06012
A*02172	B*1403	B*4005	Cw*0705	DRB1*0402	DRB1*1302	DRB4*0104	DQB1*06013
A*0218	B*1404	B*4006	Cw*0706	DRB1*0403	DRB1*13031	DRB4*0105	DQB1*0602
A*0219	B*1405	B*4007	Cw*0707	DRB1*0404	DRB1*13032	DRB4*0201N	DQB1*0603
A*0220	B*1501	B*4008	Cw*0708	DRB1*04051	DRB1*1304	DRB4*0301N	DQB1*0604
A*0221	B*1502	B*4009	Cw*0801	DRB1*04052	DRB1*1305	DRB5*01011	DQB1*06051
A*0222	B*1503	B*4010	Cw*0802	DRB1*0406	DRB1*1306	DRB5*01012	DQB1*06052
A*0224	B*1504	B*4011	Cw*0803	DRB1*0407	DRB1*1307	DRB5*0102	DQB1*0606
A*0225	B*1505	B*4012	Cw*12021	DRB1*0408	DRB1*1308	DRB5*0103	DQB1*0607
A*0226	B*1506	B*4014	Cw*12022	DRB1*0409	DRB1*1309	DRB5*0104	DQB1*0608
A*0301	B*1507	B*4015	Cw*1203	DRB1*0410	DRB1*1310	DRB5*0105	DQB1*0609
A*0302	B*1508	B*4016	Cw*12041	DRB1*0411	DRB1*1311	DRB5*0106	DQB1*0610
A*0303N	B*1509	B*4017	Cw*12042	DRB1*0412	DRB1*1312	DRB5*0107	DQB1*0611
A*0304	B*1510	B*4018	Cw*1205	DRB1*0413	DRB1*1313	DRB5*0108N	DQB1*0612
A*1101	B*1511	B*4101	Cw*1301	DRB1*0414	DRB1*1314	DRB5*0109	DQB1*0613
A*1102	B*1512	B*4102	Cw*1402	DRB1*0415	DRB1*1315	DRB5*0201	DQB1*0614
A*1103	B*1513	B*4103	Cw*1403	DRB1*0416	DRB1*1316	DRB5*0202	
A*1104	B*1514	B*4201	Cw*1502	DRB1*0417	DRB1*1317	DRB5*0203	
A*2301	B*1515	B*4202	Cw*1503	DRB1*0418	DRB1*1318	DRB5*0204	
A*2402	B*1516	B*4402	Cw*1504	DRB1*0419	DRB1*1319		
A*2402102L	B*1517	B*44031	Cw*15051	DRB1*0420	DRB1*1320		
A*2403	B*1518	B*44032	Cw*15052	DRB1*0421	DRB1*1321		
A*2404	B*1519	B*4404	Cw*1506	DRB1*0422	DRB1*1322		
A*2405	B*1520	B*4405	Cw*1601	DRB1*0423	DRB1*1323		
A*2406	B*1521	B*4406	Cw*1602	DRB1*0424	DRB1*1324		
A*2407	B*1522	B*4407	Cw*16041	DRB1*0425	DRB1*1325		
A*2408	B*1523	B*4408	Cw*1701	DRB1*0426	DRB1*1326		
A*2409N	B*1524	B*4409	Cw*1702	DRB1*0427	DRB1*1327		
A*2410	B*1525	B*4410	Cw*1801	DRB1*0701	DRB1*1328		
A*2411N	B*1526N	B*4501	Cw*1802	DRB1*0703	DRB1*1329		
A*2413	B*1527	B*4601		DRB1*0801	DRB1*1330		
A*2414	B*1528	B*4701		DRB1*08021	DRB1*1331		

(cont.)

Table 3 *(cont.)*
HLA-A, B, Cw, DRB1, DRB3, DRB4, DRB5, and DQB1 Alleles Considered for This Publication

HLA-A	HLA-B		HLA-Cw	HLA-DRB1		HLA-DRB3/4/5	HLA-DQB1
A*2501	B*1529	B*4702		DRB1*08022	DRB1*1332		
A*2502	B*1530	B*4703		DRB1*08032	DRB1*1333		
A*2601	B*1531	B*4801		DRB1*08041	DRB1*1401		
A*2602	B*1532	B*4802		DRB1*08042	DRB1*1402		
A*2603	B*1533	B*4803		DRB1*08043	DRB1*1403		
A*2604	B*1534	B*4901		DRB1*0805	DRB1*1404		
A*2605	B*1535	B*5001		DRB1*0806	DRB1*1405		
A*2606	B*1537	B*5002		DRB1*0807	DRB1*1406		
A*2607	B*1538	B*51011		DRB1*0808	DRB1*1407		
A*2608	B*1539	B*51012		DRB1*0809	DRB1*1408		
A*2609	B*1801	B*51021		DRB1*0810	DRB1*1409		
A*2901	B*1802	B*51022		DRB1*0811	DRB1*1410		
A*2902	B*1803	B*5103		DRB1*0812	DRB1*1411		
A*2903	B*1805	B*5104		DRB1*0813	DRB1*1412		
A*3001	B*2701	B*5105		DRB1*0814	DRB1*1413		
A*3002	B*2702	B*5106		DRB1*0815	DRB1*1414		
A*3003	B*2703	B*5107		DRB1*0816	DRB1*1415		
A*3004	B*2704	B*5108		DRB1*0817	DRB1*1416		
A*3006	B*27052	B*5109		DRB1*0818	DRB1*1417		
A*31012	B*27053	B*5110		DRB1*0819	DRB1*1418		
A*3201	B*2706	B*5111N		DRB1*09012	DRB1*1419		
A*3202	B*2707	B*52011		DRB1*1001	DRB1*1420		
A*3301	B*2708	B*52012		DRB1*11011	DRB1*1421		
A*3303	B*2709	B*5301		DRB1*11012	DRB1*1422		
A*3401	B*2710	B*5302		DRB1*11013	DRB1*1423		
A*3402	B*2711	B*5401		DRB1*1102	DRB1*1424		
A*3601	B*2712	B*5501		DRB1*1103	DRB1*1425		
A*4301	B*3501	B*5502		DRB1*11041	DRB1*1426		
A*6601	B*3502	B*5503		DRB1*11042	DRB1*1427		
A*6602	B*3503	B*5504		DRB1*1105	DRB1*1428		
A*6603	B*3504	B*5505		DRB1*1106	DRB1*1429		
A*68011	B*3505	B*5506		DRB1*1107	DRB1*1430		
A*68012	B*3506	B*5601		DRB1*11081	DRB1*1431		
A*6802	B*3507	B*5602		DRB1*11082	DRB1*15011		
A*6803	B*3508	B*5603		DRB1*1109	DRB1*15012		
A*68032	B*35091	B*5604		DRB1*1110	DRB1*15021		
A*6804	B*35092	B*5701		DRB1*1111	DRB1*15022		
A*6805	B*3510	B*5702		DRB1*1112	DRB1*15023		
A*6806	B*3511	B*5703		DRB1*1113	DRB1*1503		
A*6901	B*3512	B*5704		DRB1*1114	DRB1*1504		
A*7401	B*3513	B*5801		DRB1*1115	DRB1*1505		
A*7402	B*3515	B*5802		DRB1*1116	DRB1*1506		
A*7403	B*3516	B*5901		DRB1*1117	DRB1*16011		
A*8001	B*3517	B*67011		DRB1*1118	DRB1*16012		
	B*3518	B*67012			DRB1*16021		
	B*3519	B*7301			DRB1*16022		
	B*3520	B*7801			DRB1*1603		
	B*3521	B*78021			DRB1*1604		
	B*3701	B*78022			DRB1*1605		
	B*3702	B*8101			DRB1*1607		
	B*3801	B*8201			DRB1*1608		
	B*38021						
	B*38022						

2.4. Setting Up PCR

1. Method is described in the methods section. The extra ingredient required is 5U/µL *Taq* DNA polymerase (Bioline).

2.5. Gel Electrophoresis

1. Orange G loading buffer: combine 300 mL of glycerol, 250 mL of 2X Tris-borate-EDTA (TBE), 550 mL dH_2O and 0.25 g orange G (Sigma; O7629). Store at 20°C.
2. 2X and 0.5X TBE buffer: to make 2X TBE, dissolve 216 g of Tris-base and 110 g of boric acid in 9 L of dH_2O. Add 8 mL of 0.5 *M* EDTA, pH 8.3, and make up to 10 L with ddH_2O. To make 0.5X TBE, combine 750 mL of dH_2O with 250 mL of 2X TBE.
3. 1% agarose (one L): 10 g electrophoresis-grade agarose (Helena BioSciences; 8201-03), 1 L 0.5X TBE, 10 µL 10mg/mL ethidium bromide (Sigma; E1510). Dissolve, by heating in a microwave, 10 g of agarose in 400 mL 0.5X TBE. Top up to 1 L with 0.5X TBE and add 10 µL of 10 mg/mL ethidium bromide solution. Agarose solutions can be stored at 50°C for up to 1 wk.

3. Methods

3.1. DNA Extraction

1. This method is a modification of Miller's salting-out procedure ***(32)***, in which the use of proteinase K is omitted and an organic solvent extraction phase is added. This yields large quantities of good quality DNA in less than 30 min which is suitable for PCR-SSP.
2. Centrifuge 5 mL of EDTA or trisodium anticoagulated blood and aspirate buffy coat into a 15-mL polypropylene tube (Alpha; LW3075).
3. Add 10 mL of RCLB, invert several times, and leave to stand for 5 min. Centrifuge at 1000*g* for 10 min.
4. Pour off supernatant and gently rinse pellet in 2 mL of RCLB. The pellet should be white with a pink halo. If there is too much hemoglobin, resuspend the pellet in RCLB, agitate, and centrifuge. When

the pellet is homogeneously white, it can be stored at –70°C or you can continue to the next step.

5. Resuspend pellet in 3 mL of NLB+SDS (warm NLB+SDS if precipitate is visible). Add 1 mL of 6 *M* NaCl, vortex mix (precipitate should be visible). Add 2 mL of chloroform and shake until homogenous milky solution is seen. Centrifuge for 10 min at 1000 *g*.
6. Aspirate the DNA (top phase) into a 20-mL tube. If the DNA phase is not clear in appearance, transfer to a clean polypropylene tube and repeat the chloroform extraction step. Do not suck up any protein from the interface. Add 2 vol of 95% ethanol and gently rock until all of the DNA is precipitated. Centrifuge for 5 min at 700*g* and resuspend in 70% ethanol, centrifuge, and repeat this washing step.
7. Transfer the DNA precipitate into a sterile 0.5-mL microcentrifuge tube, pellet the DNA, and remove the excess ethanol either by centrifugal evaporation, lyophilization or by allowing it to dry on the bench. Resuspend the DNA in 300 µL of sterile ddH_2O. From 5 mL of blood you can expect to obtain DNA concentrations in the range of 0.2–1.0 µg/µL.

3.2. Dispensing Primer Mixes

1. Tested primer mixes (*see* **Notes 1** and **2**) should be aliquoted in 1-mL vol in 1-mL straight tubes (Integra Biosciences, Ijamsville, MD, USA; 8110-00), which are suitable for placing in standard 96-well format in a 96-well rack. These tubes and racks are suitable for use both with 8/12-channel hand-held electronic multidispensing pipets (Anachem; EP-M8-250; or Alpha; 710-310) and also with 96-well robotic dispensers such as the Robbins Hydra (Robbins Scientific, Sunnyvale, CA, USA; 1029-60-1).
2. Using a 12-channel electronic dispensing pipet, add 10 µL of mineral oil (Sigma; M8662) to 96-well PCR plates (Advanced Biotechnology, Columbia, MO, USA; AB-0600). Dispense 5 µL of each primer mixture into the appropriate wells of the PCR plates using the Robbins Hydra dispenser.
3. Completed trays may be stored for 6–12 mo at –30°C, preferably in sealed bags or with individual plate sealers (Corning Costar, Acton, MA, USA; 6524).

3.3. Setting Up PCR-SSP Using PCR Buffer

1. Thaw out plate(s) containing the primer mixes.
2. Thaw out a 13.3-mL aliquot of PCR buffer and add 64 µL of 5 U/mL *Taq* DNA polymerase. This mixture will keep at 4°C for at least 1 wk.
3. Count how many individual PCR-SSP reactions are required for each individual DNA sample (protocol given here is for 144 reactions). For each 5 µL primer mixture, 8 µL of PCR buffer/DNA/*Taq* mixture is added. It is important, for maintenance of the MgCl concentration, that the ratio of PCR buffer to all other PCR ingredients is 1:0.6. Thus, for 144 reactions add 16 µL of DNA to 1416 µL of PCR buffer/*Taq* mixture.
4. Vortex mix briefly and pour mixture into a disposable trough (Saxon Europe).
5. Place six 250-µL tips on an 8-channel electronic multidispenser and draw up an appropriate vol. Dispense 8 µL of DNA/PCR buffer /*Taq* mixture to 6 wells at a time; keep the tip at the top edge of the mineral oil meniscus, and allow the mixture to roll off the tip and through the mineral oil. On no account allow the tips to touch the primer mixture otherwise carry over and, consequently, false positive amplifications may occur. On addition of the PCR buffer mixture to the primer mixes, the cresol red will change colour from yellow to purple.
6. When the tray is complete, seal with a fresh tray sealer, centrifuge briefly (200*g* for 5 s) to ensure that all PCRs are mixed and submerged below the oil (vortex mixing completed plates is not recommended).

3.4. Setting Up PCR-SSP Using Heparin-Contaminated DNA

1. Make a 0.2 U/µL solution of heparinase II (Sigma; H6512) by adding 50 µL of ddH_2O to a 10-U vial.
2. Add 5 µL heparinase per 15 µL DNA, agitate, and incubate for 90 min at 37°C.
3. Add to PCR buffer mixture as normal (*see* **Subheading 3.3.**). Heparinase activity is destroyed by freeze-thawing.

3.5. PCR Amplification Program

1. This program is suitable for Perkin Elmer 9600 (Cetus), PTC-100™ or PTC-200 machines (MJ Research, Waltham, MA, USA). 96°C

for 60 s; 5 cycles of 96°C for 20 s, 70°C for 45 s, and 72°C for 25 s; 21 cycles of 96°C for 25 s, 65°C for 50 s, 72°C for 30 s; 4 cycles of 96°C for 30 s, 55°C for 60 s, and 72°C for 90 s. Cool by ramping to 20°C for 30 s prior to termination of the program. Program takes approx 1.5 h to run.

2. Some thermoplastics used for PCR are not an exact fit for every PCR machine and, consequently, accurate heat transfer to the PCR may be effected. Too ensure correct thermodynamics, we dip the PCR vessels into a little light paraffin oil and blot excess on tissues before placing in PCR machines.
3. Apply firm and even pressure to the top surface PCR vessels during thermal cycling. Preferably use a heated lid.

3.5. Electrophoresis

1. Use large electrophoresis tanks (Flowgen, Lichfield, Staffordshire, UK; G3-0416) utilizing gel trays accommodating gel combs with teeth spatially separated for use with multichannel pipets.
2. Pour 400 mL of 1% agarose into the taped-off gel tray, insert the combs, and allowing 20 min to set.
3. Fill electrophoresis tank with 2.2 L of 0.5X TBE (can be left in tank and reused at least 15X). Remove tape and combs and submerge gel tray in tank.
4. Using a multichannel repeating syringe (Robbins Scientific; 1021-03-5) add 5 μL of orange G loading buffer.
5. Using a multichannel pipet load 18 mL of 8 or 12 PCRs at a time to the gel (depending on tray layout).
6. Electrophorese for 20 min at 200 V or until the orange G can be seen to have traveled 3 cm.

3.6. Gel Photography

1. Visualize the PCR amplicons using 312 nm UV transillumination. Record results by gel photography using either Polaroid photography in conjunction with Wratten 22 and 2a filters or any other suitable imaging system. Two rapid exposures with a shutter speed of 4 and aperture of f5.6 are recommended.
2. To facilitate identification of positive PCRs, it is recommended that the electrophoresis lanes are labeled by using an overhead projector

(OHP) acetate with the lane numbers printed in the correct spatial orientation. The OHP acetate is laid over the gel in the correct position prior to photography. It is recommended that the OHP is laminated (to prevent wear) and that windows between one row of numbers and another (where the PCR amplicons appear) are cut out to reduce interference from the plastic fluorescing in UV light.

3.7. Interpretation of Results

PCR-SSP interpretation of HLA genotypes is relatively easy, and generally results can be interpreted with little or no prior experience. Each PCR-SSP reaction is deemed to have worked if the control amplification is present. The control amplicon in all reactions except the Bw4 and Bw6 reactions, is a 796-bp fragment from the third intron of *DRB1* , as previously used by Olerup and Zetterquist *(14)*, whereas the 256-bp control amplicon in the *Bw4* and *Bw6* reactions is from exon 15 of the adenamoutous polyposis coli gene, as previously used by Sadler et al. *(24)*. Positive allele-specific amplifications are identified by the presence of a correct sized PCR allele-specific amplicon, whereas absence of an allele-specific amplicon implies absence of the alleles identified in a given primer mixture. If a reaction has neither control nor allele-specific amplicons, the reaction has failed and is deemed "not tested." The alleles that would have been amplified in this reaction are, therefore, also not tested. Fortunately, many alleles are amplified in more than one reaction, so sporadic PCR failures do not often affect full assignation of a genotype. If all of the reactions have failed, then the whole result is not tested and must be repeated (*see* troubleshooting in **Notes 4–13**). Alleles are assigned by identifying the pattern of positive and negative reactions, and interpreted with reference to the information given in **Table 2**.

4. Notes

1. Getting the right concentration of control primers in the stock control solutions is of vital importance, as these solutions are the basis for all the primer mixes. The concentration of primers must not be so

high that the allele-specific amplicon is out-competed by the controls. On the other hand, if the concentration of control primers are too low, then the control amplicon will be difficult or impossible to visualize, and many primer mixes may appear to be not tested. To establish a good working concentration, titrate the control primers (suggested titration: 5, 2.5, 1.25 μL/mL) in the presence of a constant concentration of a pair of allele-specific primers. Test this titration against some DNA samples of varying quality and of varying genotype (some positive and some negative for the allele-specific primers). The optimal concentration of control primers is found when the control amplicon does not out-compete the allele-specific amplicon in HLA allele positive reactions and yet is present in all allele negative samples.

2. The majority of primer mixes will function well using the recommended concentrations given in **Table 2**. However, there is some variability in amplification efficiencies between batches of oligonucleotides and primer mixes, so that every new synthesis of primer mixes must be properly tested, as optimal primer concentrations will fluctuate from batch to batch. If a large volume (20–50 mL) of primer mixes are being synthesized for the first time, it is advisable to test the recommended concentrations first by making up 0.5 mL and testing on appropriate control samples. An ideal primer mixture should produce allele-specific amplicons that are easily visible in all expected positive samples and clearly negative in expected negative samples; if possible, test some DNA of poor quality as well as normal DNA, to ensure that a robust primer mixture is obtained. If a primer mixture is weak or negative with expected positive DNA samples, increasing the concentration of primers seldom fails to improve results. Similarly, false positive amplifications can be removed by titrating either both primers or one of the primers (asymmetric titration). Sense primers with the 3'- end at position 272 (e.g., primer 192) all utilize internal mismatches at positions 12 and 14 of the primer. Primer mixes involving these primers require careful testing and subsequent titration of the sense primer to avoid weak amplification of similar sequences. For example, primer mixture 73 is designed to amplify *B*1501* and some other *B*15* alleles but not *B*1502*. Because the primer 192 is only mismatched internally between *B*1501* and *B*1502*, it is possible that *B*1502* might be amplified in this mixture, and therefore, this mixture must be tested with *B*1502* as well as *B*1501* before being put into general use.

Initially, some primer mixes (especially primer mixes 152 and 76) failed to produce allele-specific and control-specific amplicons, instead producing primer-dimer. These primers also need careful reduction of the allele-specific primers to a point where they amplify alleles but do not go to primer-dimer.

Some primer mixes only detect very rare alleles (such as *HLA-A*4301* or *B*4802*) and are, consequently, difficult to positively test for unless reference DNA samples are available. If reference samples are not available, test the individual primers in different combinations to estimate the optimal concentrations. For example, if the *HLA-A*4301* primer mixture (PM12) is not positively tested, the sense primer (primer 174), which is specific for *A*2901-3* as well as *A*4301*, could be tested in conjunction with another antisense primer that also detects *A*2901-3*. Similarly, the antisense primer (primer 298) could be tested, and the optimal concentration of primers could be extrapolated to the *A*4301*-specific primer mixture. If a primer cannot be tested in this way because its sequence is unique to a rare allele, it is best to use it at the highest possible concentration that does not produce false positives.

3. Each batch of PCR buffer should be tested in comparison with the previous batch before general use. It is best to test several different DNA samples of different phenotypes to ensure that the buffer is efficient for most primer mixes. The most common defect found when testing PCR buffer is that the control bands appear strong, but the allele bands appear weak or even nonexistent. This is commonly caused by an imbalance in the MgCl:dNTP ratio; either too high a concentration of MgCl or too low a concentration of dNTPs has been used. It is thought that as dNTPs efficiently chelate MgCl, an excess of dNTPs sequesters free MgCl and, thus, deprives *Taq* of the magnesium that it requires as a cofactor.
4. If all reactions have failed (no allele, no control-specific amplicons):
   a. Poor quality or too little DNA. Test another DNA sample previously shown to work to test reagents and PCR machine. If poor quality DNA is suspected, using less DNA with 50% more *Taq* may work. If it looks like there is lots of DNA by gel electrophoresis, then it could be heparin or protein contamination. If heparin contamination is suspected, use the heparinase protocol. If protein contamination is suspected, try re-extracting by adding 20% (v/v) 6 *M* NaCl to the remaining DNA and an equal vol of

chloroform, vortex mix, centrifuge at high speed in a microfuge for 5 min, extract the aqueous DNA phase, and ethanol-precipitate as usual.

b. If DNA samples shown to previously work start failing, it is possible that one of the PCR ingredients is faulty or that the DNA sample is degrading over time. Always keep a batch of working frozen stock ingredients, so that troubleshooting can be made easier. Fluctuations can be due to variation in *Taq* supply.

5. Generally weak reactions:
   a. Usually due to insufficient or poor quality DNA. Try adding more DNA.
   b. Incorrectly made buffer or poor/dilute *Taq*. Remake buffer or try increasing *Taq* concentrations. Some laboratories use 2 or 3X the *Taq* concentration that we use.
   c. Inefficient PCR machine. Not all PCR machines work well for this PCR-SSP protocol. If reactions are always weak, try elongating some of the PCR program sections or try lower annealing temperatures at the start of the program (68°C instead of 70°C).
6. Too many allele-specific amplicons in one locus:
   a. Possible new allele. Try to confirm by PCR mapping techniques ***(33,34)*** or by using another molecular method such as sequencing or sequence-specific oligonucleotide primers (SSOP).
   b. Sample contaminated with another DNA sample. Most such contaminations would yield extra bands at all loci tested but it is possible to get a combination of alleles in two samples such that the contamination was only noticed at one locus. Most accidental contamination involves small amounts of contaminant being introduced into a larger amount, so contaminating bands are typically weak but consistent.
   c. Sample contaminated with a locus-specific amplicon from another part of the laboratory. Try to minimize contamination by maintaining good laboratory procedures and spatial separation of pre- and postamplification areas.
   d. Incorrectly made up or contaminated primer mixture. Retest suspected primer mixture and resynthesize primer mixture if faulty.
7. Too many allele-specific amplicons in all loci.
   a. PCR machine error. If the PCR program is interrupted and restarted (especially at the early stages), multiple bands are seen

due to the low stringency PCR induced. Use a PCR machine that gives error messages when programs have been interrupted.

b. Sample or PCR buffer contamination. Remake solutions if contaminated.

8. No allele-specific amplicons at one locus.
   a. Homozygous example of a new allele not detected by the given PCRs. This is unlikely.
   b. Incorrectly made primer mixes. Ensure all primer mixes are tested before use.
   c. Incorrect buffer mixture. If the dNTP to MgCl ratio is incorrect, it can effect one locus more than another, so that it appears as if there are no alleles at one particular locus. For unknown reasons, the HLA-B locus, and especially the primer mixes specific for B*44, B*08, B*51, Bw4 and Bw6 are most susceptible to this phenomenon. Classically, false negative allele amplifications due to incorrect MgCl concentrations are associated with much stronger control amplicons and much weaker than normal allele-specific amplicons.
9. Individual reaction failure. Approx 0.5–1% of PCR-SSP reactions spontaneously fail for no apparent reason. Possible causes include: incomplete PCR, PCR inhibitory contaminant in an individual well, or failure of individual PCR vessels. If a reaction has failed and no primer or primer-dimer is visible on the gel, it is likely that either the agarose well was incomplete, or the reaction was not loaded into the gel properly.
10. Allele-specific bands present, but no controls.
    a. Degraded DNA may produce only small amplicons, such as the allele-specific amplicons, but not larger amplicons such as the control amplicon.
    b. Insufficient PCR extension time. Try increasing the time the PCR program spends at the extension temperature (72°C).
    c. PCR machine needs recalibration.
    d. Concentration of control primers is too low.
11. Control-specific amplicons, but no alleles.
    a. Magnesium concentration too high; recalibrate PCR buffer.
    b. PCR program is inefficient. Try different PCR programs.
    c. Poor fit of PCR tubes/plate into PCR block. The bottom of the PCR vessel must be in direct contact with the PCR block, other-

wise, the correct temperature will not be applied to the PCRs. If the fit is suspect, dip the vessels in a little light paraffin oil to coat the exterior of the vessel before placing in the PCR machine.

d. Insufficient pressure from above. If pressure is not applied to the PCR plate the plate may lift out of the block slightly, or the thermoseal may peel off. Either way, you end up with different PCR thermodynamics, which can produce allele drop-out.

12. Part of the typing has worked well, but the remainder has failed.
    a. PCR machine failure. This is a common failure if a PCR machine is used intensively. Test block uniformity by amplifying 96 identical reactions in one plate. If a problem does exist contact a PCR service engineer.
    b. PCR plate not placed in machine properly.
    c. Uneven pressure applied during PCR.
    d. Gel artefact caused by insufficient ethidium bromide.
13. Updating the typing system. The ever increasing number of HLA alleles makes updating large PCR-SSP typing systems very difficult. Each new allele must be cross-referenced with the two (or more) *cis*-located polymorphisms identified in each primer mixture, and internal mismatches with primer lengths must also be considered. This is a difficult time-consuming task, which frequently leads to errors being made. To facilitate updating various primer mixture sets, we have developed a computer program known as "PCR-SSP Manager Program." This program allows all new HLA sequences to be aligned. Once a new allele is inserted, the specificities of all the primers and, consequently, the primer mixes is adjusted, and the new updated specificity list for a tray of reactions is produced. Furthermore, the program can assist the investigator developing new primer mixes by suggesting new primer mixes to sort out the inevitable conundrums created by new alleles. Added to this are various interpretation facilities that make the program an invaluable tool for anyone using PCR-SSP for HLA typing. This program is now available on request from the Oxford Transplantation Immunology Laboratory.

    Inevitably, further primer mixes must be included to accommodate new alleles and higher resolution requirements. Indeed, we now use 192 reactions for class I, 96 reactions for DR and DQ, and 96 reactions for DP. Many of these extra reactions will shortly be published or have already been published ***(29)*** . Most PCR-SSP primer mixes taken from papers using different PCR protocols and parameters can be successfully adapted to the universal method presented here, by ensuring that the primer temperatures are 60° or 62°C and that the working concentration of the primers is properly evaluated prior to general use.

References

1. Jordan, F., McWhinnie, A. J., Turner, S., et al. (1995) Comparison of *HLA-DRB1* typing by DNA-RFLP, PCR-SSO and PCR-SSP methods and their application in providing matched unrelated donors for bone marrow transplantation. *Tissue Antigens* **45**, 103–110.
2. Mytilineos, J., Lempert, M., Middleton, D., et al. (1997) HLA class I typing of 215 "HLA-A, B, -DR zero mismatched" kidney transplants. *Tissue Antigens* **50**, 355–358.
3. Bunce, M., Barnardo, M. C. N. M., Procter, J., Marsh, S. G. E., Vilches, C., and Welsh, K. I. (1996) High resolution *HLA-C* typing by PCR-SSP: identification of allelic frequencies and linkage disequilibria in 604 unrelated random UK Caucasoids and a comparison with serology. *Tissue Antigens* **48**, 680–691.
4. Lorentzen, D. F., Iwanaga, K. K., Meuer, K. J., Moritz, T. L., and Watkins, D. I. (1997) A 25% error rate in serologic typing of HLA-B homozygotes. *Tissue Antigens* **50**, 359–365.
5. Yu, N., Ohashi, M, Alosco, S., et al. (1997) Accurate typing of HLA-A antigens and analysis of serological deficiencies. *Tissue Antigens* **50**, 380–386.
6. Mytilineos, J., Scherer, S., and Opelz, G. (1990) Comparison of RFLP-*DR beta* and serological HLA-DR typing in 1500 individuals. *Transplantation* **50**, 870–873.
7. Barnardo, M. C., Davey, N. J., Bunce, M., et al. (1996) A correlation between HLA-C matching and donor antirecipient CTL precursor frequency in bone marrow transplantation. *Transplantation* **61**, 1420–1423.
8. Speiser, D. E., Tiercy, J. M., Rufer, N., et al. (1996) High resolution HLA matching associated with decreased mortality after unrelated bone marrow transplantation. *Blood* **87**, 4455–4462.
9. Opelz, G., Scherer S., and Mytilineos, J. (1997) Analysis of HLA-DR split-specificity matching in cadaver kidney transplantation: a report of the Collaborative Transplant Study. *Transplantation* **63**, 57–59.

10. Fanning, G. C., Bunce, M., Black, C. M., and Welsh, K. I. (1997) Polymerase chain reaction haplotyping using 3' mismatches in the forward and reverse primers: application to the biallelic polymorphisms of tumor necrosis factor and lymphotoxin alpha. *Tissue Antigens* **50**, 23–31.
11. Procter, J., Crawford, J., Bunce, M., and Welsh, K. I. (1997) A rapid molecular method (PCR-SSP) to genotype for *ABO* blood group and secretor status and its potential for organ transplants. *Tissue Antigens* **50**, 475–483..
12. Wu, D. Y., Ugozzoli, L., Pal, B. K., and Wallace, R. B. (1989) Allele-specific enzymatic amplification of beta-globin genomic DNA for diagnosis of sickle cell anemia. *Proc. Natl. Acad. Sci. USA* **86**, 2757–2760.
13. Newton, C. R., Graham, A., Heptinstall, L. E., et al. (1989) Analysis of any point mutation in DNA. The amplification refractory mutation system (ARMS). *Nucleic Acids Res.* **17**, 2503–2516.
14. Olerup, O. and Zetterquist, H. (1992) *HLA-DR* typing by PCR amplification with sequence-specific primers (PCR-SSP) in 2 hours: an alternative to serological DR typing in clinical practice including donor-recipient matching in cadaveric transplantation. *Tissue Antigens* **39**, 225–235.
15. Chien, A., Edgar, D. B., and Trela, J. M. (1976) Deoxyribonucleic acid polymerase from the extreme thermophile *Thermus aquaticus. J. Bacteriol.* **127**, 1550–1557.
16. Tindall, K. R. and Kunkel, T. A. (1988) Fidelity of DNA synthesis by the *Thermus aquaticus* DNA polymerase. *Biochemistry* **27**, 6008–6013.
17. Kwok, S., Kellogg, D. E., McKinney, N., et al. (1990) Effects of primer-template mismatches on the polymerase chain reaction: human immunodeficiency virus type 1 model studies. *Nucleic Acids Res.* **18**, 999–1005.
18. Bunce, M., Taylor, C. J., and Welsh, K. I. (1993) Rapid *HLA-DQB* typing by eight polymerase chain reaction amplifications with sequence-specific primers (PCR-SSP). *Hum. Immunol.* **37**, 201–206.
19. Olerup, O., Aldener, A., and Fogdell, A. (1993) *HLA-DQB1* and *-DQA1* typing by PCR amplification with sequence-specific primers (PCR-SSP) in 2 hours. *Tissue Antigens* **41**, 119–134.
20. Knipper, A. J., Hinney, A., Schuch, B., Enczmann, J., Uhrberg, M., and Wernet, P. (1994) Selection of unrelated bone marrow donors by PCR-SSP typing and subsequent nonradioactive sequence-based typ-

ing for *HLA DRB1/3/4/5, DQB1*, and *DPB1* alleles. *Tissue Antigens* **44**, 275–284.

21. Browning, M. J., Krausa, P., Rowan, A., Bicknell, D.C., Bodmer, J. G., and Bodmer, W. F. (1993) Tissue typing the HLA-A locus from genomic DNA by sequence-specific PCR: comparison of HLA genotype and surface expression on colorectal tumor cell lines. *Proc. Natl. Acad. Sci. USA* **90**, 2842–2845.
22. Krausa, P., Bodmer, J. G., and Browning, M.J. (1993) Defining the common subtypes of *HLA A9, A10, A28* and *A19* by use of ARMS/PCR. *Tissue Antigens* **42**, 91–99.
23. Krausa, P. and Browning, M. J. (1996) A comprehensive PCR-SSP typing system for identification of *HLA-A* locus alleles. *Tissue Antigens* **47**, 237–244.
24. Sadler, A. M., Petronzelli, F., Krausa, P., et al. (1994) Low-resolution DNA typing for *HLA-B* using sequence-specific primers in allele- or group-specific ARMS/PCR. *Tissue Antigens* **44**, 148–154.
25. Savelkoul, P. H., de Bruyn-Geraets, D. P., and van den Berg-Loonen, E. M. (1995) High resolution *HLA-DRB1* SSP typing for cadaveric donor transplantation. *Tissue Antigens* **45**, 41–48.
26. Bunce, M. and Welsh, K I. (1994) Rapid DNA typing for *HLA-C* using sequence-specific primers (PCR-SSP): identification of serological and non-serologically defined HLA-C alleles including several new alleles. *Tissue Antigens* **43**, 7–17.
27. Bunce, M., Barnardo, M. C., and Welsh, K. I. (1994) Improvements in *HLA-C* typing using sequence-specific primers (PCR-SSP) including definition of *HLA-Cw9* and *Cw10* and a new allele *HLA-"Cw7/8v."* *Tissue Antigens* **44**, 200–203.
28. Bunce, M., Fanning, G. C., and Welsh, K. I. (1995) Comprehensive, serologically equivalent DNA typing for *HLA-B* by PCR using sequence-specific primers (PCR-SSP). *Tissue Antigens* **45**, 81–90.
29. Gilchrist, F. C., Bunce, M., Lympany, P. A., Welsh, K. I., and du Bois, R. M. (1998) Comprehensive *HLA-DP* typing using polymerase chain reaction with sequence-specific primers and 95 sequence-specific primer mixes. *Tissue Antigens* **51**, 51–61.
30. Bunce, M., O'Neill, C. M., Barnardo, M. C. N. M., et al. (1995) Phototyping: Comprehensive DNA typing for *HLA-A, B, C, DRB1, DRB3, DRB4, DRB5 & DQB1* by PCR with 144 primer mixes utilising sequence-specific primers (PCR-SSP). *Tissue Antigens* **46**, 355–367.

31. Satsangi, J, Jewell, D. P, Welsh, K., Bunce, M., and Bell, J. I. (1994) Effect of heparin on polymerase chain reaction. *Lancet* **343**, 1509–1510.
32. Miller, S. A., Dykes, D. D., and Polesky, H. F. (1988) A simple salting out procedure for extracting DNA from human nucleated cells. *Nucleic Acids Res.* **16**, 1215.
33. Vilches, C., Bunce, M., de Pablo, R., Herrero, M. J., and Kreisler, M. (1995) Anchored PCR cloning of the novel *HLA-Cw*0704* allele detected by PCR-SSP. *Tissue Antigens* **46**, 19–23.
34. Krausa, P., Barouch, D., Bodmer, J. G., and Browning, M. J. (1995) Rapid characterization of *HLA* class I alleles by gene mapping using ARMS PCR. *Eur. J. Immunogenet.* **22**, 283–287.

8

HLA Typing With Reference Strand-Mediated Conformation Analysis

J. Rafael Argüello, Martha Pérez-Rodríguez, Andrea Pay, Gaby Fisher, Alasdair McWhinnie, and J. Alejandro Madrigal

1. Introduction

Comparison of HLA allelic sequences reveals a patchwork pattern in which individual alleles are defined by unique combinations of sequence motifs. Although each individual motif may be shared with other alleles, only a few alleles have specific sequences that are not present in other alleles *(1)*. Thus the main problem facing conventional methods of human leukocyte antigen (HLA) typing is their inherent inability to determine, in heterozygous individuals, to which allele a particular motif belongs *(2)*. To overcome this problem, allele separation must be performed in some samples prior to typing, making the whole procedure technically labor-intensive and time-consuming *(3,4)*.

The use of DNA conformational analysis methods relies on differences in the behavior of DNA molecules of differing sequence in polyacrylamide gel electrophoresis (PAGE). Single-strand conformation polymorphism (SSCP) *(5)*, which separates molecules of different sequence on the basis of the different stem and loop structure that they adopt, is the most widely used conformational method.

From: *Methods in Molecular Biology, vol. 210: MHC Protocols*
Edited by: S. H. Powis and Robert W. Vaughan © Humana Press Inc., Totowa, NJ

However, because single-stranded DNA fragments can adopt multiple stable conformations under identical conditions, resulting in complex banding patterns after electrophoresis *(6)*, this technique is in general restricted to fragments of around 200 bp in length *(7)*. Heteroduplex analysis separates molecules on the basis of the length and distribution of mismatched regions in a heteroduplex. Under given conditions, the conformation of such DNA duplexes will be unique for each different duplex *(8)*. Despite the increased stability of DNA duplexes as compared with single-strand DNA, the band patterns generated in conventional heteroduplex analysis are very complex and difficult to interpret. The presence of multiple duplexes can be overcome by incorporating a label (e.g., radioisotope or biotin) into one of the strands of a known reference polymerase chain reaction (PCR) product *(9)* or by isolating single strands from both the test sample and the known reference PCR product *(10)*. However, factors, such as lack of reproducibility due to inter- and intra-gel variability and the subjective interpretation of electrophoresis banding patterns, have limited the general use of all these conformational methods for HLA typing.

Reference strand-mediated conformation analysis (RSCA) utilizes simple DNA manipulations, together with laser-based instrumentation and computer software, to detect differences in DNA conformation between HLA alleles that differ by as little as one nucleotide substitution *(11,12)*. RSCA overcomes the problem of complex banding patterns generated in conventional heteroduplex analysis by utilizing a locus-specific fluorescent-labeled reference (FLR) DNA fragment, generated by PCR using a primer that is labeled at the 5' end with a fluorescent dye (**Fig. 1**). This locus-specific FLR is then hybridized with the PCR products of the *HLA* locus to be tested. Because of similarity between the reference DNA and the test sample, duplexes will form between the sense and antisense strands of all the DNA molecules present in the mixture (**Fig. 1**). The resulting molecules contain double-stranded regions in which the nucleotide sequence is complementary, but regions of nucleotide sequence mismatch lead to the formation of loops and bulges along the length of the DNA molecule. However, only

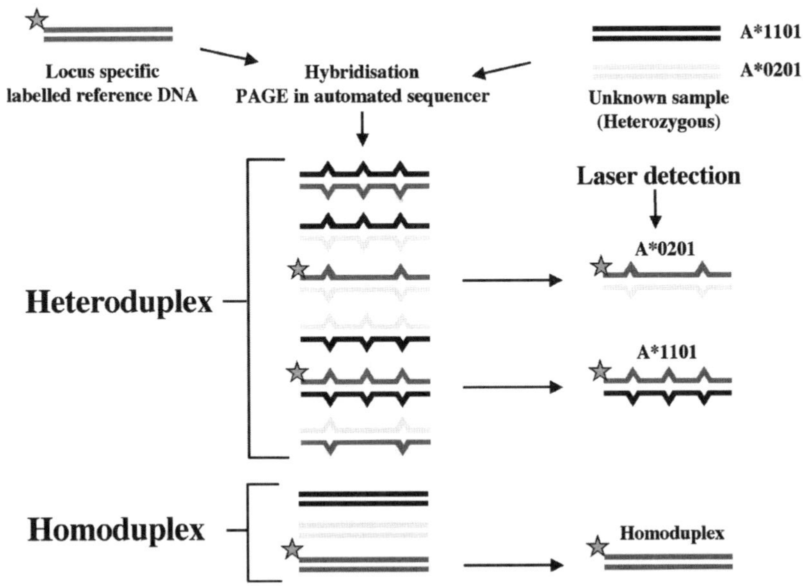

Fig. 1. DNA from the locus of interest is amplified by PCR. The amplified product is hybridized with a locus-specific FLR DNA molecule. This FLR molecule has been prepared such that the sense strand has a fluorescent label attached. Because of the similarity between the reference DNA and the sample tested, duplexes will form between the sense and antisense strands of all the DNA molecules. However as the reference sense strand is labeled, only duplexes formed with this strand will be detected after electrophoresis in an automated DNA sequencer instrument with a laser detection system. For example, as illustrated, three duplex will be detected for a heterozygous loci.

duplexes formed with the labeled reference strand will be identified by a laser detection system after nondenaturing PAGE analysis in an automated DNA sequencer instrument. The number of heteroduplexes identified will equal the number of HLA alleles in addition to the reference homoduplex band. Differences in mobility of the labeled heteroduplexes will depend entirely on the number and distribution of the mismatches with the reference strand.

Several reference strands can be tested to yield optimal resolution of alleles from a particular locus, since the molecular confor-

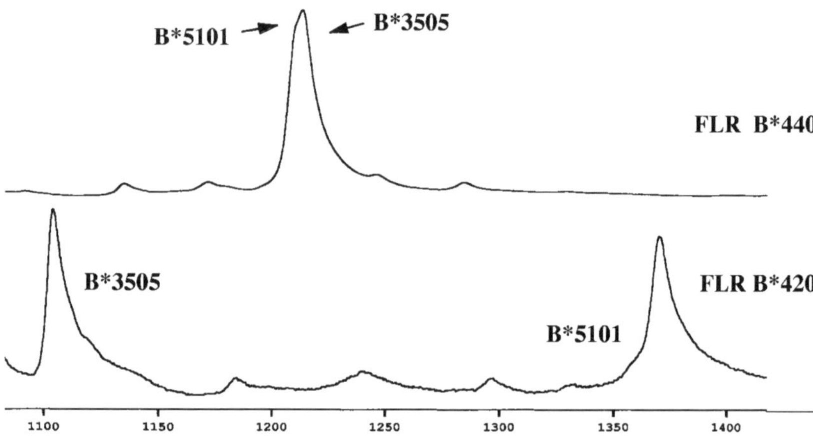

Fig. 2. The use of two FLRs per sample improves the resolution of duplexes that, despite having different conformations, migrate at a similar rate when duplexed with one of the two FLRs. In the first lane, the *HLA* alleles *B*5101* and *B*3505* have overlapping mobilities when hybridized with the FLR *B*4402*, whereas in the second lane, using a different FLR, *B*4201*, their mobilities are significantly different.

mation depends on the differences with respect to the reference strand. **Fig. 2** illustrates how the use of different FLRs under the same electrophoretic conditions alters the molecular conformation and electrophoretic mobility of two HLA-B alleles, allowing the resolution of alleles differing by a single nucleotide substitution. Unlike other conformational techniques, such as denaturing gradient gel electrophoresis (DGGE) ***(13)***, heteroduplex analysis, and SSCP, in which the velocity of migration changes during electrophoresis, in RSCA the migration for each DNA duplex remains constant under given electrophoretic conditions. Furthermore, in other conformational techniques, the resolution of DNA fragments of different mobility depends on the effective distance that is migrated during electrophoresis. This reduces the resolution of DNA strands with low mobility, because they have migrated a reduced distance compared to those with high mobility. In RSCA, this problem is overcome by having the detection system at a fixed point in the bottom of the gel where all measurements are performed. In this way, each duplex is detected after having run for different times but at the same (maximum) distance. Therefore, the resolution of RSCA is determined more by the gel length used than by the electrophoretic conditions.

2. Materials

2.1. Preparation of Genomic DNA

To obtain the best possible PCR amplification results, a pure sample of genomic DNA should be used. DNA can be extracted using the standard phenol-chloroform method or by one of the many available commercial kits.

2.2. Oligonucleotide Primers

1. Primers used for amplification of HLA class I and II test and reference alleles, modified from Cereb et al. *(14)* Fernandez-Vina and Bignon *(15)*, are shown in **Table 1**.

2.3. Preparation of Fluorescent-Labeled References

1. The following International Histocompatibility Workshop homozygous cell lines can be used to produce locus specific FLRs (*see* **Table 2**).
2. The 5' primer should be labeled with the fluorochrome Cy5™ at the time of primer synthesis (Amersham Pharmacia Biotech, Piscataway, NJ).
3. Two FLRs are utilized for each locus. For *DQB1* analysis, one FLR is the *DQB1* allele (*DQB1*0402*), and the second is the *DQB2* allele from the same sample (**Table 1**).

2.4. PCR Amplification.

1. 10X NH_4 PCR buffer (Bioline, UK).
2. dNTPs, 12.5 m*M* each (Bioline, UK).
3. $MgCl_2$ (50 m*M*).
4. Distilled water.
5. *Taq* DNA polymerase (Bioline, UK).

Table 1
Primers Used for the Amplification of HLA Class I and II Alleles for RSCA

Primer name	Locus	Anneal site (nucleotide)	Primer sequence 5' to 3'	Reference
LH: 5'AIn RH: 3'AInL	*HLA-A*	Intron 1, 21–46 Intron 3, 66–101	GAAACG/CGCCTCTGT/CGGGGAGAAGCAA TGTTGGTCCCAATTGTCTCCCCTCCTTGTGGGAGGC	14
LH: 5'BIn RH: 3'BinXL	*HLA-B*	Intron 1, 36–57 Intron 3, 37–63	GGGAGGAGCGAGGGGACCG/CCAG GGAGGCCATCCCCGGCGACCTATAGGAGATGGGG	14
LH: 5'CIn RH: 3'Cin	*HLA-C*	Intron 1, 42–62 Intron 3, 9–34	5'-AGCGAGGG/TGCCCGCCCGGCGA 5'-GGAGATGGGGAAGGCTCCCCACTGCC	14

Primer name	Locus	Anneal site (codon)	Primer sequence 5' to 3'	Reference
LH: 2DRBAMP-A RH: 2DRBAMP-B	*HLA-DRB*	Exon 2, 2–8 Exon 2, 87–94	CCCCACAGC A CGT TTC TTG CCG CTG CAC TGT GAA GCT CT A	15
LH: 2DQAAMP-A RH: 2DQAAMP-B	*HLA-DQA1*	Exon 2, 11–18 Exon 2, 80–87	AT GGT GTA AAC TTG TAC CAG T TT GGT AGC AGC GGT AGA GTT G	15
LH: 2DQBAMP-A RH: 2DQBAMP-B	*HLA-DQB1*	Exon 2, 13–20 Exon 2, 78–84	C ATG TGC TAC TTC ACC AAC GG CTG GTA GTT GTG TCT GCA CAC	15
LH: 2DPAAMP-A RH: 2DPAAMP-B	*HLA-DQA1*	Exon 2, 3–9 Exon 2, 76–82	GCG GAC CAT GTG TCA ACT TAT GC CTG AGT GTG GTT GGA ACG	15
LH: 2DPBAMP-A RH: 2DPBAMP-B	*HLA-DPB1*	Intron 1 Intron 2	GAGAGTGGCGCCTCCGCTCAT GCCGGCCCAAAGCCCTCACTC	15

Table 2
International Histocompatibility Workshop Homozygous Cell Lines

Locus	Allele	Cell line
HLA-A	*A*0101*	STEINLIN
	*A*0217*	AMALA
HLA-B	*B*4201*	RSH
	*B*4402*	SP0010
HLA-C	*Cw*0303*	AMALA
	*Cw*0701*	STEINLIN
HLA-DRB	*DRB1*08021*	SPL
	*DRB1*0101*	plasmid pBActinNeo DRB1*0101 construct
HLA-DQA1	*DQA1*0101*	KAS 116
	*DQA1*05011*	VAVY
HLA-DQB1	*DQB1*0402*	BTB
	DQB2 allele	BTB
HLA-DPA1	*DPA1*0103*	T5-1
	*DPA1*02011*	SAVC
HLA-DPB1	*DPB1*0202*	QBL
	*DPB1*1501*	PLH

2.5. Nondenaturing Polyacrylamide Gel Electrophoresis

1. 6X Ficoll® loading buffer: 15% Ficoll, 0.25% bromophenol blue.
2. Bind silane: 200 mL ethanol (96%), 300 µL acetic acid, 40 µL bind silane (Amersham Pharmacia Biotech).
3. Gel for HLA class I analysis (6% final concentration): 9.6 mL Long Ranger™ gel solution (FMC BioProducts, Rockland, ME, USA), 8 mL 10X Tris-borate EDTA (TBE) buffer (BioWhittaker, Rockland, ME, USA), 61.84 mL water, 48 µL TEMED (Amersham Pharmacia Biotech), 480 µL 10% ammonium persulphate (freshly prepared) (Amersham Pharmacia Biotech).
4. Gel for HLA class II analysis (0.5X mutation detection enhancement acrylamide formulation [MDE] solution final concentration): 20 mL MDE solution (FMC BioProducts), 8 mL 10X TBE buffer (BioWhittaker), 51.5 mL water, 48 µL TEMED, 480 µL 10% ammonium persulphate (freshly prepared).

2.6. Equipment

1. Thermal cycler.
2. Electrophoresis conditions described are for the ALFexpress™ (Amersham Pharmacia Biotech).
3. Fragment Manager™ or Allele Links™ software programs (Amersham Pharmacia Biotech).

3. Methods

3.1. Preparation of Genomic DNA

DNA prepared without the aid of a commercial kit should be resuspended and stored in 1X TE (0.01 *M* Tris, 0.1 m*M* EDTA, pH 8.0) (*see* **Note 1**). As a matter of principle, DNA should be prepared in a designated clean area and not come into contact with PCR product or any reagents, equipment, or materials exposed to PCR product.

3.2. PCR Amplification of Test and Reference (FLR) Strands

1. PCR stock solution: 169 µL 10X NH_4 PCR buffer, 12 µL dNTPs (12.5 m*M* each), 34 µL $MgCl_2$ (50 m*M*), 435 µL distilled water. Total vol is 650 µL.
2. Amplification mixture (per 25 µL mixture): 0.5 µL 5' primer (25 pmol/µL), 0.5 µL 3' primer (25 pmol/µL), 8 µL PCR stock solution, 0.25 µL *Taq* DNA polymerase, 13.25 µL distilled water, 2.5 µL DNA (200 ng/µL).
3. Place the PCR tubes in the thermal cycle and perform the following program: 95°C for 4 min, then 33 cycles of 95°C for 30 s, 65°C for 50 s, 72°C for 30 s, followed by 72°C for 8 min single-step extension.
4. After the PCR is complete, run 5 µL of the mixture on a 1.5% agarose gel to check for amplification product

3.3. Hybridization Reaction

1. In a new PCR tube, mix 1 µL of FLR with 3 µL of the respective sample PCR product.

   Set up a separate hybridization reaction for each FLR to be used.

2. Place PCR tubes in the thermal cycler and perform the following program: 95°C for 4 min, 55°C for 5 min (to allow reannealing), 15°C for at least 3 min.
3. After hybridization, 0.8 µL of 6X Ficoll loading buffer is added to each tube for HLA-Cw and HLA class II loci. Two microliters of this mixture can then loaded in PAGE.
4. For HLA-A and B analysis, duplexes are mixed with 2 µL of special loading buffer, which contain two internal DNA markers and is prepared as described below.

3.4. Preparation of Top and Bottom Marker Duplexes for HLA-A Typing

A locus internal markers used with samples generated with both A*0101 and A*0217 FLRs are prepared as follows.

1. The B*5701 allele is amplified by PCR from DNA from the cell line DEM using unlabeled primers.
2. The PCR product is hybridized to the fluorescent-labeled B*4201 reference DNA as described above in **Subheading 3.3.** at the ratio of 1:3 (*B*4201:B*5701*) and mixed with an equal volume of 6X Ficoll loading buffer as described in **Subheading 2.5.**
3. This special loading buffer (2 µL) is added to the appropriate A locus-specific sample prior to electrophoresis.
4. The final mixture (2 µL) can be loaded in the polyacrylamide gel.

3.5. Preparation of Top and Bottom Marker Duplexes for HLA-B Typing

1. B locus markers for use with samples generated with the *B*4201* FLR are prepared by amplifying the *B*4601* allele from the cell line TAB089 and hybridizing with the fluorescent-labeled *B*4402* reference DNA at a ratio of 1:3 (*B*4402:B*4601*).

   B locus markers to be used with samples generated with the fluorescent-labeled B*4402 reference DNA are prepared by amplifying the B*4501 allele from the cell line OMW and hybridizing with the fluorescent labeled B*4201 reference DNA at a ratio of 1:3 (B*4201:B*4501).

2. After these independent hybridizations, which are performed in a total vol of 160 µL, 10 µL of fluorescent-labeled Cw*0701 PCR product from the cell line STEINLIN and 170 µL 6X Ficoll loading buffer are added.
3. This special loading buffer (2 µL) is added to the appropriate B locus-specific sample prior to electrophoresis.
4. The final mixture (2 µL) can be loaded in the polyacrylamide gel.

3.6. Preparation of Nondenaturing Polyacrylamide Gel

1. Before preparing the gel, clean the plates with 10% Synperonic N detergent, scrubbing with a brush to ensure that any residual polyacrylamide is removed from the plates.
2. Dry the plates with lint-free tissue.
3. Wipe the plates (standard size), spacers (0.5 mm), and comb (0.5 mm) with 96% ethanol.
4. Apply bind silane solution to the top half of both plates.
5. The plates can now be clipped together, and the comb can be inserted and fixed with bulldog clips.
6. The TEMED and ammonium persulphate should be the final components added when preparing the gel solution. Care should be taken not to introduce any air bubbles into the solution. A 50-mL syringe should be used to mix and load the solution into the bottom of the gel plates, moving slowly from side to side.
7. Leave to polymerize for 45 min at room temperature, then lift the plates onto position on the ALFexpress instrument, add 1 L of 1X TBE buffer in the top and bottom container, and leave for 2 h at 40°C (*see* **Note 2**).

3.7. Electrophoresis

1. Electrophoresis is performed in an ALFexpress automated sequencer at 30W constant power.
2. Running times are 580 and 300 min for HLA class I and class II, respectively. Warm the running buffer to 40°C before electrophoresis, as any slight temperature difference in the buffer and the gel may alter the mobility of the heteroduplex affecting the reproducibility in the results.

3. The gel temperature is maintained at 40°C during electrophoresis by external cooling system (Amersham Pharmacia Biotech).

3.8. Analysis of Results

The apex position of each peak is assigned a value based on scales generated between the markers in each lane. For HLA-A and -B analysis, the bottom and top markers are assigned the values of 1000 and 2000, respectively. For HLA-C and class II loci analysis, the value of 1 is assigned to the high mobility fluorescent primer signal, and the value of 1000 is assigned to the fastest duplex signal, which in all cases is the fluorescent reference homoduplex. The position of sample fluorescent heteroduplexes is computed from the mobility scales generated. These marker values are also used for alignment of tracks across each gel. Detailed information concerning the use of Fragment Manager™ and Allele Links™ is given in the programs manuals (Amersham Pharmacia Biotech) (*see* **Notes 3–6**).

4. Notes

1. The resolution of duplexes is inversely proportional to the quantity of loaded DNA. Analyzing low quantities of DNA improves not only the resolution of the duplex bands but also allows very accurate discrimination of duplexes that traverse the laser with time differences of less than 1 s. RSCA requires much less DNA, because the detection system is laser-based. By using a sensitive DNA detection system, not only is sample vol reduced, but thinner gels can be run, thus allowing faster electrophoresis and higher resolution.
2. The polyacylamide gels can be reused up to 5 times, but the running buffer (1X TBE) must be changed for each run.
3. The high reproducibility of RSCA relies mainly on two important features, the first one is the use of internal control markers in each lane, and the second is that all measurements of mobility are performed after all analyzed DNA fragments have migrated the same distance. By combining these two elements, the intra- and inter-gel variability is reduced to a minimum. Variations in gel composition across a lane will affect all duplexes equally, including the internal

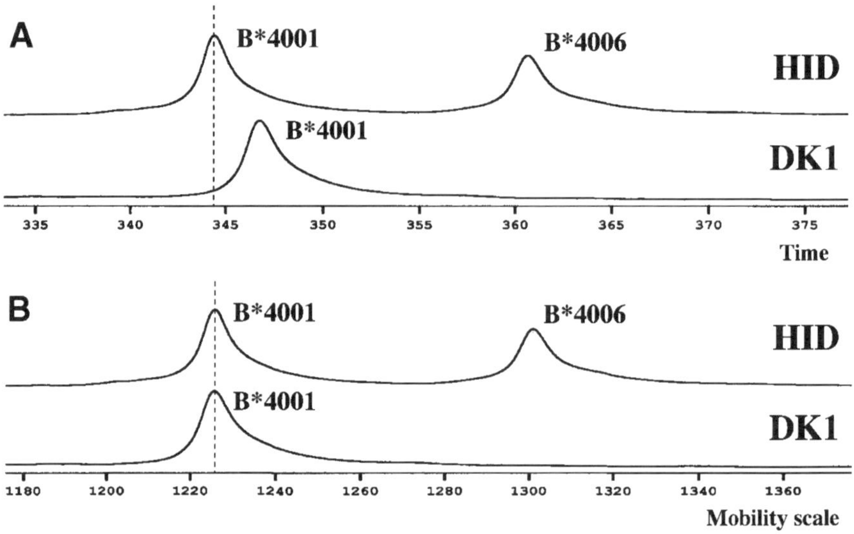

Fig. 3. Correction of gel variability. (**a**) Electropherogram showing HLA-B analysis by RSCA from two DNA samples. Sample HID has the allele *B*4001* and *B*4006*, and sample DKI is homozygous for *B*4001*. Before gel correction, the position of *B*4001* from HID appears to be different from that of the same allele present in sample DKI. (**b**) With the use of internal top and bottom markers (not shown) in each lane, an arbitrary scale can be generated. The value of 1000 is assigned to the bottom markers and 2000 to the top markers. By using a special computer program, all these internal markers can be aligned, thus resulting in the correction of lane-to-lane variability. After correction, the allele *B*4001* has the same position in both samples.

control markers. Therefore, although the absolute mobility of a particular DNA fragment may be different in distinct lanes, its electrophoretic mobility in relation to the control markers will remain constant. A "mobility scale" is generated by the use of these internal markers, which are utilized to normalize lanes across the gel, allowing direct comparison between lanes in the same gel and between different gels as is illustrated in **Fig. 3**. Based on this mobility scale, the software can compensate for lane-to-lane variation in duplex band intensities, allowing direct comparisons of bands with different intensities, which otherwise would not be possible.

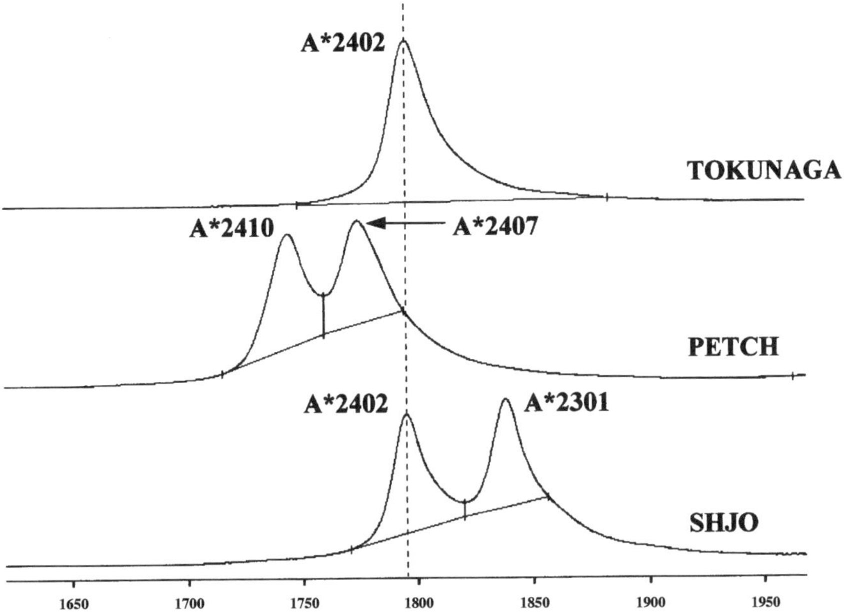

Fig. 4. HLA-A*2402 (from the DNA samples TOKUNAGA and SHJO) and A*2407 (PETCH) differ by one nucleotide in exon 2 at position 282. TOKUNAGA is homozygous for A*2402. The reference A*0101 shares a cytosine with the A*2402, whereas the A*2407 is mismatched with a guanine at position 282.

4. Using RSCA, we have been able to distinguish over 30 *HLA* class I and class II alleles that differ by single nucleotide substitutions. In two blind studies, one for *HLA-A* (unpublished data) and other for *HLA-DPB1* ***(16)***, we were able to type correctly 168 and 104 samples, respectively, that contained allele pairs that differed by a single nucleotide substitution (**Fig. 4**). There was only one pair of alleles (*A*2601* and *A*2602*) that we were unable to resolve using a 20-cm gel length and *A*0101* and *A*0217* as FLRs. However, it is anticipated that these alleles can be resolved by running longer gels and/or by using different FLRs.
5. An ambiguous combination is defined as the occurrence of two different alleles resulting in a heterozygous sequence identical to the heterozygous sequence of a different allele pair *(2)*. *HLA* typing

techniques, such as sequence-specific oligonucleotide (SSO) and sequencing-based typing (SBT) can generate ambiguous results when typing heterozygous individuals. This is due to the inability of these techniques to determine the *cis* or *trans* relationship of the multiple polymorphic motifs present in the two alleles *(**17**)*. Some ambiguous combinations may be resolved by separating the alleles using group-specific amplifications, but others (e.g., *A*0201/A*0205 = A*0202/A*0206*) will require a different strategy (e.g., cloning, allele separation by techniques such as complementary strand analysis *[**10**]* or DGGE *[**13**]*). RSCA is capable of resolving ambiguous combinations without the necessity of extra PCR amplifications or DNA manipulations and without previous knowledge of a broad HLA type *(**16**)*. **Fig. 5** shows the resolution of one sample that contains a combination of alleles that give ambiguous results with SSO and SBT if allelic separation is not performed. In RSCA, there are cases where two or more alleles can have the same mobility when hybridised with one FLR, however, the use of a second or third FLR will resolve these cases. This feature of RSCA precludes ambiguities due to the presence of identical sequence motifs in the amplified DNA fragments.

6. RSCA analysis of HLA class I alleles is performed on amplified DNA fragments that contain exon 2 and 3 together with the intervening intron 2. Consequently, any polymorphism in intron 2 will also affect the mobility of the duplexes. To investigate this potential problem, we have studied over 3800 alleles from different samples representative of various ethnic groups expressing common HLA class I alleles present on the same and different haplotypes. Only one intra-allelic intron 2 variant was found, being present in two samples. This one variant was typed by serology as HLA-A3, yet displayed a RSCA mobility value that differed from the *A*0301* and *A*0302* alleles (**Fig. 6**). Sequencing analysis of this allele demonstrated one nucleotide difference from *A*0301* located at position 19 in intron 2. Therefore, intra-allelic intron 2 polymorphism appears to be a very uncommon feature, however, where it does occur it can be defined by RSCA.

References

1. Parham, P., Lawlor, D. A., Lomen, C. E., and Ennis, P. D. (1989) Diversity and diversification of HLA-A,B,C alleles. *J. Immunol.* **142**, 3937–3950.

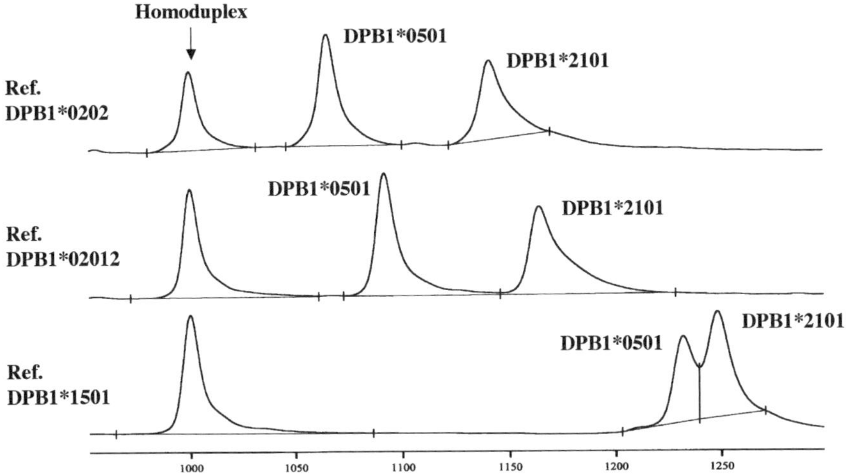

Fig. 5. Resolution of heterozygous ambiguous combination. Electropherogram showing the resolution by RSCA of one sample containing a combination of alleles that are considered ambiguous by SSO and SBT. This heterozygous ambiguous combination is *DPB1*0501/DPB1*2101* = *DPB1*2201/DPB1*3601*. In this case, the two alleles are clearly distinguished by RSCA.

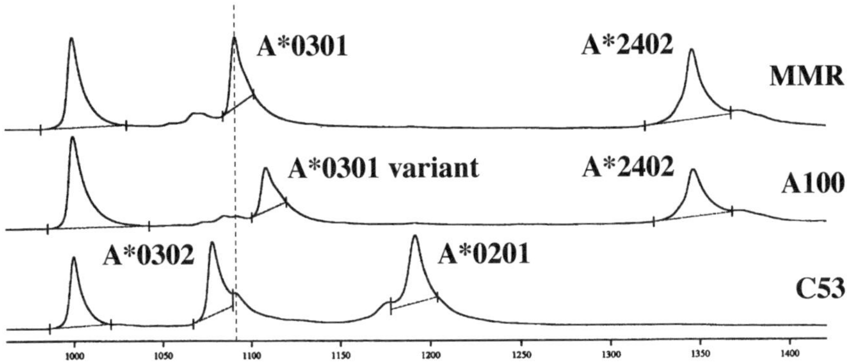

Fig. 6. RSCA resolves alleles that differ by intron polymorphism. The samples MMR and A100 were both defined as containing the allele *A*0301* and are in fact identical in their HLA-A locus exon sequences. However, sample A100 shows a peak shift, which is due to a single nucleotide difference at position 19 in intron 2 resulting in an *A*0301* variant, which can be clearly distinguished from *A*0301* and *A*0302* by RSCA

2. Rozemuller, E. H. and Tilanus, M. G. (1996) A computerized method to predict the discriminatory properties for class II sequencing based typing. *Hum. Immunol.* **46**, 27–34.
3. Arguello, R., Avakian, H., Goldman, J. M., and Madrigal, J. A. (1996) A novel method for simultaneous high resolution identification of *HLA-A*, *HLA-B*, and *HLA-Cw* alleles. *Proc. Natl. Acad. Sci. USA* **93**, 10,961–10,965.
4. Knapp, L. A., Lehmann, E., Hennes, L., Eberle, M. E., and Watkins, D. I. (1997) High-resolution *HLA-DRB* typing using denaturing gradient gel electrophoresis and direct sequencing. *Tissue Antigens* **50**, 170–177.
5. Orita, M., Iwahana, H., Kanazawa, H., Hayashi, K., and Sekiya, T. (1989) Detection of polymorphisms of human DNA by gel electrophoresis as single-strand conformation polymorphisms. *Proc. Natl. Acad. Sci. USA* **86**, 2766–2770.
6. Spinardi, L., Mazars, R., and Theillet, C. (1991) Protocols for an improved detection of point mutations by SSCP. *Nucleic Acids Res.* **19**, 4009.
7. Sheffield, V. C., Beck, J. S., Kwitek, A. E., Sandstrom, D. W., and Stone, E. M. (1993) The sensitivity of single-strand conformation polymorphism analysis for the detection of single base substitutions. *Genomics* **16**, 325–332.
8. Wood, N. A., Clay, T. M., and Bidwell, J. L. (1991) *HLA-DR/Dw* matching by PCR fingerprinting: the origin of PCR fingerprints and further applications. Eur J *Immunogenet.* **18**, 147–153.
9. Zimmerman, P. A., Carrington, M. N., and Nutman T. B. (1993) Exploiting structural differences among heteroduplex molecules to simplify genotyping the *DQA1* and *DQB1* alleles in human lymphocyte typing. *Nucleic Acids Res.* **21**, 4541–4547.
10. Arguello, R., Pay, A. L., McDermott, A., et al. (1997) Complementary strand analysis: a new approach for allelic separation in complex polyallelic genetic systems. *Nucleic Acids Res.* **25**, 2236–2238.
11. Arguello, J. R., Little, A-M., Pay, A. L., et al. (1998) Mutation detection and typing of polymorphic loci through double-strand conformation analysis. *Nat. Genet.* **18**, 192–194.
12. Arguello, J. R., Little, A-M., Bohan, E., Goldman, J. M, Marsh, S. G. E, and Madrigal, J. A. (1998) High resolution HLA class I typing by reference strand mediated conformation analysis (RSCA). *Tissue Antigens* **52**, 57–66.

13. Fischer, S. G. and Lerman, L. S. (1983) DNA fragments differing by single base-pair substitutions are separated in denaturing gradient gels: correspondence with melting theory. Proc Natl Acad Sci U S A 80, 1579-83.
14. Cereb, N., Maye, P., Lee, S., Kong, Y., and Yang, S. Y. (1995) Locus-specific amplification of HLA class I genes from genomic DNA: locus-specific sequences in the first and third introns of *HLA-A*, *-B*, and *-C* alleles. *Tissue Antigens* **45**, 1–11.
15. Fernandez-Vina, M. and Bignon, J. (1997) Primers and oligonucleotide probes (SSOP) used for DNA typing of HLA class II alleles, in *Genetic Diversity of HLA. Functional and Medical Implications. Volume 1* (Charron, D., ed.), EDK, Paris, pp. 597–632
16. Ramon, D., Arguello, J., Cox, S., et al. (1998) Application of RSCA for the typing of *HLA-DPB1*. *Hum. Immunol.* **59**, 734–747.
17. Rozemuller, E. H., Bouwens, A.G., Bast, B. E., and Tilanus, M. G. (1993) Assignment of *HLA-DPB* alleles by computerized matching based upon sequence data. *Hum. Immunol.* **37**, 207–212.

9

Sequencing Protocols for Detection of HLA Class I Polymorphism

Paul P. J. Dunn, Steven T. Cox, and Ann-Margaret Little

1. Introduction

The human leukocyte antigen (HLA) genes located within the human major histocompatibility complex on chromosome 6 are probably the most polymorphic functional genetic loci studied to date. The functional HLA genes encode protein molecules that function in antigen presentation within the immune response. Polymorphism within these genes influences diversity of the immune response against different pathogenic infections. Such polymorphism within the human population is considered an advantage, as diversity in immune responses allows some individuals to be better than others at combating certain infections and thus ensuring survival of the species. However, this polymorphism also serves as a major barrier against the transplantation of human organs and stem cells, where HLA incompatibility between donor and recipient can lead to graft rejection or, in the case of stem cells, graft vs host disease. The study of HLA polymorphism has been led by transplantation biologists because of the implication of HLA matching in improving transplant outcome. HLA polymorphism data is also utilized in anthropological studies, where the frequency of HLA alleles can be used as a marker for analysis of population genetics.

From: *Methods in Molecular Biology, vol. 210: MHC Protocols*
Edited by: S. H. Powis and Robert W. Vaughan © Humana Press Inc., Totowa, NJ

As HLA proteins function in the immune response, it is not surprising to find association between different HLA types and immune-related diseases, such as autoimmune disease, and in this field HLA typing serves as a disease marker in genetic studies.

Various approaches can be taken to determine HLA polymorphisms. Automation of DNA sequencing electrophoresis has allowed the introduction of routine HLA sequencing into clinical laboratories. The advantages of HLA sequencing over other techniques is that all possible polymorphisms within the target DNA molecule can be identified, in contrast to methods that utilize panels of oligonucleotide probes (sequence-specific oligonucleotide [SSO]) or polymerase chain reaction (PCR) primers (sequence-specific primers [SSP]), which are restricted in that they can only detect polymorphisms complementary to the range or reagents utilized. However, HLA sequencing from blind samples can be complex due to the number of heterozygous ambiguities that can be identified, and methods, such as reference strand conformation analysis (RSCA) or SSP, have to be utilized to resolve the ambiguities *(1)*.

This chapter describes current protocols that allow the determination of the nucleotide sequence of the polymorphic regions of HLA class I (A, B, C) genes. Two approaches are presented. The first involves direct sequencing of PCR products, in which the end result is a sequence typically representative of two alleles at a heterozygous loci. Due to advances in automated sequencing and sequence analysis software, this approach can be undertaken for routine HLA typings. The second approach involves the cloning of PCR products and the sequencing of individual clones to give a single allele sequence. This latter approach is usually only utilized to define new polymorphisms.

2. Materials

2.1. DNA Extraction

As for all PCR-based methodologies, the quality of DNA utilized is paramount. There are many different methods available for in-house and commercial procedures. All PCR conditions may require optimization for different types of DNA extraction methods. The authors have successfully utilized salting-out protocols and Qiagen extraction methods with the PCR protocols described in this chapter.

2.2. PCR Amplification

PCR reagents (check with **Tables 1**, **2**, and **3** before selecting reagents)

1. NH_4PCR buffer (Bioline Ltd., London, UK; M95801B) or 10X AG buffer for AmpliTaq™ Gold: 0.1 *M* $(NH_4)_2SO_4$, 0.1% (v/v) Triton® X-100, 0.5 *M* Tris-HCl, pH 8.3.
2. 5 *M* betaine (Sigma, St. Louis, MO, USA).
3. Dimethyl sulfoxide (DMSO) (Sigma).
4. 50 m*M* $MgCl_2$ (Bioline; M95801B).
5. 2 m*M* dNTPs (Bioline; S40251) or 100 m*M* each dNTP (Roche Molecular Biochemicals, Indianapolis, IN, USA; dATP 1051 440; dCTP 1051 458; dGTP 1051 466; 7-Deaza-2'-dGTP 988 537; dTTP 1051 482).
6. DNA polymerases: *Taq* DNA polymerases from Bioline (M95801B) or ABgene (AB-0192/b) or AmpliTaq Gold DNA polymerase (Applied Biosystems, Foster City, CA, USA; N8080243).
7. Expand™ High Fidelity DNA polymerase system (Roche Molecular Biochemicals; 1 732 641).
8. PCR primers (*see* **Table 1**).
9. Pre-PCR pipets (P1000, P200, or P100, P10).
10. 0.2-mL PCR tubes/strips with lids.
11. Thermal cycler machine, e.g., Model 9600 (Applied Biosystems), Tetrad DNA Engine (Genetic Research Instruments).
12. Post-PCR pipets (P200 or P100, P10).
13. Agarose gel electrophoresis equipment.
14. Agarose.
15. Ethidium bromide.
16. Molecular weight marker.
17. UV transilluminator for visualizing DNA in agarose gel.

Table 1
Primers Used for the Amplification of HLA Class I Alleles for Sequencing Based Typing

Primer name	Locus	Fig. 1a[a]	Anneal site (nuc)[b]	Primer sequence 5' to 3'
Short HLA				
LH: 5AIn1[c]	HLA-A	a	Intron1, 21-46	[d]TGTAAAACGACGGCCAGT GAAACSGCCTCTGYGGGGAGAAGCAA
RH: 3AIn3[c]	HLA-A	b	Intron 3, 66-89	[e]CAGGAAACAGCTATGACC TGTTGGTCCCAATTGTCTCCCCTC
LH: 5BIn1[c]	HLA-B	a	Intron 1 36-57	TGTAAAACGACGGCCAGT GGGAGGAGCGAGGGGACCSCAG
LH: 5BIn1CG[f]	HLA-B	a	Intron 1, 46-77	CAGGAAACAGCTATGACC GGCGGGGGCGCAGGACCCGG
LH: 5BIn1TA[f]	HLA-B	a	Intron 1, 46-77	TGTAAAACGACGGCCAGT GGCGGGGGCGCAGGACCTGA
RH: 3BIn[c]	HLA-B	c	Intron 3, 37-59	TGTAAAACGACGGCCAGTGGA GGCCATCCCCGGCGACCTAT
LH: 5CIn1[c]	HLA-C	a	Intron 1, 42-61	TGTAAAACGACGGCCAGT AGCGAGGKGCCCGCCCGGCGA
RH: 3BCIn3[c]	HLA-C	c	Intron 3, 12-35	CAGGAAACAGCTATGACC GGAGATGGGGAAGGCTCCCCACT
Long HLA				
LH: 5'UTA[g]	HLA-A	d	-24 to +6	CCCAGACGCCGAGGATGGCC
RH: 3'UTA[h]	HLA-A	e	171-202[i]	TTGGGGAGGGAGCACAGGTCAGCGTGGGAAG
LH: 5'UT[g]	HLA-A,B,C	d	-28 to –2	GGACTCAGAATCTCCCCAGACGCCGAG

Long HLA				
RH: 3'UTB[h]	HLA-B	e	171-202[i]	CTGGGGAGGAAACACAGGTCAGCATGGGAAC
LH: 5'UT[g]	HLA-A,B,C	d	-27 to –1	GGACTCAGAATCTCCCCAGACGCCGAG
LH:CLSP-23[k]	HLA-C	d	-23 to +3	TCAGATTCTCCCCAGACGCCGAGATG
RH: 3'UTC[h]	HLA-C[k]	e	171-202[i]	TCGGGGAGGGAACACAGGTCAGTGTGGGGAC
RH: CLSP1085	HLA-C	f	1085-10[l]	GGCAGCTGTCTCAGGCTTTACAAGYGA

[a]See figure 1a for approximate location of primer anneal sites within the HLA class I gene.

[b]Numbering of nucleotides within introns starts at 1 for each intron, whereas numbering of nucleotides within exons starts at nucleotide number 1 of exon 1 and continues through exon 8 (HLA-A and –C) and exon 7 (HLA-B), and excludes introns.

[c]**Ref.** ***8***.

[d]Underlined sequence = M13 –21sequence.

[e]Underlined sequence = M13 Rev sequence. The M13 sequences are included in primers to facilitate use of dye-primer chemistry with the M13 primers. For dye-terminator chemistry, the M13 sequences can be removed.

[f]These primers detect a dimorphism within HLA-B alleles ***(9)*** References: [g]***(10)*** [h]***(11)***.

[i]Annealing site is within 3' untranslated region.

[j] Ref [12].

[k]All HLA-C alleles except Cw*07.

[l]Sequence ends at nucleotide 10 of the 3' untranslated region.

Table 2
PCR Mix for Short HLA

Protocol 1		Protocol 2	
Reagents[a]	Vol/µL	Reagents	Vol/µL
dH_2O[b]	22.2	dH_2O	29.20
10X NH4 buffer	5.0	10X buffer AG	5
2 m*M* dNTPs	5.0	25 m*M* dNTPs	0.4
50 m*M* $MgCl_2$[c]	2.0	25 m*M* $MgCl_2$	3
2 pmol/µL LH primer	4.0	10 µ*M* LH primer	1
2 pmol/µL LH primer	4.0	10 µ*M* RH primer	1
Taq DNA polymerase (5 U/µL)	0.3	AmpliTaq Gold DNA polymerase or AB gene Taq (5 U/mL)	0.4
		5 *M* betaine[d]	5
DNA (50–200 ng/µL)	7.5	DNA (50 ng/µL)	5
Total Vol	50.0	Total Vol	50

Cycling Parameters:

Protocol 1	Protocol 2
1. 95°C for 4 min.	1. 96°C for 12 min.
2. 95°C for 30 s; 65°C for 50 s; 72°C for 30 s.	2. 96°C for 30 s; 65°C for 45 s; 72°C for 45 s.
3. Repeat step 2 for 32 cycles.	3. Repeat step 2 for 20 cycles.
4. 72°C for 8 min.	4. 96°C for 30 s; 60°C for 45 s; 72°C for 45 s.
5. 15°C for ever.	5. Repeat step 4 for 15 cycles.
	6. 96°C for 30 s; 55°C for 45 s; 72°C for 45 s.
	7. Repeat step 4 for 10 cycles.
	8. 72°C for 5 min.

[a]Reagents from Bioline or similar.
[b]For HLA-C locus, use 21.7.
[c]for HLA-C locus use 2.5.
[d]omitted for HLA-C.

Table 3
PCR Set Up for Long HLA

Reagents[a]	Vol/µL
dH_2O	31.5
10X buffer 1[b] (contains 17.5 m*M* $MgCl_2$)	5
25 m*M* dNTPs	0.75
10 µ*M* LH primer	1
10 µ*M* RH primer	1
Expand™ enzyme mix[b] (2.8 U)	0.75
DMSO	5
DNA (50 ng/µL)	5
Total Vol	50
1. 94° for 2 min.	
2. 94°C for 10 s; 65°C for 45 s; 72°C for 3 min.	
3. Repeat step 2 for 35 cycles.	
4. 72°C for 7 min.	

[a]Reagents from Roche Expand High Fidelity PCR System.
[b]Buffer components unknown.

2.3. PCR Product Clean-Up

1. Microcon® 100 microconcentrators (Millipore, Bedrod, MA, USA; 42413) or GFX column purification kit (Amersham Pharmacia Biotech, Piscataway, NJ, USA; 27-9602-01).
2. Distilled water (dH_2O).
3. Post-PCR pipets.
4. Microcentrifuge.

2.4. Dye Terminator Cycle Sequencing

1. Big Dye Terminator cycle sequencing ready reaction kit (ABI; 4303152).
2. 1.6 pmol/µL HLA sequencing primers (**Table 4**).
3. Distilled water.
4. Purified and diluted HLA PCR product (or cloned DNA template—*see* under **Subheading 3.4.**).

Table 4
Sequencing Primers

Primer	Fig. 1b[a]	Locus/Anneal site[b]	Primer sequence (5' to 3')
M13 -21		A, B, C Tailed PCR product/ sequencing vector	TGT AAA ACG ACG GCC AGT
M13 REV		A, B, C Tailed PCR/vector	CAG GAA ACA GCT ATG ACC
B3.6 (Reverse)[c]	g	A, B, C nuc 330 exon 2 to nuc 7 intron 2	GAC ACT CAC CGG CCT CGC TCT GG
BINT2 (Forward)[d]	h	B, C nuc 232-253 intron 2	GAC TGA CCG CGG GGS CKG GGC CAG
A5.10 (Forward)[c]	h	A nuc 203-219 intron 2	GAG GGC TCG GGG GAC YGG G
LD1S (Forward)[e]	i	A, B nuc 585-601 exon 3	GAC CTG GAG AAC GGA AGG
LD1N (Reverse)[e]	j	A, B nuc 582-606 exon 3	GAG CTA CTA CAA CCA GAG CGA GG
C1S (Forward)[f]	k	C nuc 625-642	CCA AAG ACA CAC GTG ACC
C1N (Reverse)[f]	l	C nuc 625-642	GGT CAC GTG TGT CTT TGG
C2S (Forward)[f]	k	C nuc 657-673	TGA CCA TGA GGC CAC CC
C2N (Reverse)[f]	l	C nuc 657-673	GGG TGG CCT CAT GGT CA
JD3S (Forward)[e]	o	A, B, C nuc 918-937 exon 5	GAC CCC ATC GTG GGC ATC GTT G
JD3N (Reverse)[e]	p	A, B, C, nuc 915-938 exon 5	GAG CAA CGA TGC CCA CGA TGG GGA TG
4S (Forward)[g]	m	A, B, C nuc 703-724 exon 4	GCG GAG ATC ACA CTG ACC TGG C
4N (Reverse)[g]	n	A, B, C nuc 703-724 exon 4	GCC AGG TCA GTG TGA TCT CCG C
6S (Forward)[g]	q	A, B, C, nuc 1023-1043 exon 6	AGG GAG CTA CTC TCA GGC TGC
6N (Reverse)[g]	r	A, B, C nuc 1023-1042 exon 6	TGC AGC CTG AGA GTA GCT CCC T

[a]See figure 1b for approx location of primer anneal sites within the HLA class I gene.
[b]Exon numbering is continuous from the beginning of exon 1 and exludes introns. Introns are numbered individually.
References: [c]***(13)***, [d]***(14)***, [e]***(15)***, [f]***(16)***, [g]***(10)***.

5. 0.2-mL PCR tubes and lids.
6. 75% Isopropanol.
7. AGTC spin columns (VH-Bio; 42453).
8. Centrifugal vacuum drier (Savant Instruments, Holbrook, NY, USA).
9. Formamide (ultra pure grade).
10. Blue dextran-EDTA loading buffer (ABI; 402055).
11. Thermal cycler.

2.5. Selection of Recombinant Clones

All solutions must be autoclaved (121°C, psi) before use.

1. Sterile inoculating loops (Sigma; I18263).
2. Sterile 2.0-mL cryotubes (Nalgene).
3. LB-Broth tablets (Sigma; L7275).
4. LB Agar tablets (Sigma; L7025).
5. Ampicillin, 50 mg/mL pure water (Sigma; A2804).
6. Microfuge tubes (sterile).
7. Sterile Universal containers (Sterilin).
8. Sterile dH_2O.
9. Sterile glycerol.
10. PCR reagents as in **Subheading 2.2.**

2.6. Sequencing Gel

1. 50% stock solution Long Ranger™ (Flowgen, Staffordshire, UK; 50615).
2. Sterile dH_2O.
3. Urea, ultra-pure grade (Sigma; U5378).
4. Amberlite IRN-150L Resin (Amersham Pharmacia Biotech; 17-1326-01).
5. Vacuum filtration unit (Sartorious, Edgewood, NY, USA).
6. Ammonium persulfate.
7. TEMED.
8. 10X Tris-borate-EDTA (TBE) buffer.
9. 50-mL Luer-lock syringe.

3. Methods

3.1. PCR Primer Design and Sequence

Within the HLA complex, there are many genes with related sequences. Thus the oligonucleotide primers utilized in order to achieve locus-specific PCR amplification have to be thoroughly checked for cross-reactivity with other loci. Each PCR amplification requires a 5' or left-hand (LH) primer and a 3' or right-hand (RH) primer. The LH primer is complementary to and anneals to the antisense strand of the template DNA, while the RH primer anneals to the sense strand. In the amplification process, the LH primer is a substrate to make a copy of the sense strand, whereas the RH primer allows copies of the antisense strand to be made. Each primer within a primer pair must not contain a stretch of nucleotides complementary to the other primer of the pair, otherwise dimerisation between the primers can occur, which will inhibit annealing to and amplification of the intended target. Similarly, complementary sequences within an oligonucleotide may force the primer to snap-back on itself forming a hairpin loop structure which again will inhibit optimum annealing to the target DNA.

The optimum length for PCR primers is between 18 and 23 bases in length, however longer primers have been successfully utilized for amplification of HLA alleles. Each primer in a pair should have similar annealing temperatures. The annealing temperature for an oligonucleotide is given by the equation,

T_m- 5°C, where T_m is the melting temperature or the temperature at which the primer dissociates from its target.

$$T_m = 4 \times (G+C) + 2 \times (A+T)$$

For specificity, i.e., to prevent cross-reactions, the annealing temperature should be at least 60°C.

Primers are normally supplied lyophilized and they should be dissolved in a small amount of pure water or 10 m*M* Tris-HCl, pH 8.0, to provide a high concentration stock solution from which working solutions can be made. For long term storage, they should be in a buffer rather than water, because DNA is an acid and is prone to autohydrolysis. We have found that oligonucleotides are stable if stored at high concentration (i.e., 100–500 μ*M*) at –20° or –40°C.

The molecular weight of an oligonucleotide is given by:

$$(A \times 312.2) + (C \times 288.2) + (G \times 328.2) + (T \times 303.2) - 61$$

The absorbance measured in a spectrophotometer at 260 nm of an oligonucleotide in solution is 1.0, at an approximate concentration of 20 μg/mL. For general PCR and sequencing, primers do not need to be purified.

3.2. PCR Amplification

1. Each commercially available PCR enzyme is supplied with a 10X concentrated incubation buffer. This buffer is usually ammonium sulfate or potassium chloride based, buffered with Tris-HCl (pH about 7.8). Absolute prerequisites for PCR are dNTPs (dATP, dCTP, dGTP, and dTTP), $MgCl_2$, primers, template DNA, and enzyme. Each new PCR must be optimized with respect to buffer and the concentrations of dNTPs and $MgCl_2$ that are included.
2. A PCR has 3 steps: denaturation, annealing, and extension. Denaturation (typically 94–96°C) separates the DNA strands, allowing the annealing of the primers (at 60–65°C for 30–45 s). Copying, or extension, of the DNA strands from the primers occurs for a minimum of 30 s at 72°C, although for long PCRs, this time can be increased to 3 min. The composition of each PCR, as well as the cycling parameters must be optimized for each HLA locus to be studied with a specific primer pair. PCRs must also be optimized for each different type of thermal cycler used to carry out the PCR. In our laboratories, we use Perkin Elmer 9600 or Tetrad thermal cyclers, and the PCR programs described here are interchangeable on these machines.
3. Two different PCR strategies are described, the first for amplification of a fragment containing the polymorphic exons 2 and 3 of HLA-A, B, and C loci (short HLA). This fragment also contains the intervening intron 2. The second strategy amplifies the full-length genomic fragment consisting of exon 1 to exon 7 (HLA-B) or exon 8 (HLA-A and -C) and intervening introns (long HLA).

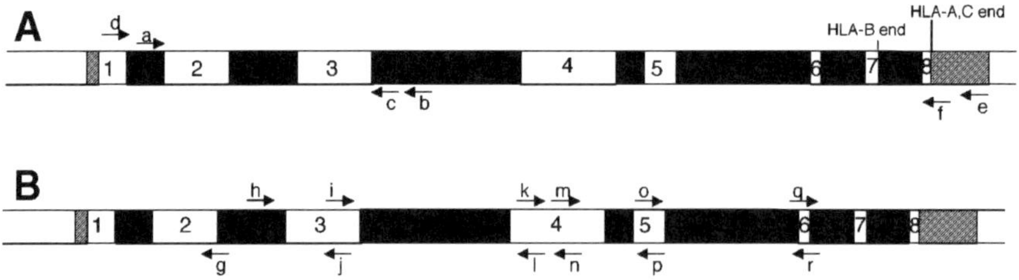

Fig. 1. Diagrammatic representation of an HLA class I (A, B, or C) gene with the exon and intron boundaries indicated. Exons are unfilled boxes, and introns are represented by black boxes. Numbers correspond to exon number. Hatched boxes at the ends of the gene represent untranslated regions. Arrows indicate location of primers from **Table 1** (PCR primers) for (**A**) and **Table 4** (sequencing primers) for (**B**).

Primer sequences for amplification of short and long fragments are given in **Table 1**. A diagrammatic representation of the annealling sites for the primers described in **Table 1** is given in **Fig. 1a**. Both long and short fragments can be sequenced directly, although the short HLA fragment is more robust for direct sequencing. Both fragments can also be subjected to cloning as described under **Subheading 3.4.**

4. Our laboratories utilize different PCR protocols, and both are given.

3.2.1. Short HLA PCR Protocol

1. Prepare the PCR reagents by mixing in a 0.2-mL PCR tube and perform PCR as detailed in **Table 2**. If multiple samples are being amplified, prepare a master mixture in an appropriate tube by multiplying the individual ingredients by the number of samples plus 2. Add the DNA after aliquoting the master mixture into the individual PCR tubes/strips.
2. Analyze 1/10 of the PCR (5 μL) using 1–2% (w/v) agarose gel electrophoresis, comparing the intensity of the DNA band with a 100-bp ladder run in parallel. The amount of DNA amplified can be approximated by comparing the intensity of the PCR product with the 1000-bp band in the ladder, which will contain a known amount of DNA based upon how the ladder was diluted.

3. A good PCR product for sequencing should give a strong fat product band and a faint or invisible primer band. Weakly amplified products are no good for sequencing, as the ratio of primer:DNA is too high—if the PCR product is not good, check genomic DNA concentration by UV spectrophotometer, and then repeat PCR. It may be necessary to re-extract the DNA or purify the DNA prior to PCR (*see* under **Notes 1–3**).
4. Store PCR products at 4° or –20°C until required.

3.2.2. Long HLA PCR Protocol

1. Prepare PCR reagents and perform PCR as described in **Table 3**.
2. Check PCR products by agarose gel electrophoresis as described in **Subheading 3.2.1.** (*see* **Notes 1–3**). An appropriate marker run for these longer fragments would be the λ *Hind*III digest (Roche Molecular Biochemicals).

3.3. Direct Sequencing of PCR products

For successful direct PCR sequencing or cloning of PCR products, excess primers and dNTPs need to be removed from the PCR. The easiest and most reliable method uses a spin column system that retains the DNA while the primers and dNTPs are eluted. The DNA is then recovered in an appropriate vol of buffer or water. This method can also be used to purify DNA prior to PCR. Two different systems have been successfully used in our laboratories.

3.3.1. GFX Column Purification

1. Place one GFX column in a collection tube (supplied with the kit) for each reaction to be purified.
2. Add 500 μL of Capture Buffer to the column.
3. Transfer the entire PCR to the column and mix thoroughly by pipeting up and down 6–8 times.
4. Centrifuge in a microcentrifuge at full speed for 30 s (add the time taken for microfuges to reach full speed).
5. Discard the flow-through by emptying the collection tube. Place the GFX column back into the collection tube.

6. Add 500 μL of Wash Buffer to the column. Centrifuge at full speed for 30 s.
7. Discard the collection tube and place the column inside a new labeled 1.5-mL Eppendorf tube.
8. Apply 50 μL of dH_2O directly on top of the glass-fiber matrix in the GFX column. Incubate at room temperature for 1 min.
9. Centrifuge at full speed for 1 min to recover the purified DNA.
10. For Dye-Terminator chemistry make a 1 in 3 dilution of the DNA, i.e., add 100 μL of dH_2O to the 50 μL of purified DNA.
11. The DNA is now pure and at an optimum concentration for sequencing.

3.3.2. Microcon Column Purification

1. Dilute PCR by addition of 150 μL dH_2O.
2. Transfer diluted DNA to Microcon column in a microcentrifuge tube and spin at 3750*g* for 35 min at 4°C.
3. Invert column into a second tube and spin again at 700*g* for 5 min to collect the concentrated and purified DNA.
4. The DNA will be in a very small vol, typically 1–3 μL. Add 10–100 μL dH_2O—the vol depends on the intensity of the post-PCR DNA band observed earlier by agarose gel electrophoresis.
5. Analyze 10% of the purified DNA by 1 to 2% (w/v) agarose gel electrophoresis comparing the size and intensity of the DNA band with suitable DNA markers as described earlier.

3.3.3. Dye Terminator Cycle Sequencing

1. DNA sequencing by chain termination using dideoxy nucleotides was first described by Sanger and colleagues in 1977 *(2)*. As the ddNTPs lack the 3'-OH group necessary for chain elongation, the growing oligonucleotide is terminated. Chain termination is still the method of choice for sequencing chemistry, 20 yr after its discovery. There have been two major changes in DNA sequencing: (i) the advent of PCR-generated templates for sequencing, rather than using templates from *Escherichia coli*, clones; and (ii) the introduction of fluorescence to label sequenced products rather than radioactivity. Fluorescent dyes can either be added to sequencing primer or to the

ddNTPs. Using dye-labeled ddNTPs makes sequencing very versatile, because any oligonucleotide can be used as a sequencing primer. Modifications to sequencing enzymes has ensured that there is even incorporation of dye-labeled ddNTPs (which means even peak height), and this is important for heterozygote detection in the analysis of polymorphic genetic loci such as HLA.

2. The protocols described here incorporate dye terminator cycle sequencing using the enzyme *Taq* DNA polymerase. DNA fragments of differing lengths are generated by *Taq* DNA polymerase-mediated extension of a primer annealed to the template DNA, by thermal cycling. Each growing chain of DNA is terminated by the random incorporation of a ddNTP, which can be A, C, G, or T, as all four reactions are carried out in the same tube. With dye terminator sequencing, each of the four ddNTPs is tagged with a different fluorescent dye. Thus, the growing chain is terminated and labeled with a dye corresponding to the terminating ddNTP. During automated sequencing, the fragments are separated according to size, smallest first, by electrophoresis. The fluorescent tag of the terminating ddNTP is detected by the laser, which is continuously scanning the read region of the gel. Sufficient fragments are generated during cycle sequencing to ensure that each base in the sequence is detected. As data collection continues, a series of bands are produced with different colors, according to the ddNTP. After the run, the computer extracts and interprets the data to produce an electropherogram showing peaks of different colors corresponding to the nature and position of each nucleotide in the DNA sequence. Primers that have been utilized for sequencing HLA alleles either directly from PCR products or from cloned DNA are described in **Table 4**. A diagrammatic representation of the annealing sites for the primers described in **Table 4** is given in **Fig. 1b**.
3. The following is a general recipe for Big Dye Terminator cycle sequencing reactions using half-strength reactions: 4 μL terminator ready reaction mix, 1.6 pmol (1 μL) sequencing primer, 3 μL sterile dH_2O, and Template (50–100 ng single-stranded DNA, 200–500 ng double-stranded DNA, 30–90 ng [2 μL diluted PCR product] PCR product DNA). Final vol 10 μL.
4. Prepare cycle sequencing reaction, fasten lids securely on PCR tubes/strips, and mix contents thoroughly. Centrifuge briefly to concentrate the reagents in the bottom of the tubes.

5. Use a Geneamp® 9600 or 2400 PCR machine (Applied Biosystems) and set the vol to 10 μL. Repeat the following for 25 cycles: rapid thermal ramp (1°C/s) to 96°C, 96°C for 10 s; rapid thermal ramp to 50°C, 50°C for 5 s; rapid thermal ramp to 60°C, 60°C for 4 min, rapid thermal ramp to 4°C, and hold until ready to purify.
6. Alternative "quick" cycle sequencing program: 98°C for 10 s, followed by 30 cycles of 96°C for 5 s and 65°C for 1 min.
7. To remove unincorporated dye-labeled terminators two methods are described in **Subheading 3.3.5.**

3.3.4. Isopropanol Precipitation

1. Remove the PCR tubes from the thermal cycler, centrifuge briefly, and remove lids.
2. To precipitate the cycle sequencing products in the PCR tubes, add 40 μL 75–80% isopropanol to each tube/reaction.
3. Seal the tubes by replacing lids or using adhesive tape, and mix the contents by inverting a few times.
4. Leave the tubes at room temperature for 15 min to precipitate the extension products.
5. Place the tray containing the PCR tubes in a centrifuge and spin at 2000–3000*g* for 30 min. It is important to proceed to the next step immediately. If not possible, then spin the tubes for 2 min more immediately before performing the next step.
6. Without disturbing the precipitates, remove the lids or adhesive tape and discard the supernatant by inverting the tray on a paper towel folded to the size of the tray.
7. Place the inverted tray with the tissue back into the centrifuge and spin at 70*g* for 1 min. Then remove the tray and discard the tissues. At this stage the pelleted products may or may not be visible (*see* **Notes 4–10**). Vacuum drying of the samples is not necessary.

3.3.5. Spin Column Purification of Sequenced Products

1. Centrifuge AGTC gel filtration columns for 750*g* for 2.5 min to remove excess buffer and preservative. Transfer columns to new labeled tube.
2. Add each cycle sequencing product to top of gel. Spin again as above.
3. Discard columns and dry down sequencing products in a centrifugal vacuum drier on medium heat for 20–25 min.

3.3.6. Addition of Loading Buffer and Denaturation of Samples

1. Prepare the sequencing loading buffer by diluting the blue dextran-EDTA 1:4 with formamide. For example, for 36 reactions, use 20-µL of blue dextran-EDTA and 100 µL of formamide, and vortex mix.
2. Add 2 µL loading buffer/formamide to each tube/reaction ensuring that the buffer drop is at the bottom of the tube.
3. Seal tubes with adhesive tape or lids and denature extension products by placing in a thermal cyler at 95°C for 2 min.
4. Place tubes on ice.
5. Do not leave samples in loading buffer for more than 2 h as the formamide can degrade DNA over time. This problem is avoided if you freeze the samples.
6. Load 1.5–2.0 µL on a sequencing gel.

3.3.7. Electrophoresis of Sequencing Reactions

1. There are several different automated DNA sequencers available, some of which utilize slab gel electrophoresis and others which utilize capillary electrophoresis. This subheading describes preparation of sequencing gels for an ABI automated slab gel sequencer.
2. Weigh 18 g urea and place in a 200-mL beaker.
3. Weigh 0.5 g Amberlite resin and add to beaker.
4. Using a sterile pipet, measure 5 mL 50% Long Ranger and add to the beaker.
5. Measure 27 mL distilled sterile water using a pipet or measuring cylinder and add to the beaker.
6. Add a clean magnetic bar to the beaker and place on a stirrer for 10 min. Note: When these reagents are mixed together, an endothermic reaction takes place.

   To ensure gel-to-gel consistency, always allow contents to warm to room temperature.
7. While the gel mixture is stirring, assemble plates and spacers in the gel cassette according to manufacturer's instructions.
8. Weigh 100 µg ammonium persulphate in a 1.5-mL Eppendorf and put to one side.
9. Attach the vacuum filtration unit to a vacuum pump and filter 5 mL 10X TBE.

10. When TBE has completely passed through the filter, add the beaker contents from **steps 1–5** above to the filtration unit.
11. Allow solution to degas for exactly 5 min (this ensures gel-to-gel consistency).
12. While the gel is degassing, prepare ammonium persulphate by adding 1 mL sterile dH_2O to the Eppendorf prepared in **step 8**.
13. Remove top part of filtration unit and discard. Place lid on lower chamber and swirl to mix.
14. When ready to cast the gel, add 250 μL ammonium persulphate and 25 μL TEMED to the gel and swirl to mix, being careful not to introduce air.
15. Take up about 40 mL of the gel in the luer-lock syringe—very SLOWLY.
16. Ensure there is no trapped air in the syringe before proceeding.
17. Attach syringe to the gel casting device if casting from the bottom of the gel.
18. Eject gel media between the plates in a slow consistant manner, tapping the gel front with four fingers to prevent bubbles forming at the same time.
19. Insert gel comb between plates flat edge first, ensuring the marker on the comb lines up with the marker on the plate.
20. Allow the gel 2 h to polymerize (*see* **Notes 4–10**).
21. Remove the gel comb postpolymerization and gently rinse away excess polymerized acrylamide and dried urea from the top of the plates using a stream of dH_2O.
22. Place the sequencing gel in the automated sequencer according to manufacturer's instructions (*see* **Notes 4–10**).
23. Insert a shark's tooth comb for 24, 32, 36, 48, or 96, depending on the capacity of the gel, such that the teeth just penetrate the top of the polyacrylamide creating wells between the teeth.
24. Pour electrophoresis buffer, 1X TBE, into the top and bottom buffer tanks and flush out the wells with a 50-mL syringe containing buffer to remove urea. Following denaturation (*see* **Subheading 3.3.6.**) place sequencing reactions on ice. Load these sequenced products into alternate wells (i.e., odd numbers first) of the sequencing gel. This can be done with a P10 pipet.
25. Run the first half of the sequencing reactions into the gel for 5 min at 32 W constant power. Flush the wells out again and load the remainder of the sequence products.

3.4. Cloning of PCR Products

1. HLA alleles can be amplified and cloned or sequenced directly from either cDNA or genomic DNA. The advantages of cloning from a cDNA template is that the full-length product can be subcloned in a variety of vectors at high efficiency. For HLA class I alleles, the full-length cDNA is around 1200 nucleotides. However, it is not always possible to work with mRNA and cDNA, and for many cases where HLA cloning and sequencing is required, the only material available is genomic DNA. As virtually all the polymorphism found within HLA class I genes is located within exons 2 and 3 with some polymorphism in exon 4, PCR strategies have been developed to amplify fragments encompassing these exons. This type of strategy generates fragments from 900–1500 bp, which again can be subcloned at high efficiency. The disadvantage of restricting the analysis to partial sequences is that other polymorphisms may be missed. Also, further studies such as making a single allele transfectant cell lines cannot be performed. The development of PCR protocols, which allow amplification of long PCR products, has permitted the full-length amplification of HLA class I genes (3.2 kb), which can be subcloned for further analysis. PCR protocol for Long HLA is given above under **Subheading 3.2.2.**
2. Prior to cloning amplified HLA DNA, it may be necessary to purify the DNA to remove excess primers and other components of the PCR that may interfere with the ligation of the DNA into the plasmid. (This is not necessary if using TA cloning systems—*see* **Subheading 3.5.2., step 2**). There are a variety of methods described in the literature that include isopropanol precipitation and agarose gel excision. There are also a number of centrifugal filtration devices that rely on a "cut-off" such that large molecules, such as DNA, are retained, while small molecules, such as dNTPs and primers, are removed. Conversely, spin columns are available which contain gel filtration media. Large molecules (DNA) are excluded from the gel matrix and quickly pass through the column when it is spun while small molecules are retained. There are also spin cartridges in which DNA binds to a silica-based surface. Other molecules in the PCR can then be washed through. All of these methods conveniently use microcentrifuges for enabling the separation of DNA from the other components of the PCR.

3. One method utilizes a membrane that has a cut off of 100 kDa. Oligonucleotides (20-mers) are approx 6–7 kDa, which means that they pass through the membrane along with the other PCR components. Amplified DNA is retained because it is too large to pass through. This method serves to concentrate as well as purify the DNA. PCR products can then either be subjected to direct sequencing or ligation for cloning and sequencing.

3.5.1 Purification of PCR Products

1. Amicon microconcentrators are used as described earlier (**Subheading 3.3.2.**).

3.5.2. Ligation of Cloned DNA to Plasmid Vector

1. Traditionally, PCR products have been cloned in plasmids called T-vectors which are linearized plasmids with a single 3'-terminal thymidine overhang at each end of the linear DNA molecule *(**3**)*. The rationale behind this approach for cloning PCR products was that DNA amplified by *Taq* DNA polymerase was reported to possess 3' adenosine overhangs caused by the terminal transferase activity of certain thermophilic DNA polymerases such as *Taq* DNA polymerase *(**4**)*. However, in the long PCR system described here, the amplified DNA may be very heterogeneous in nature, being a diverse mixture of blunt-ended molecules and molecules with a variety of 5' overhangs, including T.
2. There are various different commercial kits available that provide the necessary reagents for successful ligation. One method (Perfectly Blunt Cloning Kit, 70075-3; Novagen) blunt-ends the DNA and then ligates it to a plasmid (pT7Blue) that has been linearized by digestion with *Eco*RV. Other methods include the TA cloning system (Original TA cloning kit, K2000; Invitrogen, Carlsbad, CA, USA), which utilizes the 5' A overhangs. There are systems for optimum cloning of short and long PCR products. The advantage of this method is that purification of the PCR product is not always necessary, and some purification methods may disrupt the 5' overhangs. The TOPO- (Invitrogen; K4750) method, which is utilized for long PCR products, can be used directly on neat PCR if a good product is

achieved. However it is recommended to gel purify the PCR product if the PCR does not produce a good product.

3. An alternative approach is to incorporate different restriction enzyme sites into the 5' ends of both the LH and RH primers. After successful PCR amplification, the purified DNA is digested with the appropriate LH and RH primer enzymes. Ligation will then be performed using a vector or plasmid that contains a multicloning site and that has also been previously digested with the appropriate restriction enzymes. Orientation of the inserted DNA can be controlled by designing the PCR primers with different LH and RH enzyme sites *(**5**)*.

3.5.3. Transformation

1. Transformation protocols have been extensively described elsewhere *(**6**)*. With the current trend to make use of T vectors, most commercial kits include competent *E. coli* derivative cells, which are used following the manufacturer's instructions. Most protocols are based on the disruption of the plasmid's β-galactosidase gene after correct ligation of template DNA. This system enables blue-white color selection of bacteria colonies after transformation, with the white colonies possessing the recombinant plasmid. DNA can be extracted from bacterial colonies by PCR, and the amplified product can then be sequenced as described earlier.
2. In the TA and pT7Blue cloning systems, agar plates are pretreated with an antibiotic such as ampicillin for selective growth of antibiotic-resisitant plasmid-containing cells. A color selection method is used to identify colonies that contain the DNA of interest (recombinants). The plasmid contains a lacZα fragment, which encodes the first 146 amino acids of β-galactosidase. The plates are also treated with X-gal, a substrate for β-galactosidase, that turns colonies blue. However, the DNA to be cloned interrupts the lacZα fragment resulting in nonexpression of β-galactosidase and colonies appear white. White colonies that are well separated from each other are selected and cultured in medium containing antibiotic.
3. The pCR®-XL-TOPO® vector (Invitrogen) allows the direct selection of recombinants via disruption of a lethal gene, *ccd*B. The vector contains this gene fused to the C-terminus of the lacZα fragment. Ligation of a long PCR product disrupts expression of the lacZα-

*ccd*B gene fusion, therefore, only positive recombinants can be cultured on the plate. Cells that contain nonrecombinant vector are killed upon plating and blue/white screening is not required.

4. For both methods, more recombinant colonies than needed should be screened to ensure that enough clones are produced containing the HLA allele of interest, as most samples will be heterozygous for the HLA loci being studied. Typically, 20 clones are screened for each ligation reaction. Also, false positives can occur with both TA and TOPO cloning by PCR artefacts that were ligated into the vector (this can be avoided by gel purification of PCR products). In addition, frameshift mutations within the reading frame for either the *ccd*B lethal gene or the lacZα fragment can cause inactivation leading to false positive colonies.

3.5.4. Selection of Recombinant Clones

1. Pick individual white colonies into labeled microcentrifuge tubes containing 25 µL H_2O (*see* **step 4**) and also into sterile L-broth containing 50 µg/mL ampicillin (minimum 0.85 mL) in a labeled sterile universal.
2. Grow the L-broth cultures at 37°C for 12–24 h with shaking (200 rpm, New Brunswick).
3. In a labeled cryotube containing 0.15 mL sterile glycerol, add 0.85 mL culture, vortex mix, then plunge into a dry-ice/ethanol bath to quickly freeze the culture. Store cultures at –80°C for long-term storage of clones. Ensure you keep an accurate record of each culture you have in such a culture collection. The DNA insert in each clone is characterized by PCR (*see* **step 4** and following).
4. Place microcentrifuge tubes in a heating block, set at 99°C for 5 min to degrade cells and DNases, and then spin at top speed for 5 min in a microfuge to pellet debris.
5. Transfer 5 µL supernatant to 200-mL PCR strip tubes on ice. Use a 50-µL PCR to analyze the DNA insert in each recombinant clone as described earlier (**Subheading 3.3.**). These PCR products can also be sequenced using methods described earlier.

3.5.5. Dye Terminator Cycle Sequencing of Clones

As described under **Subheading 3.3.**

3.6. Sequence Analysis

3.6.1. Description of Analysis

1. Following an automated sequencing run, raw data is extracted from the run by the sequencing software and a raw sequence is generated. Where the data is ambiguous, the computer calls the base 'N' (aNy base). Base calls may also be missed, for example, a mononucleotide repeat may not contain the correct number of nucleotides. Data can also be very 'noisy' at the beginning and end of a sequence. To eliminate these errors, the same template should be sequenced in the sense and antisense directions and then compared to each other. A program such as Sequence Navigator is used to do this. Sequence files generated by the ABI sequencer can be imported into the program to be manipulated manually. Antisense sequences can be reverse-complemented so that sequence from both strands are the same. The noisy data is then removed from the beginning and end of the sequence (**Fig. 2**). Any ambiguities between the two sequences can be highlighted using the program, then checked by eye, by calling up the electropherograms for the particular sequences. Most errors can be easily eliminated this way, but difficulty can sometimes be encountered when sequencing heterozygous HLA locus-specific PCR products. When sequencing clones, there should be no heterozygous positions—if there are, this is indicative of contamination between different clones. A heterozygous sequence will contain two bases at certain polymorphic positions that are represented by two peaks imposed on each other. Before analyzing heterozygous sequences, codes are assigned to denote a position as heterozygous. This is a standard nomenclature assigned by the International Union of Biochemists. For example a heterozygous g and c peak is called s and is termed IUB coding (**Table 5**) (*see* **Notes 4–11**).
2. Heterozygous peaks can sometimes be clearly seen in one direction but not the other. Experience in our laboratory has shown that some positions in the sequence show constant sequence anomalies such as this, and experience allows the identification of such problems. Once sequences have been compared and edited, a unanimity sequence is created, which is a consensus of all the sequences. In the case of HLA typing, this can then be compared to a library of HLA alleles using a computer program that calculates all the possible combinations of alleles that are possible and produces a report (**Fig. 3**).

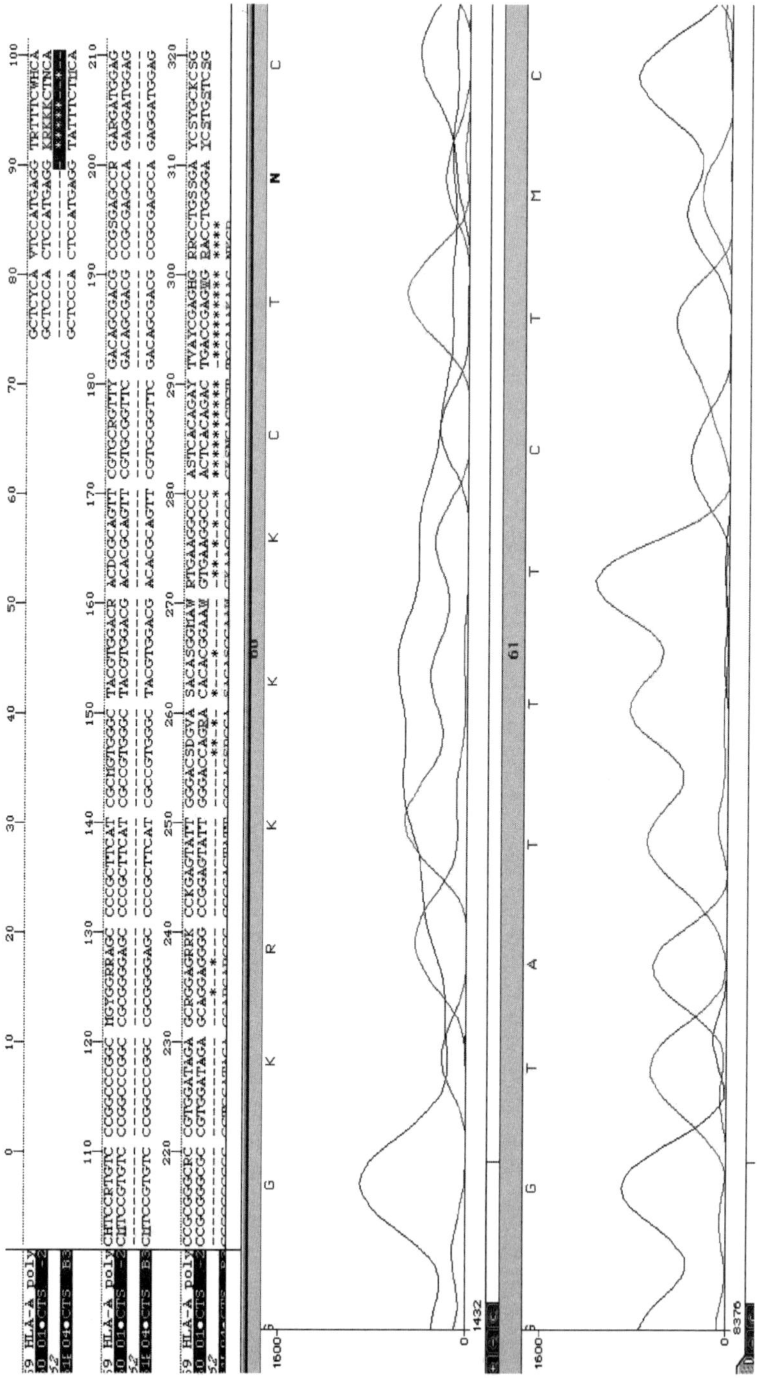

Table 5
IUB Codes

A = Adenosine	R = A or G	B = C, G or T
C = Cytosine	Y = C or T	D = A, G or T
G = Guanosine	K = G or T	H = A, C or T
T = Thymidine	M = A or C	V = A, C or G
N = any base	S = G or C	
	W = A or T	

3.6.2. Misincorporations

1. When sequencing cloned DNA it is necessary to sequence several clones to verify the sequence, as misincorporations can be introduced during the original PCR that generated the template for cloning. It is essential to sequence multiple clones when identifying a novel HLA sequence to ensure that the new sequence is not an artefact. Typically, three clones out of many should be sequenced to ensure identity.

3.7. Conclusion

As the number of defined HLA sequences continues to increase *(7)*, the number of ambiguities will continue to increase such that even common HLA types will not be assigned immediately. Cloning of PCR products resolves the ambiguity issue, but such procedures are not very useful for routine laboratory typings. All the laboratory manipulations can be automated for high sample processing, but the time required to analyze the sequencing data limits

---

Fig. 2. *(opposite page)* The base miscalls have been caused by dye blobs near the beginning of the sequence present in the top electropherogram, the true sequence is obtained from the opposite strand which is the lower electropherogram. To rectify this situation, highlight each letter on the DNA sequence table and correct in lower-case by following the good sequence. By using lower-case letters for corrections, it is easy to track manual intervention in the sequence analysis.

```
MatchTools v1.0            For Research Use Only.  Not for use in diagnostic          31/10/00 4:50 pm  pg 1
PE Applied Biosystems                          procedures.                                    A. N. Other
Sample: A. N. Other                            Library: A-Ex23.V180
Allele Report: Exact match to: A-02011/A-2902 A-0209/A-2902 A-0243N/A-2902.  See Warnings Below.

Files : A. N. Other
____________________________________________________________

Warning #5: Only one sequence orientation is present.

____________________________________________________________

Polymorphic Position Report

     11111 1111111122 2222222222 2222222222 3333333333 3333333333 4444444444 4444444444 4555555555 5555555555 66 | r
7899900222 2446678900 1233344555 6667789999 0000111113 4566678999 0011111112 4566788899 9001222233 4555567789 11 |
8127826136 7240360403 9838903679 1580120249 1278134791 1523865179 2412346893 8338705937 8267146789 5357900135 68 |

YCAWYAGCCG GGMGAGYCAG RAAGGKGSKS GCAWGSCSCY RAGGCCTGGC CrkGAGTKTC CGCCRSAKAC YmCTGGTrCY CMRGCrGTTT CGGGACGTTG AG <>
.......... .......... .......... .......... .......... .......... .......... .......... .......... .......... .. >

.......... .......... .......... .......... .......... .......... .......... .......... .......... .......... .. A-
.......... .......... .......... .......... .......... .......... .......... .......... .......... .......... .. A-
.......... .......... .......... .......... .......... .......... .......... .......... .......... .......... .. A-
.......... .......... .......... .......... .......... .....S.... .......... .......... .......... .......... .. A-
.......... .......... .......... ..M....... .......... .......... .......... .......... .......... .......... .. A-
C......... .......... .......... .......... .......... .......... .......... .......... .......... .......... .. A-
.......... .......... .......... ........Y. .......... .......... .......... .......... .......... .......... .. A-
.......... .......... .......... .......... .......... .......... .......... ....R..... .......... .......... .. A-
.......... .......... .......... .......... .......... ..T....... .......... .......... .......... .......... .. A-
.......... .......... .......... .......... .......... ....R..... .......... .......... .......... .......... .. A-
.......... .......... .......... .......... .......... .....S.... .......... .......... .......... .......... .. A-
.......... .......... .......... ..M....... .......... .......... .......... .......... .......... .......... .. A-
.......... .......... .......... .......... .......... ....R..... .......... .......... .......... .......... .. A-
.......... .......... .......... ...T...... .......... .......... .......... .......... .......... .......... .. A-
.......... .......... .......... ...Y...... .......... .......... .......... .......... .......... .......... .. A-
.......... .......... .......... .......... .......... .......... .......... .......... .........K .......... .. A-
.......... .......... .......... .......... .......... .......... .......... .C........ .......... .......... .. A-
.......... .......... .......... .......... .......... .......... .......... .......... ...R...... .......... .. A-
.......... .......... .......... .......... .......... .......... .......... .......... .......W.. .......... .. A-
.......... .......... .......... ..M....... .......... .......... .......... .......... .......... .......... .. A-
.......... .......... .......... ..C....... .......... .......... .......... .......... .......... .......... .. A-
```

the throughput of sequencing for routine typings. This limitation is likely to be resolved by the introduction of more robust sequencing chemistries and translation from gel and capillary systems to microarray technology with improvements in software for handling the sequencing data.

4. Notes

1. No bands are seen for products but primer bands are present. (*i*) PCR failure may be caused by DNA concentration being too low (or too high). Check OD and concentration and repeat PCR; (*ii*) dNTPs and/or other reagents may have been missed. Repeat PCR; and (*iii*) DNA may be impure, if repeating doesn't work, try purifying the DNA with a GFX column (*see* **Subheading 3.3.1.**) but only if you have sufficient vol of DNA (>40 μL).
2. A stronger than usual primer band is seen. This can be caused by primer dimerization that is then amplified in the reaction. This can cause a bad sequencing result if the PCR primer sequence is utilized as a sequencing primer—the strong stop signal caused by sequencing to the end of the primer dimer may result in the rest of the sequence being noisy. If after purification of the PCR product, a strong primer band is still seen, you may need to gel purify the sample or repeat the PCR after optimizing primer concentration.
3. Extra bands are seen that are larger than the required product, but are much fainter after agarose gel electrophoresis. This can sometimes occur if the concentration of DNA is too high. If the DNA appears viscous and/or stringy when it is pipeted then dilute the DNA to the appropriate concentration. If a strong PCR product is present, these extra bands will not interfere with sequencing.

Fig. 3. *(opposite page)* Example of a sequencing result from analysis of a heterozygous HLA-A type by direct sequencing of PCR product. Exon 2 and exon 3 have been included in this analysis. Three HLA types share sequence identity with the sample (AN Other): A*02011, *2902 or A*0209, *2902 or A*0243N, *2902. The A*02 alleles cannot be distinguished by sequencing exon 2 and exon 3 alone, as their differences are found within exon 4.

4. The sequencing gel has not polymerized. The sequencing gel mixture was made up incorrectly, check reagents, and freshness of ammonium persulphate.
5. Gel has lots of air bubbles. The gel was not poured carefully enough. Bubbles in the gel will ruin the data and unless there are only a few bubbles affecting lanes, a fresh gel will have to be prepared.
6. Gel leaks from loading device. The plates were not aligned correctly. A little leakage is acceptable but if too much leaks out and there is not sufficient gel mixture to fill the space between the plates, the procedure will have to be repeated.
7. No pellets can be seen after precipitating cycle sequencing products. Do not worry. A pellet is not normally seen, but if the tube containing the sequencing products is held up to the light very tiny strands should be seen in the bottom of the tube.
8. The plate check gel image has red/green vertical stripes. There may be some acrylamide solution on the outside surfaces of the plates. Remove plates from machine and wipe the read region with a moist lint-free tissue. If the problem persists, it may be cleared during the prerun, if not, avoid the affected lane(s) if possible.
9. The plate check gel image shows a green haze over the entire gel. Plates have a build-up of fluorescent contaminants. Prepare 2 L of 0.2 *M* nitric acid and 2 L of 0.2 *M* NaOH. Soak the plates in the acid for 2–3 min with the inside (gel side) facing up. Rub the plates with a gloved hand. Rinse in copious amounts of dH_2O and immerse in the NaOH; repeat procedure, and allow to air-dry.
10. Sequence data is very noisy/unusable. DNA template is of poor quality, therefore prepare it again.
11. Examples of sequencing anomalies are given in **Fig. 4**.

Fig. 4. *(opposite page)* Common sequencing anomalies found in HLA analysis. **(A)** HLA-B, nucleotide 419. A heterozygous G/C = S is not automatically detected, as the G signal is lower than the C signal. This heterozygous position has to be corrected manually. **(B)** HLA-B, nucleotide 317. A heterozygous G/T = K is not automatically detected, as the G signal is lower than the T signal. This heterozygous position has to be corrected manually. *Figure and legend continued on pages 220–221.*

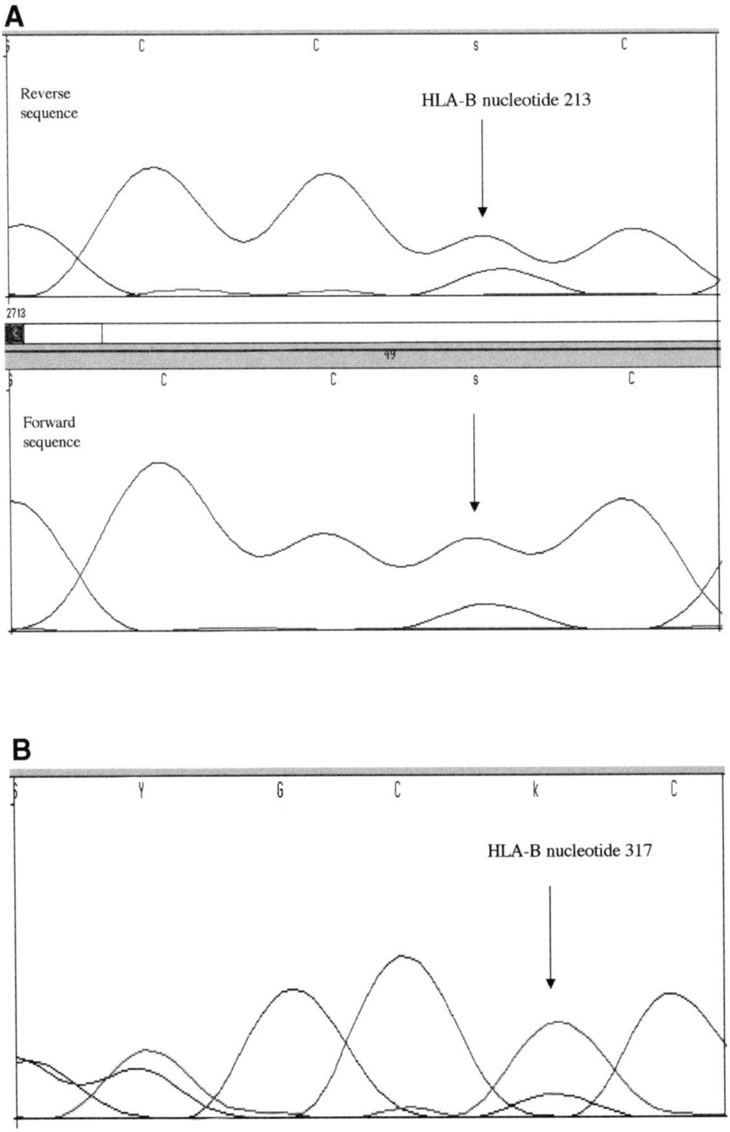
A
Reverse sequence
HLA-B nucleotide 213
Forward sequence
B
HLA-B nucleotide 317

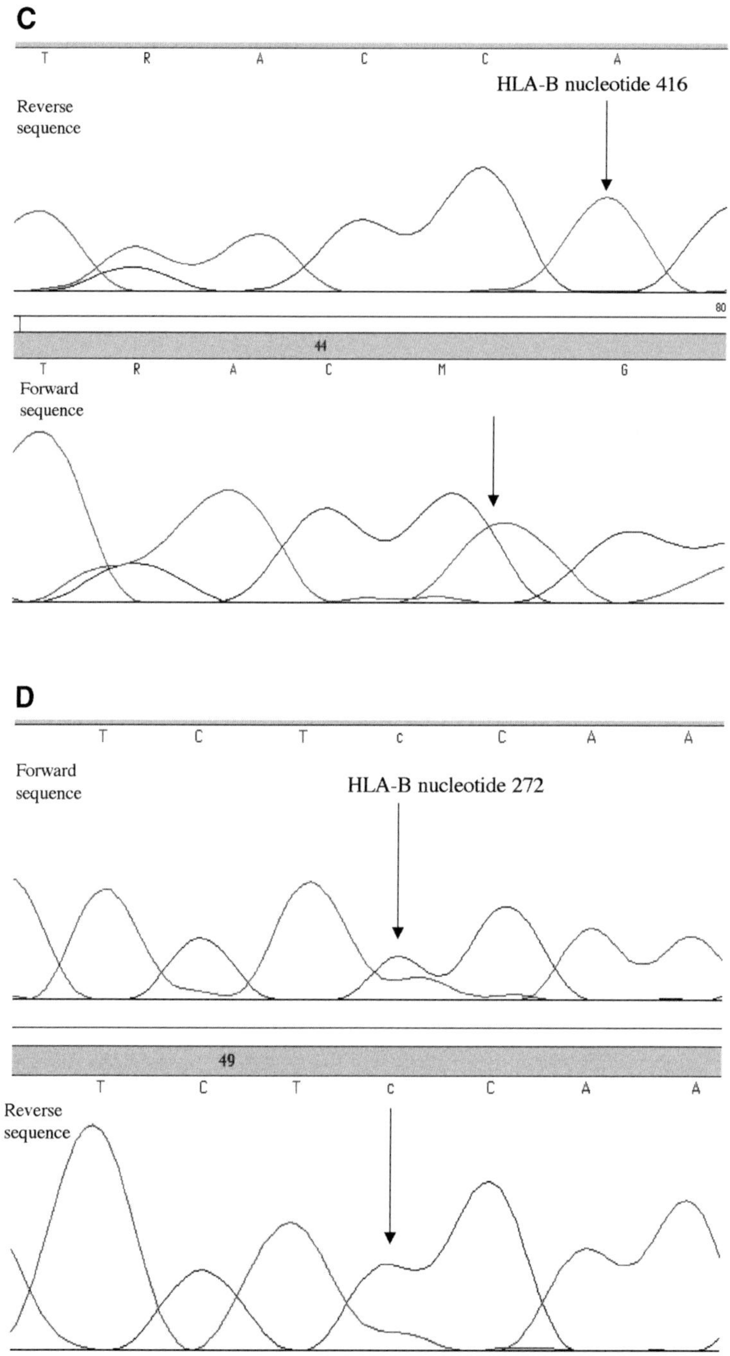
C
T R A C C A
HLA-B nucleotide 416
Reverse sequence
80
44
T R A C M G
Forward sequence
D
T C T c C A A
Forward sequence
HLA-B nucleotide 272
49
T C T c C A A
Reverse sequence

Figure and legend continued from pages 218–219.

Fig. 4. *(opposite page)* Common sequencing anomalies found in HLA analysis. **(C)** HLA-B, nucleotide 416. A compression (C/A) is always seen in the forward direction but not in the reverse direction, leading to a miscall of M and deletion of the A. **(D)** HLA-B, nucleotide 272. Often a false Y is called in the forward sequence with a shoulder or small peak in the reverse sequence. True heterozygosity at this position shows more typical heterozygous peaks.

References

1. Argüello, J. R., Little, A.-M., Bohan, E., Goldman, J., Marsh, S., and Madrigal, J. (1998) High resolution HLA class I typing by reference strand mediated conformation analysis (RSCA). *Tissue Antigens* **52**, 57–66.
2. Sanger, F., Nicklen, S., and Coulson, A. (1977) DNA sequencing with chain terminating inhibitors. *Proc. Natl. Acad. Sci. USA* **74**, 5463–5471.
3. Mezei, L. M. and Storts, D. R. (1994) Cloning of PCR products, in *PCR Technology: Current Innovations* (Griffin, H. G. and Griffin, A. M., eds.), CRC Press, Boca Raton.
4. Clark, J. M. (1988) Novel non-templated nucleotide addition reactions catalyzed by procaryotic and eucaryotic DNA polymerases. *Nucleic Acids Res.* **16**, 9677–9686.
5. Little, A.-M. and Parham, P. (1993) HLA class I gene and protein sequence polymorphism, in *Histocompatibility Testing: A Practical Approach* (Dyer, P. and Middleton, D., eds.), IRL Press, Oxford, pp. 159–190.
6. Hanahan, D. (1985) Techniques for transformation of *E. coli*, in *DNA Cloning, Vol 1* (Glover, D., ed.) IRL Press, Oxford, pp. 109–135.
7. Robinson, J., Malik, A., Parham, P., Bodmer, J. G., and Marsh, S. G. E. (2000) IMGT/HLA database—a sequence database for the human major histocompatibility complex. *Tissue Antigens* **55**, 280–287.
8. Cereb, N., Maye, P., Lee, S., Kong, Y., and Yang, S.Y. (1995) Locus-specific amplification of HLA class I genes from genomic DNA: locus-specific sequences in the first and third introns of HLA-A, -B, and -C alleles. *Tissue Antigens* **45**, 1–11.

9. Cereb, N. and Yang, S. Y. (1997) Dimorphic primers derived from intron 1 for use in the molecular typing of HLA-B alleles. *Tissue Antigens* **50**, 74–76.
10. Ennis, P., Zemmour, J., Salter, R., and Parham, P. (1990) Rapid cloning of HLA-A,B cDNA by using the polymerase chain reaction: frequency and nature of errors produced in amplification. *Proc. Natl. Acad. Sci. USA* **87**, 2833–2837.
11. Zemmour, J., Little, A.-M., Schendel, D., and Parham, P. (1992) The HLA-A,B "negative" mutant cell line C1R expresses a novel HLA-B35 allele, which also has a point mutation in the translation initiation codon. *J. Immunol.* **148**, 1941–1948.
12. Little, A.-M., Mason, A., Marsh, S. G. E., and Parham, P. (1996) HLA-C typing of eleven Papua New Guineans: identification of an HLA-Cw4/Cw2 hybrid allele. *Tissue Antigens* **48**, 113–117.
13. Scheltinga, S. A., Johnston-Dow, L. A., White, C. B., et al. (1997) A generic sequencing based typing approach for the identification of HLA-A diversity. *Hum. Immunol.* **57**, 120–128.
14. Dunn, P. P., Day, S., Harvey, J., Fuggle, S. V., and Ross, J. (1998) Identification of an HLA-C variant allele, Cw*0805, by sequencing based typing. *Tissue Antigens* **52**, 587–589.
15. Domena, J., Little, A.-M., Madrigal, J. A., et al. (1993) Structural heterogeneity in HLA-B70, a high-frequency antigen of black populations. *Tissue Antigens* **42**, 509–517.
16. Parham, P., Arnett, K., Adams, E., et al. (1997) Episodic evolution and turnover of HLA-B in the indigenous human populations of the Americas. *Tissue Antigens* **50**, 219–232.

10

HLA-E and HLA-G Typing

Jorge Martinez-Laso, Eduardo Gomez-Casado, and Antonio Arnaiz-Villena

1. Introduction

Major histocompatibility complex (MHC) class I human genes have been divided into two different subgroups. The classical class I loci, human leukocyte antigen (HLA)-A, -B and -C, encode highly polymorphic membrane glycoproteins. These proteins present intracellular-derived peptides to cytotoxic T cells *(1)* and are expressed in all nucleated cells. In contrast, the nonclassical class I genes encode molecules (class Ib molecules) that have only recently been defined and their functions are not fully understood (HLA-E, -F, and -G) *(2–10)*.

Early studies suggested that HLA-G is predominantly expressed in the cytotrophoblast, and thus, the molecule may play a role in the immunological interaction between the fetus and the mother, perhaps by inhibiting maternal natural killer cells allo-reactivity against foetal tissue *(11–15)*. However, other authors have demonstrated HLA-G in eye tissues *(16)* and fetal liver *(17)*, and low levels have also been found in a wide variety of cells in both fetal and adult tissues *(18)*. HLA-E and -F transcripts have also been found in a variety of different tissues *(11,17,19–21)*.

From: *Methods in Molecular Biology, vol. 210: MHC Protocols*
Edited by: S. H. Powis and Robert W. Vaughan © Humana Press Inc., Totowa, NJ

HLA-E binds to leader peptides derived from class Ia and HLA-G molecules, and can inhibit natural killer cell lysis through an interaction with CD94/NKG2 receptors *(22)*. HLA-G and HLA-F have recently been shown to bind with ILT2 and ILT4, two members of the immunoglobulin-like receptor (ILT) family *(21,23,24)*. X-ray crystallography shows that, unlike the highly promiscuous peptide binding groove of the class Ia molecules, the peptide binding groove of HLA-E is highly adapted to form a specific binding site for the class Ia leader sequence *(25)*.

In contrast to the classical HLA class I loci, HLA-E, -F, and -G exhibit relatively limited polymorphism. HLA-G molecules exhibit polymorphism that does not affect either the T-cell receptor or the peptide binding site *(26–28)*. The very few productive allelic changes found in HLA-E *(26)* only affect the T-cell receptor binding site *(29–32)*.

Here we describe a polymerase chain reaction restriction fragment length polymorphism (PCR-RFLP) method for typing HLA-G and a PCR sequence-specific oligonucleotide (PCR-SSO) method for typing HLA-G and HLA-E.

2. Materials

2.1. Preparation of Genomic DNA

1. To obtain the best possible PCR amplification results, a pure sample of genomic DNA should be used. DNA can be extracted using the standard phenol-chloroform method or by one of the many available commercial kits.

2.2. PCR Amplification

1. PCR amplification mixture: GeneAMP® 10X PCR Buffer (100 m*M* Tris-HCl pH 8.4 at 24°C), 50 m*M* KCl, 0.1 mg/mL gelatin, 0.02% NonideP® (NP40), 1.5 m*M* $MgCl_2$ (final concentrations), 125 m*M* dNTP mix, 1 μ*M* of each primer and 0.025 U/μL AmpliTaq® Polymerase (Applied Biosystems, Foster City, CA, USA).
2. Specific primer sequences are listed in **Subheading 3.**

2.3. Digestion of PCR Products for PCR-RFLP-DNA Typing

1. Restriction endonucleases: *Msp*I, *Hin*f, *Acy*I, *Ppu*MI, and *Bse*RI (available from a variety of manufacturers).
2. A 6% acrylamide gel or 2% NuSieve® agarose gel.

2.4. Oligonucleotide Labeling for PCR-SSO Typing

We use a digoxigenin (dig)-ddUTP oligonucleotide labeling kit (Roche Molecular Biochemicals, Indianapolis, IN, USA).

1. 10X Tailing Buffer: 1.4 m*M* Na-Cacodylate, 300 m*M* Tris-HCl, pH 7.2, 10 m*M* $CoCl_2$.
2. 1 m*M* dig-11-ddUTP.
3. Terminal deoxynucleotidyl transferase.
4. Distilled water (dH_2O).

2.5. PCR-SSO Typing

1. 0.5 *M* NaOH
2. 1 *M* and 2 *M* Ammonium acetate
3. 30x SSPE: 4.5 *M* NaCl, 0.3 *M* NaH_2PO_4, 30 m*M* EDTA, adjusted with NaOH to pH 7.4.
4. 50X Denhardt's Solution: prepare 100 mL of 2% polyvinylpirolidone (PVP)40, and 2% Ficoll® 400. Autoclave for 10 min at 120ºC and cool down to room temperature. Dissolve 2 g of bovine serum albumin (BSA) (Fraction V; Roche Molecular Biochemicals)) in approximately 80 mL of dH_2O. Adjust to pH 3.0 with 2 N HCl. Boil gently for 15 min and cool down in ice at 4ºC. Add dH_2 O up to 100 mL. Mix both solutions well and store at –20ºC.
5. TMAC Solution: 50 m*M* Tris-HCl, pH 8.0, 3.0 *M* tetramethylammonium chloride (TMAC), 0.5 *M* EDTA, 0.1% sodium dodecyl sulfate SDS.
6. Hybridization Buffer: 6X SSPE, 5X Denhardt's, 0.1% lauryl sarcosine, 0.02% SDS.
7. Buffer 0: 0.1 *M* Tris-HCl, pH 7.5, 150 m*M* NaCl.
8. Buffer 1: Buffer 0, 0.1% Tween® 20.

9. Buffer 2: 2% Blocking reagent (Roche Molecular Biochemicals) diluted in Buffer 0.
10. Buffer 3: 0.1 *M* Tris-HCl, pH 9.5, 0.1 *M* NaCl, 0.1 *M* $MgCl_2$.
11. AMPDD.
12. NBT Stock: 100 mg/mL nitroblue tetrazolium salt in 70% dimethyl formamide.
13. X-Phosphate stock: 50 mg/mL 5-Bromo-4-chloro-3-indolyl phosphate in dimethyl formamide.
14. Nylon filters.
15. Baking oven.
16. Whatman 3MM paper.
17. Plastic bags.
18. X-ray film.

3. Methods

3.1. Preparation of Genomic DNA

DNA prepared without the aid of a commercial kit should be resuspended and stored in 1X TE (0.01 *M* Tris, 0.1 m*M* EDTA, pH 8*M*). As a matter of principle, DNA should be prepared in a designated clean area and not come into contact with PCR product or any reagents, equipment, or materials exposed to PCR product.

3.2. HLA-G PCR-RFLP DNA Typing

3.2.1. PCR Amplification

1. The primer sequences used to amplify the second and third exons of the HLA-G gene are: for exon 2, 5'-TCCATGAGGTATTTC AGCGC-3' (Primer G-25') and 5'-CTGGGCCGGAGTTACTCACT-3' (Primer G-23'); for exon 3, 5'-ACACCCTCCAGTGGATGAT-3' (Primer G-35') and 5'-GGTACCCGCGCGCTGCAGCA-3' (Primer G-33').
2. Amplification conditions using the reaction mixture under **Subheading 2.2.** are: 30 cycles of 96°C for 15 s, 55°C for 15 s, and 72°C for 1 min in a programmable heat block (e.g., 9600 Perkin Elmer Cetus).

Table 1
HLA-G Allele Specificities Recognized by PCR-RFLP

	Enzymes	Band size(s)	Allele specificity
EXON 2	*MSP*I	125 + 113	01011, 0103
		238	01012, 01013
	*HINF*I	79 + 27	0103
		106	01011, 01012, 01013
EXON 3	*ACY*I	45 + 231	01011, 01012, 0103
		276	01013
	*Ppu*MI	108 + 168	01011, 01012, 01013, 0103, 0104
		276	0105N
	*Bse*RI	40 + 236	01011, 01012, 01013, 0103, 0105N
		276	0104

For further interpretation of individual alleles see **Table 2**.

3.2.2. Digestion of PCR Products

1. Following amplification of exon 2, 10-µL aliquots of the reaction mixture are digested with the restriction endonucleases *Msp*I and *Hinf*I, according to the manufacturer's recommendations.
2. Following amplification of exon 3, 10-µL aliquots of the reaction mixture are digested with the restriction endonucleases, *Acy*I, *Ppu*MI, and *Bse*RI, according to the manufacturer's recommendations.
3. Digestion products are sized-resolved by standard electrophoresis in a 6% acrylamide gel or 2% NuSieve agarose gel and detected by staining with ethidium bromide.

3.2.3. Interpretation of Results

Alleles can be determined by reference to **Tables 1** and **2**.

Table 2
HLA-G Polymorphic Restriction Motifs

	HLA-G*01011	HLA-G*01012	HLA-G*0103	HLA-G*01013	HLA-G*0104	HLA-G*0105N
*Msp*I	+	–	+	–	–	–
	(30+5+125+113+8)	(30+5+238+8)	(30+5+125+113+8)	(30+5+238+8)	(30+5+238+8)	(30+5+238+8)
*Hinf*I	–	–	+	–	–	–
	(106+175)	(106+175)	(79+27+175)	(106+175)	(106+175)	(106+175)
*Acy*I	+	+	+	–	+	+
	(45+231)	(45+231)	(45+231)	(276)	(45+231)	(45+231)
*Ppu*MI	+	+	+	+	+	–
	(108+168)	(108+168)	(108+168)	(108+168)	(108+168)	(276)
*Bse*RI	+	+	+	+	–	+
	(40+236)	(40+236)	(40+236)	(40+236)	(276)	(40+236)

Presence (+) or absence (-) of the polymorphic restriction motif is shown together with the expected sizes (in bp) of the digested product.

3.3. PCR-SSO DNA Typing for HLA-E and HLA-G

3.3.1. PCR Amplification

1. The primer sequences used to amplify the second and third exons of the HLA-E gene are: for exon 2, 5'-TGTGAATTCTCTACCGGA GTAGAGAGG-3' (Primer E-25') and 5'-TGTAAGCTTGGG CGGGTTCCGCAGCCTT-3' (Primer E-23'); for exon 3, 5'-CGGG ACTGACTAAGGGGC-3' (Primer E-35') and 5'-AGCCCTGTGGA CCCTCTT-3' (Primer E-33').
2. The primer sequences used to amplify the second and third exons of the HLA-G gene are: for exon 2, 5'-TCCATGAGGTATTTC AGCGC-3' (Primer G-25') and 5'-CTGGGCCGGAGTTACTCACT-3'; for exon 3, 5'-CACACCCTCCAGTGGATGAT-3' (Primer G-35') and 5'-GGTACCCGCGCGCTGCAGCA-3' (Primer G-33')
3. Amplification conditions using the reaction mixture under **Subheading 2.2.** are: 30 cycles of 94°C for 30 s, 65°C for 1 min, and 72°C for 1 min in a programmable heat block (e.g., 9600 Perkin Elmer Cetus).

3.3.2 Oligotyping

1. Oligonucleotide probes used are shown in **Table 3** for HLA-E and **Table 4** for HLA-G.
2. For oligonucleotide labeling, mix the following in a sterile prethawed microcentrifuge tube: 100 pmol SSO, 2.5 µL 10X Tailing Buffer, 1.0 µL 1 m*M* dig-11-ddUTP, and dH_2O to a final volume of 25 µL.
3. Add 50 U of terminal deoxynucleotidyl transferase (1 µL)
4. Incubate at 37°C for 30 min.
5. Use labeled SSO directly as a probe (*see* **Note 1**.)

3.3.3. Dot/Slot-Blotting

1. To denature, take 80 µL of amplification reaction and add 320 µL of 0.5 *M* NaOH. Mix thoroughly and maintain at room temperature for 10 min. Add 400 µL of 2 *M* Ammonium acetate, mix well, and put in ice.
2. Transfer 50 µL from this final mixture to a spot of the nylon filter previously soaked in 1 *M* ammonium acetate. Once the sample has passed through the filter add 500 µL of 1 *M* ammonium acetate.
3. Dry the filter at room temperature.
4. Dry the filter to completion at 80°C for 2 h by baking. Alternatively, illuminate the filter with 254 nm UV lamp for 5 min.

Table 3
Oligonucleotide Probes Used for HLA-E PCR-SSO DNA Typing

Probe Name	DNA sequence	Specificities
EXON 2		
E2011	5'-GGCTCCCACTCCTTG	0101,0102,01031, 01032
E2012	5'-GGCTCTCACTCCTTG	0104
E2021	5'-CGAGTGAACCTGCGG	0101,0102,01031,0104
E2022	5'-CGAGTGAATCTGCGG	01032
E2031	5'-CTGCGCGGCTACTAC	0101,01031,01032, 0104
E2032	5'-CTGCGGCGCTACTAC	0102
EXON 3		
E3011	5'-CCCGACAGGCGCTTC	0101,0102
E3012	5'-CCCGACGGGCGCTTC	01031,01032,0104
E3021	5'-CACCAGAGAGCCTAC	0101,0102,01031, 01032
E3022	5'-CACCAGGGAGCCTAC	0104

3.3.4. Hybridization

1. To prehybridize the filter: place the filter in a plastic bag with hybridization buffer (0.2 mL/cm^2) for 30 min at 42°C.
2. For hybridization with the oligonucleotide probe: remove the hybridization solution used in prehybridization and add hybridization solution (0.2 mL/cm^2) containing the dig-ddUGTP-labeled oligonucleotide probe.
3. Incubate for 3 h at 42ºC.
4. With constant gentle agitation, rinse the filter twice in 100 mL per filter of 2X SSPE, 0.1% SDS at room temperature for 5 min.
5. Wash the filter in 100 mL of TMAC solution at room temperature for 10 min.
6. Wash the filter twice for 15 min in 100 mL of preheated oligoprobes (at 50°C for HLA-E or at 58°C for HLA-G oligoprobes).

Table 4
Oligonucleotide Probes Used for HLA-G PCR-SSO DNA Typing

Probe Name	DNA sequence	Specificities
EXON 2		
G-2.1	GTGGACGACTCGCAGTTC	0103
G-2.2	GTGGACGACACGCAGTTC	01011, 01012, 01013, 0102,0104, 0105N
G-2.3	GAGGGGCCAGAGTATTGG	01012, 01013,0104, 0105N
G-2.4	GAGGGGCCGGAGTATTGG	01011, 0102, 0103
EXON 3		
G-3.1	TCCGACGGTCGCCTCCT	01013
G-3.2	TCCGACGGACGCCTCCT	01011, 01012, 0102, 0103,0104, 0105N
G-3.3	GAACGAGGACTGCGCTCC	0105N
G-3.4	AACGAGGACCTGCGCTCC	01011, 01012, 01013, 0102,0103, 0104
G-3.5	CGCCTCATCCGCGGGTAT	0104
G-3.6	CGCCTCCTCCGCGGGTA	01011, 01012, 01013, 0102,0103, 0105N

3.3.5. Chemiluminescent Detection

1. After the last washing step, the excess of solution should be removed by putting the filter on a filter sheet, avoiding complete dryness.
2. Immediately wash in buffer 1 (approx 100 mL/100 cm^2 of filter) for 5 min at room temperature with constant agitation.
3. Wash in buffer 2 (approx 100 mL/100 cm^2 of filter) for 30 min at room temperature with constant agitation.
4. Incubate in buffer 2 (approx 100 mL/100 cm^2 of filter) containing AP-antibody anti-dig (75mU/mL: dilution 1/10,000) for 30 min at room temperature with constant agitation.
5. Wash in 20 mL of buffer 3 for 5 min at room temperature with constant agitation.

6. Incubate for 5 min in 10 mL of buffer 3 containing AMPPD (0.1mg/mL: dilution 1:100).
7. Let excess AMPPD drip off the membrane before blotting for a few seconds on a sheet of dry Whatman 3 MM paper, but not to complete dryness (*see* **Note 2**).
8. Place the filter into a new heat-sealed plastic bag.
9. Incubate at 37°C for 15 min.
10. Expose for 30 min at room temperature to X-ray film.

3.3.6. Alkaline Phosphatase Detection

1. This is the same protocol used for chemiluminiscent detection in **Subheading 3.2.5.**, but at **step 6**, instead of adding Buffer 3 containing AMPPD, add exactly the same volume (10 mL) of Buffer 3 containing nitroblue tetrazolium (30 µL of NBT stock per 10 mL of Buffer 3) and X- Phosphate (30 µL of X-phosphate stock per 10 mL of Buffer 3).
2. Incubate in complete darkness.
3. The color will develop in a few minutes and is completed in 1 day (Do Not Shake).
4. Wash the filter in 50 mL of TE 10/1, pH 8.0, to halt the color development.

3.3.7. Interpretation of Alleles

1. HLA-E alleles can be determined by reference to **Table 5**, and HLA-G alleles can be determined by reference to **Table 6**.

4. Notes

1. Dig-ddUTP-labeled probe is stable at –20°C.
2. The diluted solutions of AMPPD can be stored at 4°C in complete darkness and reused.

Preliminary data presented at the 13th International Histocompatibility Workshop suggested that the HLA-E alleles *0102 and *0104 occur very infrequently. Readers are advised to consult the final workshop when it becomes available.

Table 5
HLA-E Allele Specificities Recognized by PCR-SSO

Oligoprobe	Allele Specificity
EXON 2	
E*2011	0101,0102,01031,01032
E*2012	0104
E*2021	0101,0102,01031,0104
E*2022	01032
E*2031	0101,01031,01032,0104
E*2032	0102
EXON 3	
E*3011	0101, 0102
E*3012	01031,01032,0104
E*3021	0101, 0102, 01031,01032
E*3022	0104

Table 6
HLA-G Allele Specificities Recognized by PCR-SSO

Oligoprobe	Allele Specificity
EXON 2	
G-2.1	0103
G-2.2	01011, 01012, 01013, 0102, 0104,0105N
G-2.3	01012, 01013,0104, 0105N
G-2.4	01011, 0102, 0103
EXON 3	
G-3.1	01013
G-3.2	01011, 01012, 0102, 0103, 0104,0105N
G-3.3	0105N
G-3.4	01011, 01012, 01013, 0102, 0103,0104
G-3.5	0104
G-3.6	01011, 01012, 01013, 0102, 0103,0105N

References

1. Townsend, A. R. M., Rothbard, J., Gotch, F. M., Bahador, G., Wraith, D., and McMichael, A. J. (1986) The epitopes of influenza nucleoprotein peptides. *Cell* **44**, 959–968.
2. Geraghty, D. E., Koller, B. H., and Orr, H. T. (1987) A human histocompatibility complex class I gene that encodes a protein with a shortened cytoplasmatic segment. *Proc. Natl. Acad. Sci. USA* **84**, 9145–9149.
3. Geraghty, D. E., Pei, J., Lipsky, B., et al. (1992) Cloning and physical mapping of the HLA class I region spanning the HLA-E to HLA-F interval by using yeast artificial chromosomes. *Proc. Natl. Acad. Sci. USA* **89**, 2669–2673.
4. Geraghty, D. E., Koller, B. H., Pei, J., and Hansen, J. A. (1992) Examination of four HLA class I pseudogenes. Common events in the evolution of HLA genes and pseudogenes. *J. Immunol.* **149**, 1947–1956.
5. Ragoussis, J., Bloemer, K., Pohla, H., Masser, G., Weiss, E. H., and Ziegler, A. (1989) A physical map including a new class I gene (cda 12) of juman major histocompatibility complex (A2/B13 haplotype) derived from a monosomy 6 mutant cell line. *Genomics* **4**, 301–308
6. Schmidt, C. M. and Orr, H. T. (1991) A physical linkage map of HLA-A, -G, -7.5p and -F. *Hum. Immunol.* **31**, 180–185.
7. Chimini, G., Boreto, J., and Marget, G. (1990) Molecular analysis of the human Mhc class I region using yeast artificial chromosome clones. *Immunogenetics* **32**, 419–426
8. Messer, G., Zemmour, J., Orr, H. T., Parham, P., Weiss, E. H., and Girdlestone, J. (1992) HLA-G and HLA-A. Implications for the evolution of the HLA-A related genes. *J. Immunol.* **148**, 4043–4053.
9. Carosella, E. D., Dausset, J. and Kirszenbaum, M. (1996) HLA-G revisited. *Immunol. Today* **17**, 407–409.
10. Pook, M., Woodocock, V., Tassabehji, M., et al. (1991) Characterization of an expressible nonclassical class I HLA gene. *Hum. Immunol.* **29**, 102–109
11. Wei, X. and Orr, H. T. (1990) Differential expression of HLA-E, HLA-F and HLA-G transcripts in human tissue. *Hum. Immunol.* **29**, 131–142.

12. Ellis, S., Palmer, M., and McMichael, A. (1990) Human trophoblast and the choriocarcinoma line BeWo express a truncated HLA class I molecule. *J. Immunol.* **144**, 731–735.
13. Kovats, S., Main, E., Librach, C., Stubblebine, M., Fisher, S., and DeMars, R. (1990) A class I antigens, expressed in human trophoblasts. *Science* **248**, 220–223.
14. Chumbley, G., King, A., Holmes, N., and Loke, Y. M. (1993) In situ hybridization and northern blot demostration of HLA-G mRNA in human trophoblast populations by locus-specific oligonucleotide. *Hum. Immnol.* **37**, 17–22.
15. O'Callaghan, C. A. and Bell, J. I. (1998) Structure and function of the human MHC class Ib molecules HLA-E, HLA-F and HLA-G. *Immunol. Rev.* **163**, 129–138.
16. Shukla, H., Swaroop, A., Srivatav, R., and Weissman, S. (1990) The mRNA of a human class I gene HLA G/HLA 6.0 exhibits a restricted pattern of expression. *Nucleic Acids Res.* **18**, 2189.
17. Houlihan, J. M., Biro, P. A., Fergar-Payne, A., Simpsons, K. L., and Holmes, C. H. (1992) Evidence for the expression of non-HLA-A, -B, -C class I genes in the human fetal liver. *J. Immunol.* **149**, 668.
18. Onno, M., Guillaueux, T., Amiot, L., et al. (1994) The HLA-G gene is expressed at a low mRNA level in different human cells and tissues. *Hum. Immunol.* **41**, 79–86.
19. Geraghty, D. E., Wei, X., Orr, H. T., and Koller, B. H. (1990) HLA-F: an expressed HLA gene composed of a class I coding sequenced liked to a novel transcribed repetitive element. *J. Exp. Med.* **171**, 1–18.
20. Koller, B. H., Geraghty, D. E., Shimuzu, Y., DeMars, R., and Orr, H. (1998) T. HLA-E: a novel HLA class I gene expresed in resting T lymphocytes. *J. Immunol.* **141**, 897–904.
21. Lepin, E. J. M., Bastin, J. M., Allan, D. S. J., et al. (2000) Functional characterization of HLA-F and binding of HLA-F tetramers to ILT2 and ILT4 receptors. *Eur. J. Immunol.* **30**, 3552–3561.
22. Braud, V., Allan, D. S. J., O'Callaghan, C. A., et al. (1998) HLA-E binds to natural killer cell receptors CD94/NKG2A, B and C. *Nature* **391**, 795–799.
23. Colonna, M., Nakajima, H., and Cella, M. (2000) A family of inhibitory and activating Ig-like receptors that modulate function of lymphoid and myeloid cells. *Semin. Immunol.* **12**, 121–127.

24. Allan, D. S., Colonna, M., Lanier, L. L., et al. (1999) Tetrameric complexes of human major histocompatibility leukocyte antigen (HLA)-G bind to peripheral blood myelomonocytic cells. *J. Exp. Med.* **189**, 1149–1156.
25. O'Callaghan, C. A., Tormo, J., Willcox, B. E., et al. (1998). Structural features impose tight peptide binding specificity in the non-classical 'MHC molecule HLA-E. *Mol. Cell* **1**, 531–541.
26. Arnaiz-Villena, A., Martinez, J., Alvarez, M., et al. (1997) *Immunogenetics* **46**, 251–266.
27. Castro M. J, Morales P, Fernández-Soria, V., et al. (1996) Allelic diversity at the primate Mhc-G locus: exon 3 bears stop codons in all *Cercophitecinae* sequences. *Immunogenetics* **43**, 327–336.
28. Suárez, B., Morales, P., Castro. M. J., et al. (1997) A new HLA-G (HLA-G*0105N) and its distribution in the spanish population. *Inmunogenetics* **45**, 464–465.
29. Bjorkman, P. J., Saper, M. A., Samraoui, B., Bennet, W. S., Strominger, J. L., and Wiley, D. C. (1987) Structure of the human class I histocompatibility antigen, HLA 2. *Nature* **329**, 506–512.
30. Alvarez, M., Martínez-Laso, J., Varela, P., et al. (1997) High polymorphism of Mhc-E locus in non-human primates: alleles with identical exon 2 and 3 are found in two different species. *Tissue Antigen* **49**, 160–167.
31. Gómez-Casado, E., Martínez-Laso, J., Vargas-Alarcón, G., et al. (1997) Description of a new HLA-E (E* 01031) allele and its frequency in the spanish population. *Hum. Inmunol.* **54**, 69–73.
32. Suárez, B., Morales, P., Castro. M. J., et al. (1997) Mhc-E polymorphism in Pongidae primates: the same allele is found in two different species. *Tissue Antigens* **50**, 695–698.

11

Typing Alleles of HLA-DM

Hélène Teisserenc

1. Introduction

Human leukocyte antigen DM (HLA-DM) is a major histocompatibility complex (MHC) class II-like molecule that facilitates antigen processing by catalyzing the exchange of invariant chain-derived peptides (CLIP) from class II molecules for antigenic peptides. The genes encoding HLA-DM, *HLA-DMA* and *-DMB*, are located in tandem in the class II region of the MHC, between *HLA-DP* and *HLA-DQ* loci *(1)*. HLA-DMA and -DMB products co-localize with HLA-DR in the specialized acidic MHC class II compartment (MIIC) of antigen presenting cells, but no expression of these molecules is detectable at the cell surface. The interaction of DM with class II also aids in the subsequent rapid loading of high-affinity antigen-derived peptides into the MHC class II groove. There is now accumulating evidence that HLA-DM functions as a peptide editor that removes low-stability ligands and skews the class II peptide repertoire toward high-stability ligands *(2–4)*. Because of their role in antigen processing and their location within the class II region of the MHC, these genes have attracted attention as possible disease susceptibility loci.

Limited polymorphism of HLA-DMA and -DMB has been described within the third exon *(5–7)*. **Table 1** lists polymorphic

From: *Methods in Molecular Biology, vol. 210: MHC Protocols*
Edited by: S. H. Powis and Robert W. Vaughan

Table 1
The World Health Organization (WHO)
HLA Nomenclature for HLA-DMA and -DMB Alleles

DMA	140	155	184	DMB	144	179
0101	val	gly	arg	0101	ala	ile
0102	ile	gly	arg	0102	glu	ile
0103	val	ala	his	0103	ala	thr
0104	ile	gly	cys	0104	val	thr

residues, and **Table 2** lists the corresponding reference cell lines exhibiting polymorphic variants *(8)*. We have found polymerase chain reaction sequence-specific oligonucleotide (PCR-SSO) typing an appropriate method to type DM polymorphism *(6)*. With this technique, practically all the HLA-DMA and -DMB coding regions can be analyzed for all polymorphic sites using specific oligoprobes (**Tables 3** and **4**).

2. Materials

2.1. Preparation of Genomic DNA

To obtain the best possible PCR amplification results, a pure sample of genomic DNA should be used. DNA can be extracted using the standard phenol-chloroform method or by one of the many available commercial kits.

2.2. PCR Amplification

1. Primers used to amplify DMA and DMB are listed in **Tables 5** and **6**.
2. For the reaction mixture: 200 µ*M* dNTPs, 1X *Taq* DNA polymerase buffer, $MgCl_2$, and *Taq* DNA polymerase.
3. A 1.2% agarose gel and ethidium bromide for staining DNA.

Table 2
Reference Cell Lines from the Xth Histocompatibility Workshop

Alleles	Cell lines
DMA codon 140	
val	AZH-BM9-BM16-BM21
ile	AZL-DEU-KT3
val+ile	KOZE
DMA codon 155	
gly	JY-MANN-AZL-BM21
ala	HOM2
DMA codon 184	
arg	JY-MANN
his	HOM2
cys	BM21
DMB codon 144	
ala	AZL-BRIP-BM16-COX-CB6B
glu	KAS116-YAR
val	CEPH 23-01
DMB codon 179	
ile	AZL-BM9-BM21-CB6B-KAS116
thr	BM16-BRIP-COX

Table 3
Polymorphic Sites Within the Coding Regions of HLA-DMA[a]

Oligoprobe name	Codon sequence	Codon position	Amino acid
DMA			
EXON 3			
S1	GTC	140	VAL
S2	ATC	140	ILE
S3	GGA	155	GLY
S4	GCA	155	ALA
S5	CGC	184	ARG
S6	CAC	184	HIS
S7	TGC	184	CYS

[a]The names of the oligoprobes used for the PCR-SSO typing are listed.

Table 4
Polymorphic Sites Within the Coding Regions of HLA-DMB

Oligoprobe name	Codon sequence	Codon position	Amino acid
DMB			
EXON 2			
V6	GAT	31	ASP
V7[a]	GTT	31	VAL
EXON 3			
	CTG	115	LEU
	CTA	115	LEU
V1	GCG	144	ALA
V2	GAG	144	GLU
V3	GTG	144	VAL
V4	ATT	179	ILE
V5	ACT	179	THR

[a]Variant described in only one individual of Hispanic origin.

2.3. Dot blotting of Amplified DNA

1. A suitable nylon membrane filter, for example Hybond® N+ from Amersham Pharmacia Biotech (Piscataway, NJ, USA).
2. Dot blotting apparatus with a vacuum source.
3. 0.4 *N* NaOH.
4. 10X sodium solution phosphate EDTA (SSPE): 1.5 *M* NaCl, 0.1 *M* NaH_2PO_4, 10 m*M* EDTA, pH 7.4 (adjust with NaOH).

2.4. Probe Labeling

1. Oligonucleotide primers used as probes are listed in **Tables 5** and **6**.
2. For nonradioactive labeling of probes, we recommend the digoxigenin (dig)-ddUTP kit from Roche Molecular Biochemicals (Indianapolis, IN, USA). In addition to the oligonucleotide probe, the labeling mixture contains: 10X tailing buffer, dig-11-ddUTP, terminal deoxynucleotidyl transferase, and distilled water (dH2O).

Table 5
Primers and Oligonucleotides Used to Type the Polymorphic Sites Within the Third Exon of HLA-DMA

Oligonucleotide primers	Name
Forward primer	PL DMAex3 5'-GGGTTTCCTATCGCTGAAGTG-3'
Reverse primer	PR DMAex3 5'-CCAATAGGCAATTGCTGTGTA-3'
Oligoprobes[a]	**Name**
S1	5'-ATCATTCCGTCCCTGTGG-3'
S2	5'-ATCATTCCATCCCTGTGG-3'
S3	5'-TGTCGATGGACTCAGCTT-3'
S4	5'-TGTCGATGCACTCAGCTT-3'
S5	5'-AATTGACCGCTACACAGC-3'
S6	5'-AATTGACCACTACACAGC-3'
S7	5'-AATTGACTGCTACACAGC-3'

[a]Washing temperature (conditions should be optimized in each laboratory.

Table 6
Primers and Oligonucleotides Used to Type the Polymorphic Sites Within the Third Exon of HLA-DMA

Oligonucleotide primers	Name
Forward primer	PL DMBex3 5'-TGCAAGTAGCCAAAACCACTCC-3'
Reverse primer	PR DMBex3 5'-CCAGTCCCGAAGGATGGGCT-3'
Oligoprobes[a]	**Name**
V1	5'-ACAGCAGTGCGCACAAGA-3'
V2	5'-ACAGCAGTGAGCACAAGA-3'
V3	5'-ACAGCAGTGTGCACAAGA-3'
V4	5'-TAGAGCACATTGGGGCTC-3'
V5	5'-TAGAGCACACTGGGGCTC-3'

[a]Washing temperature conditions have to be optimized in each laboratory.

2.5. Hybridization

1. A 50 mL Falcon® tube.
2. Hybridization buffer: 5X SSPE, 5X Denhardt's solution, 0.1% sodium dodecyl sulfate (SDS), 3 *M* tetramethylammonium chloride (TMAC).
3. 50X Denhardt's solution: prepare 100 mL of 2% polyvinylpirolidone(PVP) 40, and 2% Ficoll® 400. Autoclave for 10 min at 120°C and cool down to room temperature. Add 2 g of bovine serum albumin (BSA) (fraction V; Roche Molecular Biochemicals) and sterile water to make up to 200 mL of solution. Store in 50-mL aliquots at –20°C.

2.6. Washing

1. 10X SSPE, 0.1% SDS.
2. Washing solution: 50 m*M* Tris-HCl, pH 8.0, 2 m*M* EDTA, 0.1% SDS, 3 *M* TMAC.
3. 20X SSC stock solution: 3 *M* NaCl, 0.3 *M* sodium citrate.

2.7. Detection of Hybridized Probes

1. Buffer 1: 100 m*M* Tris-HCl pH 7.2, 150 m*M* NaCl.
2. Buffer 2 (blocking reagent buffer): buffer 1 containing 1% blocking reagent (Roche Molecular Biochemicals). Dissolve blocking reagent at 70°C.
3. Buffer 3: 100 m*M* Tris-HCl pH 9.5, 100 m*M* NaCl, 50 m*M* $MgCl_2$.
4. Luminigen PPD (Roche Molecular Biochemicals) 10 mg/mL.
5. Alkaline phosphatase conjugated anti-dig antibody (Roche Molecular Biochemicals) 750 U/mL.

2.8. Other Stock Solutions

1. 5 *M* TMAC.
2. 0.5 *M* EDTA, pH 8.0.
3. 10% SDS.
4. 1 *M* Tris-HCl, pH 8.0.

3. Methods

3.1. Preparation of Genomic DNA

DNA prepared without the aid of a commercial kit should be resuspended and stored in 1X TE (0.01 *M* Tris-HCl, 0.1 m*M* EDTA, pH 8.0). As a matter of principle, DNA should be prepared in a designated clean area and not come into contact with polymerase chain reaction (PCR) product or any reagents, equipment, or materials exposed to PCR product.

3.2. PCR Amplification

1. DNA samples (0.5 µg) are amplified in a 50-µL reaction mixture containing 100 ng of each oligonucleotide primer, 200 µ*M* dNTPs, 1X *Taq* DNA polymerase buffer ($MgCl_2$ at 1.5 m*M* for DMA and 2.5 m*M* for DMB), and 2 U of *Taq* DNA polymerase with sterile water to make up to the final reaction vol.
2. Reaction conditions are: 94°C for 3 min, 30 cycles of 92°C for 1 min, an annealing temperature of 65°C (DMA) or 57°C (DMB) for 30 s, 72°C for 30 s, and one cycle of 72°C for 5 min. These conditions should be optimized for individual thermal cyclers.
3. Reaction products (10 µL) are verified on a 1.2% agarose gel stained with ethidium bromide.

3.3. Dot Blotting of Amplified DNA

1. Equal amounts of DNA should be spotted on the nylon membrane filter.
2. Prepare one membrane filter for every probe.
3. Prewet the nylon filter in dH_2O.
4. Soak the nylon filter in 10X SSPE for 15 min.
5. Dry the filter at 60°C.
6. Use 50 ng of amplified DNA per spot (approx 3–5 µL of PCR product). Always spot amplified DNA from HLA-DM typed cells as controls.
7. Dry the filter at room temperature.

8. Rewet the filter with 0.4 *N* NaOH for 5 min.
9. Soak the filter in 10X SSPE for 10 min.
10. Place the membrane on a filter paper (e.g., Whatman 3 MM; Whatman, Clifton, NJ) to remove excess fluid or dry the membrane completely by baking.

3.4. Probe Labeling

1. Set up the labeling mixture as follow: 60 pmoles probe, 2.5 L of 10X tailing buffer, 1 µL of 1 m*M* dig-11-ddUTP, 1 L containing 50 U of terminal deoxynucleotidyl transferase and dH_2O to 30 µL.
2. Incubate at 37°C for 30 min.
3. Use the labeled probe directly for hybridisation of the dot blot (*see* **Note 1**).

3.5. Hybridization

1. Place the filter in a 50-µL Falcon tube with 10 mL of hybridization buffer.
2. Add 20 pmole (10 µL) of the dig-ddUTP-labeled oligonucleotide probe.
3. Incubate overnight at 54°C with constant agitation.

3.6. Washing

1. With constant gentle agitation, rinse the filters twice in 100 mL per filter of 2X SSPE, 0.1% SDS at room temperature for 10 min to remove excess probe.
2. With constant gentle agitation, wash the filters in 200 mL of washing solution preheated at appropriate temperature for 10 min (*see* **Note 2**). Washing temperatures must be optimized for each probe; start at 58°C for 10 min and adjust the temperature according to the results given by the amplified control DNAs.
3. With constant gentle agitation, wash the filters in 2X SSC at room temperature for 10 min.

3.7. Detection of Hybridized Probes

1. The following **steps 2–6** are all performed at room temperature with gentle shaking.
2. Wash the filters in buffer 1 for 5 min.
3. Incubate the filters in buffer 2 (i.e., blocking reagent buffer) for 30 min.
4. Dilute the alkaline phosphatase-conjugated anti-dig antibody 1/10,000 in buffer 2 (200 mL for 10 filters). Incubate the filters for 30 min in this solution.
5. Wash 2 times for 15 min in buffer 1.
6. Equilibrate for 5 min in buffer 3.
7. Dilute the luminigen PPD (10 mg/mL) to 1/100 in buffer 3 (i.e., substrate solution).
8. Incubate the filters in the substrate solution for 5 min (100 mL for 10 filters).
9. Let excess liquid drip off the filters, and leave them on dry Whatman 3 MM paper.
10. Before complete dryness, cover the filters on the Whatman paper with cling film and expose for 10 min to 2 h at room temperature to X-ray film (e.g., Hyperfilm™ MP from Amersham Pharmacia Biotech).
11. Immerse the filters in 10 m*M* Tris-HCl, pH 8.0, 0.5% SDS at 70°C for 20 min.
12. Remove excess fluid on a filter paper.
13. Put the filters in an envelope and store them at room temperature (*see* **Note 3**).

4. Notes

1. If the labeled probe is not used immediately, it can be stored at –20°C.
2. When washing the dot blots, we have found that a vol of 500 mL is sufficient for 10 filters.
3. Following exposure, the filters can be stored by immersing them in 10 m*M* Tris-HCl (pH 8.0), 0.5% SDS at 70°C for 20 min, removing the excess fluid on with filter paper, and putting the filters in an envelope and storing them at room temperature.

References

1. Kelly, A., Monaco, J., Cho, S., and Trowsdale J. (1991) A new human HLA class II-related locus, DM. *Nature* **353**, 571–573.
2. Denzin, L. and Cresswell, P. (1995) HLA-DM induces CLIP dissociation from MHC class II alpha beta dimers and facilitates peptide loading. *Cell* **82**, 155–165.
3. Sherman, M., Weber, D., and Jensen, P. (1995) DM enhances peptide binding to class II MHC by release of invariant chain-derived peptide. *Immunity* **3**, 197–205.
4. Sloan, V., Cameron, P., Porter, G., Gammon, M. M. A., Mellins, E., and Zaller, D. (1995) Mediation by HLA-DM of dissociation of peptides from HLA-DR. *Nature* **375**, 802–806.
5. Carrington, M., Yeager, M., and Mann, D. (1993) Characterization of *HLA-DMB* polymorphism. *Immunogenetics* **38**, 446–449.
6. Powis, S. H. and Teisserenc, H. (1996) *TAP*, *LMP*, and *HLA-DM* typing protocols, in *Genetic Diversity of HLA: Functional and Medical Implication* (Charron, D., ed.), EDK, Paris, pp. 226–235.
7. Sanderson, F., Powis, S. H., Kelly, A. and Trowsdale, J. (1994) Limited polymorphism in HLA-DM does not involve the peptide binding groove. *Immunogenetics* **39**, 56–58.
8. Bodmer, J. G., Marsh, S. G. E., Albert, E. D., et al. (1996) Nomenclature for factors of the HLA system, in *Genetic diversity of HLA: Functional and Medical Implication* (Charron, D., ed.), EDK, Paris, pp. 505–532.

III

Polymorphism in Non-HLA MHC Genes

12

Typing Alleles of *TAP1* and *TAP2*

Stephen H. Powis

1. Introduction

TAP1 and *TAP2* (for transporter associated with antigen processing) are two ATP-binding bassette (ABC) transporter genes located within the class II region of the human major histocompatibility complex (MHC) *(1)*. Their protein products form a heterodimer within the membrane of the endoplasmic reticulum (ER), which transports oligopeptides from the cytoplasm into the lumen of the ER. Once within the ER, peptides are bound by MHC class I molecules and then transported to the cell surface where they can be recognized by cytotoxic T cells *(2)*. Because of their role in antigen processing and their location within the class II region of the MHC, these genes have attracted attention as possible disease susceptibility loci *(3–8)*. In the rat, alleles of *TAP2* transport different subsets of peptides *(9)*, although this does not appear to be the case in humans.

A number of laboratories have defined alleles of *TAP1* and *TAP2*, which consist of various combinations of dimorphisms *(10,11)*; (*see* **Tables 1** and **2** and **Notes 1** and **2**). Several methods have been described for TAP typing, but most involve typing individual dimorphisms *(10,11)* (*see* **Note 3**). This chapter describes methodology for typing the common dimorphisms of *TAP1* and

From: *Methods in Molecular Biology, vol. 210: MHC Protocols*
Edited by: S. H. Powis and Robert W. Vaughan © Humana Press Inc., Totowa, NJ

Table 1
TAP1 Nomenclature

Allele Nomenclature		Codon					
WHO	Unofficial	333	370	458	637	648	661
*TAP1*0101*	*TAP1A1*	ATC	GCT	GTG	GAC	CGA	CCG
		Ile	Ala	Val	Asp	Arg	Pro
	TAP1A2	ATC	GCT	GTG	GAC	CAA	
		Ile	Ala	Val	Asp	Gln	
*TAP1*02011*	*TAP1B1*	GTC	GCT	GTG	GGC	CGA	CCG
		Val	Ala	Val	Gly	Arg	Pro
*TAP1*02012*	*TAP1B2*	GTC	GCT	GTG	GGC	CGA	CCA
		Val	Ala	Val	Gly	Arg	Pro
*TAP1*0401*	*TAP1B3*	GTC	GCT	TTG	GGC	CAA	CCG
		Val	Ala	Leu	Gly	Gln	Pro
	TAP1B4	GTC	GTT	GTG	GGC	CGA	
		Val	Val	Val	Gly	Arg	
	TAP1B5	GTC	GTT	GTG	GGC	CAA	
		Val	Val	Val	Gly	Gln	
*TAP1*0301*	*TAP1C1*	GTC	GCT	GTG	GAC	CGA	CCG
		Val	Ala	Val	Asp	Arg	Pro
	TAP1C2	GTC	GCT	GTG	GAC	CAA	
		Val	Ala	Val	Asp	Gln	
	TAP1D1	ATC	GCT	GTG	GGC	CGA	
		Ile	Ala	Val	Gly	Arg	

The WHO HLA nomenclature for *TAP1* alleles is shown in the first column. An unofficial nomenclature used in the 12th HLA Workshop is shown in the second column *(**10,11**)*. Polymorphic nucleotides are underlined.

TAP2 *(**12**)* by a modification of the amplification refractory mutation system polymerase chain reaction (ARMS-PCR) *(**13**)*. This technique permits rapid genotyping of any known mutation or polymorphism and is based upon the observation that, under appropriate conditions, oligonucleotides with a mismatched 3'-residue will not permit PCR amplification. The method described permits typing of the common dimorphisms at *TAP1* codons 333 and 637 and *TAP2* codons 379, 565 and 665.

Table 2
TAP2 Nomenclature

Allele Nomenclature		Codon									
WHO	Unofficial	163	379	386	387	436	565	604	651	665	687
*TAP2*0101*	*TAP2A1*	GTC Val	GTA Val	GGG Gly	GTGl Val	AAC Asn	GCT Ala	GGA Gly	CGT Arg	ACA Thr	TAG Stop
	TAP2A2		GTA Val	GGT Gly			GCT Ala		TGT Cys	ACA Thr	TAG Stop
*TAP2*0201*	*TAP2B1*	GTT Val	GTA Val	GGG Gly	GTG Val	AAT Asn	GCT Ala	GGG Gly	CGT Arg	GCA Ala	CAG Gln
TAP2B2			GTA Val	GGG Gly			GCT Ala		TGT Cys	GCA Ala	CAG Gln
TAP2C			ATA Ile	GGG Gly			GCT Ala		CGT Arg	ACA Thr	TAG Stop
TAP2D			ATA Ile	GGG Gly			ACT Thr		CGT Arg	ACA Thr	TAG Stop
*TAP2*0102*	*TAP2E*	GTT Val	GTA Val	GGT Gly	GTG Val	AAC Asn	ACT Thr	GGA Gly	CGT Arg	ACA Thr	TAG Stop
TAP2F			ATA Ile				ACT Thr		CGT Arg	GCA Ala	CAG Gln
TAP2G			GTA Val	GGT Gly			ACT Thr		CGT Arg	GCA Ala	CAG Gln
TAP2H			ATA Ile	GGT Gly			GCT Ala		CGT Arg	GCA Ala	CAG Gln

The WHO HLA nomenclature for *TAP2* alleles is shown in the first column. An unofficial nomenclature used in the 12th HLA Workshop is shown in the second column (***10,11***). Polymorphic nucleotides are underlined.

2. Materials

2.1. Preparation of Genomic DNA

1. In order to obtain the best PCR amplification results, a pure sample of genomic DNA should be used. Many commercial kits are now available for purifying DNA, or standard phenol-chloroform extraction of genomic DNA can be used.

2.2. DNA Amplification

1. Clean micropipets and aerosol-resistant micropipet tips should be used exclusively for PCR setup to avoid contamination with other reagents.
2. Genomic DNA as prepared under **Subheading 2.1.**
3. Oligonucleotide primers are as shown in **Tables 3** and **4**.
4. dNTPs (can be obtained from a variety of manufacturers).
5. *Taq* DNA polymerase buffer (1.5 m*M* $MgCl_2$) (can be obtained from a variety of manufacturers).
6. *Taq* DNA polymerase (can be obtained from a variety of manufacturers).
7. Sterile double-distilled water.
8. A thermal cycler (can be obtained from a variety of manufacturers).

2.3. Gel Electrophoresis

1. TE: 10 m*M* Tris-HCl, pH 7.5, 1 m*M* EDTA, pH8.0.
2. Ethidium bromide. Ethidium bromide is a powerful mutagen and care should be exercised in its use. Gloves should be worn when handling solutions containing it, and when weighing out the powder, a mask should also be worn.
3. A 100-bp DNA ladder or similar markers (can be obtained from a variety of manufacturers).

2.4. Interpretation

1. A polaroid or charge-coupled device (CCD) camera.

3. Methods

3.1. DNA Amplification

1. Four oligonucleotides are included in each PCR mixture. Two of these, one sense and one antisense, are designed such that their 3' terminal nucleotides are complementary to one of the variants of the dimorphic nucleotide being typed (*see* **Tables 3** and **4**). Other deliberate mismatches are introduced one or two nucleotides from the 3' end of each oligonucleotide in order to enhance specificity. The remaining two oligonucleotides, one sense and one antisense, are complementary to flanking sequences located at asymmetrical distances on either side of the dimorphic nucleotide.
2. A PCR mixture is prepared containing DNA (0.1–1 µg), 0.25–1 µg of the appropriate oligonucleotide primers (as shown in **Tables 3** and **4**), 200 m*M* dNTPs, 1X *Taq* DNA polymerase buffer (1.5 m*M* $MgCl_2$) and 2 U *Taq* DNA polymerase.
3. The reaction conditions are 95°C for 5 min; 35 cycles of 94°C for 1 min, the appropriate annealing temperature for 1 or 2 min (*see* **Tables 3** and **4**), 72°C for 1 or 2 min; 72°C for 10 min (*see* **Note 4**).

3.2. Gel Electrophoresis

1. Digestion products are sized-resolved by standard electrophoresis in a 2% agarose gel and detected by staining with ethidium bromide.

3.3. Interpretation

1. Each PCR amplification should produce a control product irrespective of which alleles are present in the DNA sample. The expected size of this control product is shown in the last column of **Tables 3** and **4**. If this control product is not visible, the reaction has failed and should be repeated (*see* **Note 5**).
2. In addition to the control product, one or two allele-specific products should be observed depending on whether the individual is homozygous or heterozygous. The expected size of these products is also shown in the last column of **Tables 3** and **4** (*see* **Note 6**).

Table 3
Oligonucleotide Primers Used for ARMS PCR Typing of *TAP1*

Polymorphic Site	Name	Oligonucleotide Sequence	Specificity	Product Size
TAP1 Position 333	*TAP1*/ARMS1	5'-CCCTGCACTGAGATTTGCAGACCTCTGGAG-3'	5' flanking sequence	Control–533bp
(58°C)[a]	*TAP1*/ARMS2	5'-GATCAGTGTCCCTCACCATGGTCACCCGGA-3'	*TAP1* Ile-333	Ile-333 specific–241bp
	TAP1/ARMS3	5'-GGGCAGAAGGAAAAGCAGAGGCAGGGTCAC-3'	*TAP1* Val-333	Val-333 specific–351bp
	TAP1/ARMS4	5'-ACCTGGGAACATGGACCACAGGGACAGGGT-3'	3' flanking sequence	
TAP1 Position 637	*TAP1*/ARMS5	5'-CATCTTCCCAGAATCTCCCCTATCCAGCTA-3'	5' flanking sequence	Control–429 bp
(58°C)[a]	*TAP1*/ARMS6	5'-CATCTTGGCCCTTTGCTCTGCAGAGGTACA-3'	*TAP1* Asp-637	Asp-637specific–307bp
	TAP1/ARMS7	5'-ACCCCCTGACAGCTGGCTCCCAGCCTCCC-3'	*TAP1* Gly-637	Gly-637 specific–180bp
	TAP1/ARMS8	5'-TGGGGAGGCATCCAATGGAACTGGATTTGG-3'	3' flanking sequence	

The expected sizes of the three amplified products (one constant, two variable) that can be produced in each reaction are listed in the final column.

[a]The annealing temperature used in each reaction.

Table 4
Oligonucleotide Primers Used for ARMS PCR Typing of *TAP2*

Polymorphic Site	Name	Oligonucleotide Sequence	Specificity	Product Size
TAP2 Position 379	*TAP2*/ARMS 1	5'-TTGGAGGGCTGCAGACCGTTCGCAGTTTTG-3'	5' flanking sequence	Control–427bp
(58°C)[a]	*TAP2*/ARMS2	5'-GAGACCTGGAACGCGCCTTGTACCTGCGCG-3'	*TAP2* Val-379	Val-379 specific–328bp
	TAP2/ARMS 3	5'-ACAACCACTCTGGTATCTTACCCTCCTGAT-3'	*TAP2* Ile-379	Ile -379 specific–158bp
	TAP2/ARMS 4	5'-ACATAGCTCCCCACGCTCTCCTGGTAGATC-3'	3' flanking sequence	
TAP2 Position 565	*TAP2*/ARMS 5	5'-CTCACAGTATGAACACTGCTACCTGCACAG-3'	5' flanking sequence	Control–400bp
(61°C)[a]	*TAP2*/ARMS 6	5'-TGTTCTCCGGTTCTGTGAGGAACAACAGTA-3'	*TAP2* Thr-565	Thr-565 specific–161bp
	TAP2/ARMS 7	5'-ATCATCTTCGCAGCTCTGCAGCCCATAAAC-3'	*TAP2* Ala-565	Ala-565 specific–298bp
	TAP2/ARMS 8	5'-GGAGCAAGCTTACAATTTGTAGAAGATACC-3'	3' flanking sequence	
TAP2 Position 665	*TAP2*/ARMS 9	5'-TTGGGGAATGGAATCCGGTGGTGTGAGGGC-3'	5' flanking sequence	Control–408bp
(58°C)[a]	*TAP2*/ARMS 10	5'-CAGTGCTGGTGATTGCTCACAGGCTGCAAA-3'	*TAP2* Thr-665	Thr-665 specific–141bp
	TAP2/ARMS 11	5'-CACCAGGATCTGGTGGGCGCGCTGAACTAC-3'	*TAP2* Ala-665	Ala-665 specific–326bp
	TAP2/ARMS 12	5'-TCAGCCGCTGCTGCACCAGGCGGGAATAGA-3'	3' flanking sequence	

[a]The expected sizes of the three amplified products (one constant, two variable) that can be produced in each reaction are listed in the final column.

Annealing temperature used in each reaction.

4. Notes

1. Different laboratories have used a variety of names for the same combinations of *TAP* dimorphisms. The World Health Organization human leukocyte antigen (WHO HLA) nomenclature committee has named *TAP* alleles using nomenclature similar to that used for class I and II alleles. A current list of these alleles is included in **Tables 1** and **2**. The nomenclature committee only gives names to nucleotide sequences, but current *TAP* typing methodology predominantly types nucleotide dimorphisms at known polymorphic sites. Thus, the WHO nomenclature does not completely match the nomenclature used in **Tables 1** and **2**. For example, no WHO nomenclature exists for *TAP2C*, because the existence of this allele has only been deduced by typing specific polymorphic residues in homozygous typing cell lines (HTCs) and not by nucleotide sequencing.
2. The dimorphisms described in this chapter are the most commonly studied *TAP* polymorphisms. However, other TAP polymorphisms have been described. Further information can be found in the report of the 12th International Histocompatibility Workshop *(**10,11**)*.
3. PCR-restriction fragment length polymorphism/sequence-specific oligonucleotide (RFLP/SSO) methodology for *TAP1* and *TAP2* and PCR-RFLP/SSO methodology for *TAP2* has also been described *(**11**)*.
4. PCR amplification conditions should be optimized for the particular thermal cycler in use. It may be possible to vary concentrations of DNA, *Taq* DNA polymerase, and oligonucleotides. Annealing temperatures and $MgCl_2$ concentrations may similarly need to be optimized.
5. It is also possible to perform the PCR amplification using only two of the four oligonucleotide primers described for each dimorphism, typing each of the two potential alleles in separate reactions. This is particularly useful if there is difficulty in obtaining the control PCR product. Using *TAP1* position 333 as an example, *TAP1*/ARMS1 and *TAP1*/ARMS3 would be used in one reaction to determine whether the DNA sample contained Val-333, and *TAP1*/ARMS2 and *TAP1*/ARMS4 would be used in another reaction to determine whether the sample contained Ile-333. The disadvantage of this approach is the absence of an internal control, but this can be overcome by including primers from another gene that will produce a product of an appropriate known size.

6. ARMS-PCR does not permit the unequivocal designation of alleles in individuals who are heterozygous at more than one dimorphic residue. For example, heterozygote individuals whose *TAP1* type is Ile/Val-333, Asp/Gly-637 could either be *TAP1A/B* or *TAP1C/1D* heterozygotes.

References

1. Van Edert, P. M. (1999) Genes regulating MHC class I processing of antigen. *Curr. Opin. Immunol.* **11**, 82–88
2. Koopman, J. O., Hammerling, G. J., and Momburg, F. (1997) Generation, intracellular transport and loading of peptides associated with MHC class I molecules. *Curr. Opin. Immunol.* **9**, 80–88
3. Powis, S. H., Rosenberg, W. M. C., Hall, M., et al. (1993) TAP1 and TAP2 polymorphism in coeliac disease. *Immunogenetics* **38**, 345–350.
4. Wordsworth, B. P., Pile, K. D., Gibson, K., Burnley, R. O., Mockridge, I., and Powis, S. H. (1993) Analysis of the MHC encoded transporters TAP1 and TAP2 in rheumatoid arthritis: linkage with DR4 accounts for the association with a minor TAP2 allele. *Tissue Antigens* **42**, 153–155.
5. Powis, S. H., Cooper, M. D., Trowsdale, J., Zhu, Z.-B., and Volanakis, J. E. (1994) Major histocompatibility haplotyes associated with immunoglobulin A deficiency and common variable immunodeficiency: analysis of the peptide transporter genes *TAP1* and *TAP2*. *Tissue Antigens* **43**, 261–265.
6. Jackson, D. G. and Capra, J. D. (1995) TAP2 association with insulin-dependent diabetes mellitus is aecondary to HLA-DQB1. *Hum. Immunol.* **43**, 57–65
7. Ma, L., Penfornis, A., Wang, X., et al. (1997) Evaluation of TAP1 polymorphisms with insulin dependent diabetes mellitus in Finnish diabetic patients. The Childhood Diabetes in Finland (DiMe) Study Group. *Hum. Immunol.* **53**, 159–166
8. Takeuchi, F., Nakano, K., Matsuta, K., et al. (1997) Polymorphism of TAP1 and TAP2 in Japanese patients with rheumatoid arthritis. *Tissue Antigens* **49**, 80–282
9. Powis, S. J., Deverson, E. V., Coadwell, W. J., et al. (1992). Effect of polymorphism of an MHC-linked transporter on the peptides assembled in a class I molecule. *Nature* **357**, 211–215.

10. Teisserenc, H., Besnault, L., Briaud, I., et al. (1997) TAP, LMP and HLA-DM polymorphism: 12th International Histocompatibility Workshop study, in *Genetic Diversity of HLA. Functional and Medical Implications. Volume 1* (Charron, D., ed.), EDK, Paris, pp. 159–170.
11. Powis, S. H. and Teisserenc, H. (1997) TAP, LMP and HLA-DM typing protocols, in *Genetic Diversity of HLA. Functional and Medical Implications. Volume 1* (Charron, D., ed.), EDK, Paris, pp. 226–235.
12. Powis, S. H., Tonks, S., Mockridge, I., Kelly, A., Bodmer, J. G., and Trowsdale, J. (1993) Alleles and haplotypes of the MHC encoded ABC transporters TAP1 and TAP2. *Immunogenetics* **37**, 373–380.
13. Newton, C. R., Graham, A., Heptinstall, L. E., et al. (1989) Analysis of any point mutation in DNA. The amplification refractory mutation system (ARMS). *Nucleic Acids Res.* **17**, 2503–2516.

13

Determining Alleles of the *C2* Gene by Southern Blotting

Zeng-Bian Zhu and John E. Volanakis

1. Introduction

The second component of human complement, C2, is a 102-kDa single-chain glycoprotein. It provides the catalytic subunit for the C3/C5 convertases of the classical and lectin pathways of complement activation. The single gene encoding C2 spans about 18 kb of DNA and is composed of 18 exons *(1)*. It has been mapped to the class III region of the major histocompatibility complex (MHC), 421 bp telomeric of the complement *factor B* gene *(2)* and about 92 kb centromeric of the gene encoding the major heat-shock protein HSP 70 *(3)*.

C2 protein is moderately polymorphic among Europeans and Japanese, with 94–98% of all chromosomes carrying the gene for the common allotype C, and about 2–3% of all chromosomes carrying the gene for the basic B allotype. A number of additional rare acidic (A) and basic (B) variants have also been described by using PAGIEF *(4)*. A higher allelic variation can be detected at the gene level by using restriction fragment length polymorphism (RFLP) analysis of genomic DNA *(5,6)*. There are three dimorphic RFLPs and a multiallelic polymorphism of the *C2* gene, which combine to give rise to at least nine haplotypes *(6)*.

From: *Methods in Molecular Biology, vol. 210: MHC Protocols*
Edited by: S. H. Powis and Robert W. Vaughan © Humana Press Inc., Totowa, NJ

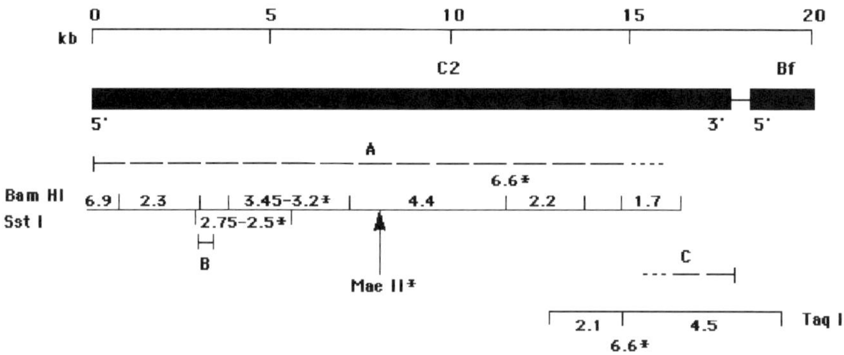

Fig. 1. Schematic representation of the RFLPs of the *C2* gene. A partial restriction map of the gene is shown with the relative positions and sizes of polymorphic fragments indicated by asterisks. Probes A and C are the 5' 2-kb and 3' 1-kb EcoRI fragments, respectively, of the full-length cDNA for *C2*, C2HL5-3 *(11)*. Probe B is a 300-bp fragment derived from the 5' region of the gene *(5)*. Probe A detects two RFLPs: (a) a multiallelic BamH1 RFLP with fragments of 3.45, 3.40, 3.35, 3.30, or 3.20 kb and (b) a second, dimorphic BamH1 RFLP with fragments of either 6.6 or 4.4 kb. Probe B detects the multiallelic SstI RFLP with fragments of 2.75, 2.70, 2.65, 2.60, or 2.50 kb. Probe C detects a dimorphic TaqI RFLP with fragments of either 6.6 or 4.5 kb. The arrow indicates the position of a polymorphic MaeII site, which can be detected by agarose gel electrophoresis of digests of a polymerase chain reaction (PCR) amplified fragment, containing exon 7.

Two of the dimorphic RFLPs map to the 3' end of the gene and are detected by Southern blotting using BamH1 and TaqI digests of genomic DNA (**Fig. 1**). The BamH1 dimorphic site gives rise to either 6.6- or 4.4- and 2.2-kb fragments, while the TaqI site gives rise to 6.6- or 4.5- and 2.1-kb fragments. The third dimorphic RFLP maps to about the middle of the gene and is detected by digesting a 241-bp PCR-amplified DNA fragment containing exon 7 with MaeII. This polymorphic site was described recently *(7)* and is caused by a G to C transversion at cDNA nucleotide 954, which generates the functionally silent E298D mutation. In the presence of G at position 954, digestion of the 241-bp fragment with MaeII gives rise to 205- and 36-bp fragments, detected by agarose gel electrophoresis. When C is present at position 954, the 205-bp fragment is split into 155- and 50-bp fragments.

The multiallelic polymorphism can be detected by Southern blotting using BamH1 or SstI digests of genomic DNA. It is located in the third intron and is caused by a human-specific variable number of tandem repeats (VNTR) locus, termed short interspersed nucleotide element-type retroposon of the *C2* gene (SINE-R.C2), which apparently was derived from the human endogenous retrovirus (HERV)-K10 *(8)*. VNTRs are thought to be the result of deletions/ insertions of the repeated nucleotide sequence through a mechanism of slipped-strand mispairing or unequal crossing over during meiosis. Most VNTRs are highly polymorphic, but only five alleles have been detected for SINE-R.C2. They can be detected by using SstI which gives fragments of 2.75, 2.70, 2.65, 2.60, or 2.50 kb or by BamHI which gives the corresponding fragments of 3.45, 3.40, 3.35, 3.30, or 3.20 kb (**Fig. 1**). A sixth size class, characterized by a 2.725-kb SstI fragment, was recently detected in our laboratory (unpublished observation), by using an enhanced resolution Southern blot as described *(9)*. The distribution of these alleles among 143 unrelated black and white Americans from the Southeastern United States was found to be skewed with 87% of the chromosomes carrying 2.7/3.4 kb SstI/ BamH1 fragments (allele a) and 6.6% 2.5/3.2 kb SstI/ BamH1 fragments (allele b), while the remaining 3 alleles were rare *(6)*. Sequencing data demonstrated that allele a and allele b contain 23 and 17 tandem nucleotide repeats, respectively.

2. Materials

2.1. Genomic DNA Digestion and Agarose Gel Electrophoresis

1. Genomic DNA is isolated from 20–30 mL blood collected in ethylenediaminetetraacetic acid (EDTA) as described *(10)*.
2. Restriction enzymes, e.g., BamHI, SstI, or TaqI are available commercially and are used according to manufacturer's instructions.

3. 10X Tris-borate-EDTA (TBE): dissolve 109 g Tris base, 55.6 g boric acid, and 9.3 g Na_2EDTA in distilled water. Make the vol to 1 L by adding distilled water. Store in a brown bottle at room temperature.
4. Ethidium bromide (10 mg/mL): add 1 g ethidium bromide to 100 mL distilled water. Stir on a magnetic stirrer for a few hours to make sure the dye has dissolved. Keep the solution in a brown bottle and store at room temperature (see **Note 1**).
5. Gel loading buffer: 0.25% bromophenol blue, 0.25% xylene cyanole FF, 15% Ficoll® (Type 400) in distilled water. Store at room temperature.
6. 100 mM Spermidine: 29 mg spermidine-HCl, 100 µL 0.1 *M* NaOH. Add distilled water to 2 mL.

2.2. Southern blotting

1. Denaturation solution: 1.5 M NaCl, 0.5 M NaOH. Mix equal vol of 3 M NaCl with 1 N NaOH. Store at room temperature.
2. Neutralization solution: 1.5 M NaCl, 0.5 M Tris-HCl, pH 7.0. Mix equal vol of 3 M NaCl with 1 M Tris-HCl, pH 7.0. Store at room temperature.
3. 20X SSC: 3 M NaCl, 0.3 M sodium citrate, pH 7.0. Autoclave and store at room temperature.
4. Deionized formamide: add 20 g fully regenerated mixed-bed resin AG 501-X8(D) (20–50 mesh) to 100 mL formamide. Mix for 1 h at room temperature. Filter and store in 50-mL fractions at –20°C.
5. 100X Denhardt's solution: combine 2 g bovine serum albumin, 2 g Ficoll (Type 400), and 2 g polyvinylpyrrolidone with 90 mL distilled water. Adjust final vol to 100 mL by adding distilled water. Keep the solution at 4°C and mix before use.
6. 1 M Sodium phosphate buffer, pH 6.5.
7. Single-stranded DNA: add 1 g salmon sperm DNA to 100 mL of 10 mM Tris-HCl, 0.1 mM EDTA, pH 7.4. Soak overnight at 4°C. Split into three parts, and sonicate for 1 min, 2 min, or 3 min, respectively. Mix the three parts and denature in boiling water for 10 min. Store in 25-mL aliquots at –20°C.
8. 50% Dextran sulfate (w/v): 50 g dextran sulfate dissolved in distilled water to a final vol of 100 mL. Store the solution at 4°C (see **Note 2**).
9. Prehybridization solution: mix 5 mL deionized formamide, 2.5 mL 20X SSC, 0.5 mL 100X Denhardt's solution, 0.5 mL 1 M sodium

phosphate buffer, pH 6.5, 0.8 mL sonicated salmon sperm DNA, and 0.7 mL distilled water.

10. Hybridization solution: mix 5 mL deionized formamide, 2.5 mL 20X SSC, 0.1 mL 100X Denhardt's solution, 0.2 mL 1 M sodium phosphate buffer, pH 6.5, 2.0 mL 50% dextran sulfate, and 0.3 mL sonicated salmon sperm DNA.
11. Washing buffer: 2X SSC, 0.2% sodium dodecyl sulfate (SDS): Mix 200 mL 20xSSC with 1760 mL distilled water first, then add 40 mL 10% SDS (see **Note 3**).

2.3. Preparation and Labeling of DNA Probes and Molecular Size Markers

1. Full-length *C2* cDNA clone, e.g., C2HL5-3 *(**11**)*.
2. *C2* genomic DNA clone, e.g., pG850 *(**5**)*.
3. 10X Klenow buffer: 100 mM Tris-HCl, pH 7.5, 150 mM $MgCl_2$, 75 mM dithiothreitol (DTT).
4. dNTPs mixture for labeling using random oligonucleotide primers: 0.5 mM dATP, 0.5 mM dGTP, and 0.5 mM dTTP.
5. dNTPs mixture for end-labeling: 330 µM dATP, 330 µM dGTP, and 330 µM dTTP.
6. Hexanucleotide mixture, $pd(N)_6$ available commercially.
7. Klenow large fragment of DNA polymerase and [^{32}P]dCTP (3000 Ci/mmol, 10 mCi/mL) available commercially.
8. Sephadex® G25 spin columns available commercially.
9. Spin column washing buffer: 10 mM Tris-HCl, pH 7.5, 100 mM NaCl, 1 mM EDTA, 0.2% SDS.
10. λDNA available commercially.

2.4. Equipment

1. Horizontal agarose gel electrophoresis system and power supply.
2. Microcentrifuge for 1.5- and 0.5-mL tubes.
3. Rocker shaker platform.
4. Incubator shaker capable of regulating temperatures between 30° and 65°C.
5. Water baths and heater blocks capable of regulating temperatures between 30° and 90°C.

6. Long-wave hand-held UV light.
7. Photography equipment with UV light box.
8. Impulse sealer capable of sealing plastic.
9. Automatic pipets capable of dispensing 0.5–20, 10–100, and 50–100 mL.

3. Methods

3.1. Genomic DNA Digestion and Agarose Gel Electrophoresis

1. Genomic DNA (10 µg) is digested to completion with BamHI, SstI or TaqI according to the manufacturer's instructions (see **Note 4**).
2. A 6-mm thick 0.8% agarose gel in 1X TBE buffer containing 0.5 µg/mL ethidium bromide is cast in a 20 x 25 cm electrophoretic tray with a 24-well (1-mm thick) analytical comb. Allow the gel to solidify at room temperature for 30 min, then move to 4°C for 30 min.
3. At the end of digestion, add 1/5 vol of gel loading buffer to DNA digests and load them into the wells of the gel. ^{32}P end-labeled HindIII, and HindIII + EcoRI digests of λDNA are used as molecular-size markers loaded into wells at either side of the samples.
4. Electrophoresis is carried out at room temperature at 65 V for 15 h using 1X TBE running buffer containing 0.5 µg/mL ethidium bromide (see **Note 5**).

3.2. Southern Blotting

1. After electrophoresis, place the gel in a clean plastic dish containing 10 gel vol (300–500 mL) of 0.25 N HCl. Shake gently on a rocker platform for 20 min at room temperature. This step results in partial depurination and cleavage of DNA fragments into smaller fragments for improved transfer.
2. Pour off the liquid and rinse the gel with distilled water. Add 10 gel vol of denaturation solution and shake gently as before for 20 min. Replace with same vol of denaturation solution and shake for an 20 additional min. Denaturation gives rise to single-stranded DNA fragments, which are suitable for subsequent hybridization with labeled probes.
3. Pour off the denaturation solution and rinse the gel with distilled water. Add 10 gel vol of neutralization solution. Shake as before for

20 min. Replace with fresh neutralization solution and shake for an additional 20 min. Neutralization solution will reduce the pH to 9.0. If the pH is still >9.0, a third neutralization step should be carried out.

4. Add 20X SSC to a plastic dish to soak two sponges about half-submerged in the solution.
5. Cut two sheets of Whatman 3 MM paper to the same length as the sponges. Wet with 20X SSC and place them on the top of the sponges; these sheets should dip into the tray and act as wicks.
6. Place the treated gel on the top of the Whatman paper and remove any air bubbles under the gel by rolling a glass pipet over the gel surface.
7. Cover the entire plastic dish with plastic wrap. Cut a window of slightly smaller size than the gel and remove the plastic wrap from the gel surface.
8. Cut a sheet of nitrocellulose membrane to the same size as the gel. Wet it with distilled water and place it on top of the gel. Remove air bubbles under the membrane by rolling a glass pipet over the surface (see **Note 6**).
9. Stack paper towels on top of the Whatman 3 MM paper to a height of about 6 cm. Place a glass plate on the top and add a weight (400–500 g) on the glass plate. Leave overnight at room temperature (see **Note 7**).
10. Remove the paper towels and Whatman paper and recover the nitrocellulose membrane. Rinse the membrane with 6X SSC, then place it on a sheet of Whatman 3 MM paper DNA side up for few minutes to dry. Mark the top-bottom and back-front orientations with pencil.
11. To immobilize DNA on the membrane, bake the membrane between two sheets of Whatman 3 MM paper under vacuum at 80°C for 2 h. At this stage, the membrane can be used for hybridization or can be stored at room temperature for several months.
12. Wet the baked membrane in 2X SSC and place in a plastic bag. Add 20 mL of the prehybridization solution into the bag and remove all air bubbles before sealing. Incubate at 42°C with shaking in a thermostat-controlled shaker for 4 h.
13. Cut a corner of the plastic bag and pour off solution. Add 20 mL of hybridization solution containing $0.5–1.0 \times 10^6$ counts per min (cpm)/mL denatured radioactive probe. Hybridize for 16 h at 42°C with shaking.
14. Pour the radioactive solution to a radioactive waste container and remove the membrane into a plastic dish with a forceps. Add 500 mL

washing buffer and agitate at room temperature for 10 min. Pour off the washing buffer into a radioactive waste container and replace with fresh washing buffer. Repeat 3X. Change the buffer to 500 mL pre-warmed washing buffer (68°C). Agitate at 68°C in water bath for 20 min. Repeat once and rinse the membrane with 500 mL fresh 2X SSC at room temperature, repeating once.

15. Blot the membrane on a Whatman 3 MM paper until almost dry (5–10 min). Wrap it with plastic wrap, place in an autoradiography cassette with intensifying screen and expose to X-ray film at –70°C.
16. Develop the film after a 1- to 3-d exposure, depending on the specific radioactivity of the probe.

3.3. DNA Probe Preparation and Labeling

1. Probe A and probe C (**Fig. 1**): a full-length C2 cDNA clone, C2HL5-3 ***(11)***, is digested with EcoRI to yield a 2-kb 5' fragment (probe A) and a 1-kb 3' fragment (probe C), which are fractionated by electrophoresis in a 0.8% low melting temperature agarose gel. The two bands are excised under a long-wave UV light and extracted with phenol.
2. Probe B (**Fig. 1**): a 300-bp C2 genomic DNA fragment derived from plasmid pG850 ***(5)*** by BamHI/KpnI double digestion, is gel-purified as described above.
3. Probe labeling using random oligonucleotide primers ***(12)***: make DNA mixture by mixing 60–100 ng DNA probe, 3 µL pd(N)$_6$, and distilled water to 13 µL. Denature DNA in boiling water for 3 min, then keep on ice for 3 min. After a brief spin, add to the DNA mixture 2.5 µL of 0.5 mM dNTPs mixture, 2.5 µL of 10X Klenow buffer, 50 µCi ^{32}P-dCTP, and 2.5 U Klenow large fragment of DNA polymerase. Add distilled water to adjust the final vol to 25 µL. Incubate the reaction mixture at 37°C for 30 min, and pass through G-25 spin column, which is in equilibrium with spin column washing buffer. The specific radioactivity of the labeled probes should vary between 0.5 and 1.0×10^9 cpm/µg. Before use, denature probe in boiling water for 3 min, then keep on ice for 3 min.
4. DNA molecular size marker preparation and end-labeling: 4 µg λDNA is digested with HindIII or HindIII/EcoRI to completion in a final vol of 10 µL according to the manufacturer's instructions. Add 10 µCi ^{32}P-dCTP, 2 µL 10X Klenow buffer, 5 µL dNTP mixture

(330 mM each), 2.5 U of Klenow large fragment of DNA polymerase to each λDNA digest. Adjust the final vol to 20 µL by adding distilled water. Incubate the reaction mixture at room temperature for 25 min, then add 1 µL 1 mM dCTP, and incubate an additional 7 min. Add 1 µL 0.5 M EDTA, pH 8.0, to end the reaction. Store labeled λDNA molecular size markers in –20°C for 1 to 2 mo. Before use, labeled DNA size markers are diluted 20–100X with 1X TBE running buffer, depending on the efficiency of the labeling.

4. Notes

1. Ethidium bromide is a powerful mutagen and is moderately toxic. Gloves should be worn when working with solutions that contain this dye.
2. Dextran sulfate in solution occupies a large vol. Therefore, distilled water should be added carefully when preparing the solution. Mix the suspension with a magnetic stirrer until it becomes a clear solution. Add distilled water to adjust the vol to 100 mL.
3. Do not mix 20X SSC with 10% SDS first, since a heavy white precipitate will form.
4. If digestion is not complete, prewarm 10 µg of genomic DNA sample to 37°C first, then add 4 mM spermidine, and appropriate restriction enzyme to DNA sample. Digest DNA samples same as above. Spermidine helps to remove unknown inhibitors present in DNA samples. Be sure that DNA samples are prewarmed to 37°C to avoid DNA precipitation when adding spermidine.
5. To increase the resolution of the electrophoresis, an enhanced resolution electrophoresis should be carried out ***(9)***. In this case, the electrophoresis is performed at 4°C with 1X TBE running buffer containing 0.5 µg/mL ethidium bromide under low voltage (30 V) for 96 h or until 2927 λ marker (from an EcoRI/HindIII double digest) has migrated approximately 16 cm from origin (examined under UV light).
6. Once placed on top of the gel, the membrane should not be moved anymore, because DNA fragments transfer can take place almost immediately.
7. Excessive weight should be avoided, because it will squeeze the liquid out of the gel and retard transfer.

References

1. Ishii, Y., Zhu, Z.-B., Macon, K.J ., and Volanakis, J. E. (1993) Structure of the human C2 gene. J. Immunol. **151**, 170–174.
2. Carroll, M. C., Campbell, R. D., Bentley, D. R., and Porter, R. R. (1984) A molecular map of the human major histocompatibility complex class III region linking complement genes C4, C2, and factor B. Nature **307**, 231–241.
3. Sargent, C. A., Dunham, I., Trowsdale, J., and Campbell, R. D. (1989) Human major histocompatibility complex contains genes for the heat shock protein HSP 70. Proc. Natl. Acad. Sci. USA **86**, 1968–1972.
4. Jahn, I., Uring-Lambert, B., Arnold, D., Clemenceau, S., and Hauptmann, G. (1990) C2 reference typing report. Complement Inflamm. **7**, 175–182.
5. Bentley, D. R., Campbell, R. D., and Cross, S. J. (1985) DNA polymorphism of the C2 locus. Immunogenetics **22**, 377–390.
6. Zhu, Z.-B., and Volanakis, J. E. (1990) Allelic associations of multiple RFLPs of the gene encoding complement protein C2. Am. J. Hum. Genet. **46**, 956–962.
7. Zhu, Z.-B., Atkinson, T. P., and Volanakis, J. E. (1998) A novel type II complement C2 deficiency allele in an African-American family. J. Immunol. **161**, 578–584.
8. Zhu, Z. B., Hsieh, S. L., Bentley, D. R., Campbell, R. D., and Volanakis, J. E. (1992) A variable number of tandem repeats locus within the human complement C2 gene is associated with a retroposon derived from a human endogenous retrovirus. J. Exp. Med. **175**, 1783–1787.
9. Ratanachaiyavong, S., Campbell, R. D., and McGregor, A. M. (1989) Enhanced resolution of the SstI polymorphic variants of the C2 locus: description of a new size class. Hum. Immunol. **26**, 310–320.
10. Sykes B. C. (1983) DNA in heritable diseases. Lancet **2**, 787–788.
11. Horiuchi, T., Macon, K. J., Kidd, V. J., and Volanakis, J. E. (1989) cDNA cloning and expression of human complement component C2. J. Immunol. **142**, 2105–2111.
12. Feinberg, A. P. and Vogelstein, B. (1983) A technique for radiolabeling DNA restriction endonuclease fragments to high specific activity. Anal. Biochem. **132**, 6–13.

14

Complement C4 Protein and DNA Typing Methods

Peter M. Schneider and Gottfried Mauff

1. Introduction

Complement C4 is the only component coded for by two nearly identical isotypic genes, C4A and C4B. The C4 genes are located tandemly arranged each with a steroid 21-hydroxylase (CYP21) gene at its 3' end, together with the genes for C2 and factor B in the major histocompatibility complex (MHC) class III gene region between HLA-B and human leukocyte antigen (HLA)-DR *(1)*. C4 polymorphism can be defined at three levels: (*i*) at the genomic level regarding the gene structure and the number of C4 genes; (*ii*) at the DNA sequence level by determining variability in the coding sequence, as well as by analyzing restriction fragment length polymorphisms (RFLPs) in introns and flanking regions; (*iii*) at the protein level by electrophoretic and serological methods to distinguish C4 isotypes and allotypes.

Based on the almost identical genomic sequence in the coding and noncoding portions of the C4A and B genes, it can be assumed that the C4 isotypes have originated from ancestral gene duplication. The tandemly arranged gene region seems to facilitate unequal recombination events generating haplotypes with a single copy, as well as three and even four copies of the C4/CYP21 gene pair. Furthermore,

From: *Methods in Molecular Biology, vol. 210: MHC Protocols*
Edited by: S. H. Powis and Robert W. Vaughan © Humana Press Inc., Totowa, NJ

a difference in C4 gene size (22 or 16 kb) can be detected, which is based on the presence or absence of a 6-kb insertion in intron 9 *(2,3)*. It has been revealed that this insertion contains a complete human endogenous retrovirus of the HERV-K family which is coded for on the opposite strand relative to the C4 coding sequence (for review, *see* **ref.** ***4***). Recently, it has been shown that this retroviral insertion may regulate the expression of retroviral sequences thus demonstrating the importance of C4 genotyping methods in the context of MHC disease association studies. *(4a)*. Southern blot analysis using the restriction enzyme *Taq*I is most useful to analyze the variation in gene size, as well as to detect gene deletions and duplications *(5)*. Population data are also available regarding the frequency of C4/CYP21 DNA haplotypes *(6)*. Furthermore, a number of conventional RFLPs have been defined either correlating with protein variants and serological determinants or dividing them into further subtypes (for review, *see* **refs.** ***7,8***).

It has been agreed that the two C4 isotypes should be defined at the functional level, i.e. on the basis of their hemolytic activity *(9)*, which is 3- to 4-fold higher for C4B than for C4A. This inherent isotype characteristic is also reflected by the presence of the Chido 4 antigenic determinant in C4B, as well as the apparent molecular weight difference of the α-chain after sodium dodecyl sulfate polyacrylamide gel electrophoresis (SDS-PAGE) (C4A: 96 kd, C4B: 94 kd). The molecular basis for these well-defined differences can be attributed to residues 1101–1106 of the C4d fragment of the α-chain *(10)*. The definition of electrophoretic variants using high-voltage agarose gel electrophoresis (HVAGE) can be facilitated by determining their relative migration (RM) based on the distance between the most common allotypes A3 and B1 *(11)*.

The functional polymorphism of human C4 is characterized by striking differences regarding the binding preferences of the reactive thiolester bond within the C4d region of the α-chain. It has been observed that the C4A thiolester preferentially transacylates onto amino group nucleophiles, whereas the C4B thiolester prefers hydroxyl groups for ester formation. This preferential binding could explain the higher hemolytic activity of C4B vs C4A and the

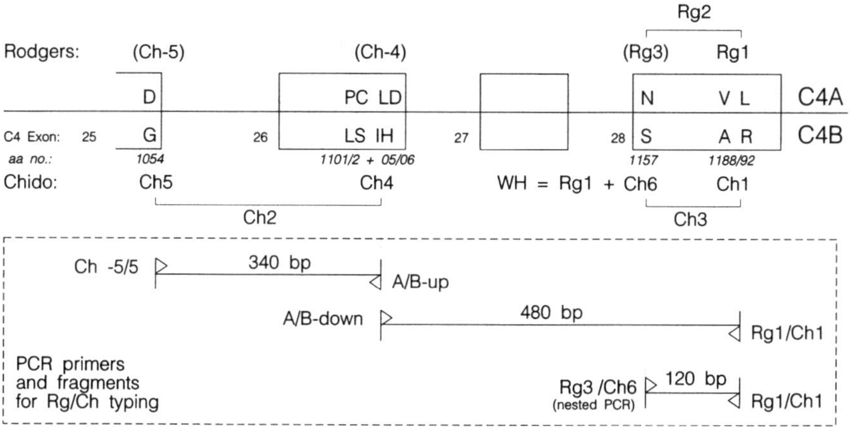

Fig. 1. Structural model for the Rodgers and Chido antigens of C4 and PCR typing strategy. Above: C4 exons 25 to 28 are schematically represented with the positions of relevant amino acid exchanges shown above for C4A and below for C4B. Except for Ch -4 (C4A) and Ch 4 (C4B) at positions 1101–1106, which define the isotype, all other determinants may occur in various combinations in rare hybrid alleles or alleles with reversed antigenicity. For determinants shown in brackets (e.g., Ch -5, Rg3) alloantisera have not been described, but have to be postulated due to the structural model. Conformational epitopes (Rg2, Ch2, Ch3) are created as indicated by the presence of two determinants in the same molecule. Below: Location of sequence-specific PCR primers and amplification strategy for PCR typing of the Rg/Ch determinants. The isotype-specific primers A-up/down and B-up/down allow the separate typing of C4A and C4B genes (modified after **ref. *21***).

higher capacity of C4A vs C4B to bind to immune complexes ***(12,13)***. Again, these isotype properties are determined only by the sequence differences in residues 1101–1106 coding for the Chido (Ch) 4 antigenic determinant ***(10)***. Furthermore, the antigenic determinants Rodgers and Chido, which are associated with C4A and C4B, respectively, can be explained by well-defined polymorphisms in exons 25–28 of the C4d region, and a structural model has been developed to explain the molecular basis of these epitopes ***(14)*** (*see* **Fig. 1**).

For the complete definition of C4 polymorphism, a combination of the following methods should be applied (modified after **ref. *15***):

1. Electrophoretic separation of allotypes using high-voltage gel electrophoresis (HVAGE) of neuraminidase- and carboxypeptidase-treated plasma followed by hemolytic overlay and immunofixation *(**16,17**)*;
2. Western blot of HVAGE-separated samples using polyclonal and/or monoclonal antibodies *(**18**)*;
3. Determination of the RM values of newly described allotypes *(**11**)*;
4. SDS-PAGE of C4 α-chains *(**19**)*;
5. *Taq*I southern blot analysis of genomic DNA using C4- and CYP21-specific probes *(**5**)*;
6. DNA typing of the Rodgers and Chido antigenic determinants based on the structural model *(**14**)* by direct polymerase chain reaction PCR analysis using sequence-specific primers *(**20,21**)*.

A separate polymorphism of the C4β-chain, linked to C4α-chain polymorphism, has been described which may be useful for segregation analysis in families and recognition of nonexpressed alleles at the protein level *(**22**)*.

2. Materials

For buffers *see* **Table 1**.

2.1. High-Voltage Immunofixation AGE (HV-AGE)

1. Electrophoretic equipment: Conventional flat bed agarose gel electrophoresis with cooling system and anti-condensation lid, if available (e.g. Pharmacia Multiphor II; power supply e.g., Pharmacia EPS 3500 XL).
2. Samples are either ethylenediaminetetraacetic acid-sodium salt (EDTA)-plasma or serum with 0.02 *M* EDTA, treated with *Clostridium perfringens* neuraminidase (NANA'se) (e.g. Sigma type VI).
3. Buffers for electrodes: Tris-glycine/barbital (buffer no. 15, **Table 1**); as gel buffer, electrode buffer 1:4 diluted is used.
4. Agarose: Seakem® ME (FMC, Rockland, ME, USA).
5. Gelbond® film (FMC), glass plates (2 mm, 12.5 x 25 cm) and slot former (Teflon teeth 0.5 x 1 mm).

Table 1
Buffers

No.	Name	Composition
1.	Alserver's solution	114 m*M* D(+) glucose, 27 m*M* Na_3-citrate, 72 m*M* NaCl; adjust to pH 6.1 with citric acid.
2.	10% Ammonium persulfate (APS)	0.1 g/1 mL H_2O (make fresh daily).
3.	Blocking buffer	Buffer no. 9, 1 g/L gelatine.
4.	EDTA-VBS	Buffer no. 17, 10 m*M* EDTA.
5.	Gelatine-VBS (GVBS)	Buffer no. 17, 1 g gelatine/L.
6.	GVBS-Suc	Buffer no. 5: 80 mL, 120 mL 9% sucrose stock solution.
7.	Mg/Ca-GVBS	Buffer no. 5, 0.83 m*M* $MgCl_2$, 0.25 m*M* $CaCl_2$, pH 7.2.
8.	PAGE solutions/buffers	1. 4 *M* KCl: 5.96 g KCl to 20 mL with H_2O. 2. 0.25 *M* EDTA: 1.86 g EDTA in 15 mL distilled H_2O, pH to 7.4 with 2 *M* NaOH, then to 20 mL with H_2O. 3. 0.14 *M* PMSF (EXTREMELY CAUSTIC): 25 mg PMSF in 1 mL ethanol (phenylmethylsulfonylfluoride is a selective enzyme inhibitor). 4. 0.01 *M* EDTA: 0.37 g EDTA in 90 mL H_2O, pH to 7.4 with 2 *M* NaOH, then to 100 mL with H_2O. 5. Disintegration buffer: a. 0.75 *M* Tris-HCl: 0.91 g tris in 9 mL H_2O, pH to 6.8 with HCl, then to 10 mL with H_2O. b. For 10 mL disintegration buffer: 0.6 g SDS, 3 mL glycerine, 10 µL 3% bromphenol blue, 1.5 mL mercaptoethanol to 10 mL with 0.75 *M* Tris-HCl, pH 6.8. Store in freezer in 0.5 mL aliquots.

(cont.)

Table 1
Buffers *(cont.)*

No.	Name	Composition
9.	Phosphate-buffered saline (PBS)-Tween	137 m*M* NaCl, 2.7 m*M* KCl, 1.5 m*M* KH_2PO_4, 8.1 m*M* Na_2HPO_4, 0.03% Tween, pH 7.3.
10.	Running gel buffer (RGB)	2 *M* Tris-HCl: 48.4 g Tris in 180 mL H_2O, pH 8.8; store at room temperature.
11.	SDS 10%	5 g SDS to 50 mL with distilled H_2O, store at room temperature.
12.	Stacking gel buffer (SGB)	1 *M* Tris-HCl: 6.06 g Tris in 30 mL H_2O, to pH 6.9 with concentrated HCl, then to 50 mL with H_2O; store at room temperature.
13.	Tris-borate-EDTA (TBE)	89 m*M* Tris, 89 m*M* boric acid, 2 m*M* EDTA, pH 7.4 with HCl.
14.	Tris-glycine	100 m*M* Tris, 200 m*M* glycine, pH 8.9.
15.	Tris-glycine-barbital	31.8 m*M* sodium barbital, 5.6 m*M* barbituric acid, 374.3 m*M* glycine, 186.6 m*M* Tris, pH 8.9, conductivity $3.4 \times 10^3\ \mu S \times cm^{-1}$.
16.	Tris-glycine-SDS buffer	Buffer no. 14, 8 g SDS to 2 L with distilled H_2O; dilute 250 mL to 1 l for use, pH 8.8.
17.	Veronal-buffered saline (VBS)	145 m*M* NaCl, 0.9 m*M* sodium barbital, 3.1 m*M* barbituric acid.

Note: Addition of EDTA and Ca-/Mg-ions is best achieved by preparing stock solutions of 0.2 *M* EDTA, 0.3 *M* $CaCl_2$, and 2.0 *M* $MgCl_2$.

6. Antiserum: polyclonal anti-C4 serum (goat, from Biogenesis or Quidel).
7. Staining reagent: Coomassie® brilliant blue R (Sigma, St. Louis, MO, USA) or Serva blue R (Serva, Heidelberg, Germany).
8. Staining/destaining solution: 100 mL acetic acid, 450 mL methanol, 450 mL distilled water.

2.2. Functional Hemolytic Detection After HV-AGE

1. Guinea pig serum (GPS): freshly bled, serum stored at –70°C; alternatively commercial serum may be used, e.g., from Behring Diagnostics.
2. C4-depleted (R4)-serum: 19 parts of GPS treated for 2 h at 37°C with 1 part 1 *M* hydrazin sulfate stock solution (e.g., Sigma) with 1 *M* NaOH adjusted to pH 8.0. R4 to be aliquotted and stored at –70°C, must not be thawed more than once. Alternatively, C4-deficient GPS may be used (Quidel).
3. Sheep red blood cells (RBCs): from freshly bled male sheeps (or commercially available, e.g., Behring Diagnostics) to be aged at 4°C for approx 2 wk in Alserver's solution (*see* buffer, **Table 1**).
4. Amboceptor: antibody from rabbits against sheep erythrocytes (e.g., from Behring Diagnostics).
5. 1% Glutaraldehyde in PBS for fixation of erythrocyte lysis patterns.
6. Two glass plates of 12.5 x 25 cm size with U-shaped rubber or plastic frame (0.5-mm thick), gel dimensions 7 × 25 cm.

2.3. Immunoblot

1. Electrophoretic equipment: *see* **Subheading 2.1.**
2. Rotating water bath.
3. Antibodies: first antibodies should be monoclonal C4A- and C4B-antibody (mouse IgG, if available) or polyclonal C4-antibody (e.g., from Biogenesis or Quidel) if isotype identification is not relevant. Second antibody: peroxidase-conjugated anti-mouse-IgG (species-specific, e.g., Sigma) for monoclonal antibodies; peroxidase-conjugated anti-goat or anti-rabbit-IgG (e.g., Sigma) for polyclonal antibody.
4. Nitrocellulose: 0.45 µm (e.g., Bio-Rad [Herculea, CA, USA] or Schleicher & Schüll, Keene, NH, USA).
5. Substrate: O-dianisidine from Sigma.

2.4. High Resolution HV-AGE

1. Electrophoretic equipment: as in **Subheading 2.1.**
2. Samples: as in **Subheading 2.1.**
3. Buffers for electrodes: Tris-glycine (buffer no.14, **Table 1**), gel electrode buffer plus 0.2 m*M* EDTA.
4. Agarose: Seakem ME.

2.5. Relative Migration (RM) Values

1. Electrophoretic equipment: For HV-AGE or for high resolution HV-AGE *see* **Subheading 2.1.**
2. Samples: as in **Subheading 2.1.**
3. Laser/video scanning densitometer (e.g., Pharmacia 2202 Ultrascan;. Gel-Doc 1000; Bio-Rad), software molecular analyst/DC.

2.6. C4 α- and C4 β-chain SDS-PAGE

1. Electrophoretic equipment: 17 x 15 x 0.3 cm vertical electrophoresis (e.g., Pharmacia; Bio-Rad) or alternative models.
2. UV-lamp: for acrylamide polymerization.
3. Samples: serum or EDTA-plasma as in **Subheading 2.1.**
4. Anti-C4: e.g., from Biogenesis or Quidel.
5. Acrylamide, bisacrylamide: e.g., from Serva, Sigma, or alternatives (research grade), TEMED, 10% ammonium persulfate (APS) in distilled water.

2.7. Southern Blot for TaqI RFLP Analysis

1. Standard equipment and TBE buffer (buffer 13, **Table 1**) for horizontal electrophoresis and Southern blotting of 20 x 20 cm agarose gels, hybridization oven with roller glass tubes, and/or rotating water bath.
2. *Taq*I restriction enzyme, 10X restriction buffer (usually supplied by the enzyme manufacturer), paraffin oil.
3. 5X sample loading buffer: 10% glycerol, 0.1% bromphenole blue, 0.1% xylene cyanol, 0.25 *N* HCl for depurination, alkaline buffer for denaturation of restriction fragments: 0.5 *M* NaOH, 1.5 *M* NaCl, 20X standard saline citrate (SSC) stock solution for blotting: 175.3 g NaCl, 88.2 g NaCitrate per liter, adjust to pH 7.0 with NaOH.
4. Neutral or positively charged nylon membrane (e.g., Hybond® N/N+, Amersham Pharmacia Biotech).
5. 5' C4 probe: 500-bp fragment of the C4 cDNA probe pAT-A ***(23)***, generated by a *Bam*HI/*Kpn*I double digestion, and purification of the 500-bp fragment by elution from agarose gel. Alternatively, a probe can be generated from human genomic DNA by PCR using the prim-

ers 3 and 4 specific for exons 5 to 9 of the human C4 genes and amplifying a 1.2-kb genomic fragment ***(24)***.

primer 3: 5'-GACCTCTGCCTGTGACCTACTTC-3'

primer 4: 5'-GTTAGCTCAGAGGTCAGAGGCAA-3'

6. CYP21 probe: 900-bp *Bgl*I fragment from the genomic subclone p21-K4 from the C4 cosmid clone 1E3 ***(25)*** or a 650-bp fragment generated by PCR using the following primers specific for exons 6 to 8 of the human CYP21A gene ***(26)***:

primer XC5: 5'-GCCATAGAGAAGAGGGACCACAA-3'

primer XC2: 5'-AGTCCGTGGTCTAGCTCCTCCTA-3'

7. Standard ^{32}-P radioisotope or chemiluminescent DNA labeling, hybridization and detection materials and reagents.

2.8. PCR-Based Rodgers and Chido Typing

1. Thermocycler GeneAmp 9600 (Applied Biosystems, Foster City, CA, USA), agarose minigel apparatus, PCR reagents.
2. PCR primers (25 μ*M* each, allele-specific bases are printed in bold face):

Isotype-specific C4A/C4B primers:

A-up	GCATG CTCCT **G**T**C**TA ACACT GGA**C**
A-down	AGGAC C**C**CT**G** TCCAG TG**T**TA **G**A**C**
B-up	TGCTC CT**A**T**G** TA**T**CA CTGGA **G**AG**A**
B-down	AGGAC C**T**CT**C** TCCAG TG**A**TA **C**A**T**

Sequence-specific Rg/Ch primers:

Ch -5	TGCGG CTTGG TTGTC ACGGG **A**
Ch 5	TGCGG CTTGG TTGTC ACGGG **G**
Rg 1	AGGTT GTTGT GGGCA ACACC **GA**
Ch 1	AGGTT GTTGT GGGCA ACACC **CC**
Rg 3	AGCCT CCATC TCAAA GGCAA **A**
Ch 6	AGCCT CCATC TCAAA GGCAA **G**

Note: The sequences of primers Ch5/Ch-5 and Rg3/Ch6 were not printed correctly in the original publication ***(20)***. The correct sequences are shown above.

PCR control (780-bp fragment from C4 gene—although any other similar fragment may be used as well):

7b	TGAGG GGACC AGCTG GAAGA GTC
8	CAAGC GCCGC CACCT GTGCC CTA

3. Methods

3.1. HV-AGE

1. Digest samples with NANA'se with 5 U/mL for 18 h at room temperature (e.g., 45 µL EDTA plasma plus 5 µL NANA'se).
2. Place plastic film on a 2-mm glass plate supported by a cooling block (20 × 25 cm); prepare 50 mL 0.5% agarose gel solution in gel buffer (buffer, no. 15, **Table 1**), and pour carefully and evenly on plastic film; submerge slot former 3 cm from cathodal edge of gel into the fluid hot agarose before solidifying.
3. For electrophoretic separation, carefully apply 3-µL samples from edge of slots; separate samples across width of cooling block for 3 h at 500 V, 80 mA, 40 W, at 10°C until a hemoglobin marker has migrated 7 cm; use a 1 to 4 dilution of gel buffer as electrode buffer (*see* **Note 1**).
4. Immunofixation: use anti-C4 serum 2–8 µL/cm^2, depending on the titer, roll carefully on agarose gel with a glass rod or pipet 1–7 cm from application slots, incubate the gel in a moist chamber for 90 min at room temperature, wash subsequently in PBS overnight, and 2X in distilled water.
5. Conserve the gel by pressing with filter paper (e.g., Whatman no. 3) and paper towels under a glass plate with approx 2 × 300 g weight, elution of unprecipitated protein in PBS, repeated pressing, and drying.
6. Stain the dry gel, fixed on plastic film, with staining solution for approx 20 min, and subsequently wash 2–3X in destaining solution until the background is clear.

3.2. Functional Hemolytic Detection after HV-AGE

1. Carry out electrophoretic separation by HV-AGE as described in **Subheading 3.1.**
2. Prepare hemolytic overlay gel with 0.8% agarose in Mg/Ca GVBS-sucr buffer (buffer no. 7, **Table 1**), amboceptor-sensitized sheep

RBCs 1.66 x 10^8/mL, 3% hydrazine-treated GPS or 2% C4-deficient GPS, and pour the gel into a 0.5-mm chamber between to glass plates (size approx 10 × 25 cm), separated by U-shaped frame.
3. Incubate the solid overlay, carefully placed on electrophoretic gel avoiding air bubbles (1–7 cm as in **Subheading 3.1.**), for approx 45 min at 37°C until lytic bands become clearly visible.
4. Conserve the overlay after development of lytic bands by submersing for approx 20 min in 1% glutardialdehyde/PBS, with subsequent washing twice in distilled water.
5. *See* also **Notes 2–6**.

3.3. Immunoblot

1. Carry out electrophoretic separation by HV-AGE (*see* **Subheading 3.1.**) or by high-resolution HV-AGE (*see* **Subheading 3.4.**).
2. Blotting is best achieved passively from the electrophoretic gel onto a dry nitrocellulose sheet for 5 min, avoid the trapping of air bubbles; cover the nitrocellulose sheet with filter paper (e.g., Whatman no. 3), paper towels, glass plate, and 2 × 300 g weight (e.g., 2 × 300 mL water in Erlenmeyer flask).
3. Block by incubating membrane for 30 min at 37°C in blocking buffer (buffer no. 3, **Table 1**) and subsequent washing 3X in PBS-Tween.
4. Incubate with monoclonal C4A- or C4B-antibody (both antibodies to be titrated before final use, undiluted to >1:5) or polyclonal C4 antiserum (dilution 1:1000–1:5000) in blocking buffer (buffer no. 3, **Table 1**). Use antibody vol of 0.5 mL/cm^2 nitrocellulose, seal in plastic bag or submerge in mini-tray, and incubate in a rotating water bath (*see* **Note 7**).
5. Color development: prepare stock solution with O-dianisidine 0.01 g/mL methanol, and develop with 500 µL stock solution plus 7 mL 0.3% H_2O_2 in 200 mL PBS, interrupt the process with tap water when bands are clearly visible.

3.4. High-Resolution HV-AGE

1. Place 2-mm glass plate on a cooling block (20 × 25 cm), and pour 0.5% agarose on plastic film (0.5-mm thick), submerge slot former 3 cm from cathodal edge of gel into the fluid agarose before solidifying.
2. For electrophoretic separation, cover the gel with plastic wrap after sample application: 3 µL of diluted plasma (10 µL of treated EDTA-

plasma plus 3 μL EDTA-veronal-buffered saline (VBS) [buffer no. 4, **Table 1**]), 3 cm from cathodal edge of gel. The separation takes place across length of cooling block for 10 min, 30 W limit; 3.5 to 4 h at 50–60 W, 1.400 V power limit.
3. Blotting: first blot: for 2 min on dry nitrocellulose membrane (e.g., BA85, Schleicher & Schüll); second blot: for 8 min on wet membrane, covered with one layer of wet Whatman no. 3 filter paper, and several layers of dry paper towels.
4. Carry out immunoblotting as described in **Subheading 3.3.**
5. Alternatively, immunofixation may also be applied as described in **Subheading 3.1.**
6. *See* **Note 8**.

3.5. RM Values

1. Electrophoretic separation should be done as described in **Subheading 3.1.** or, alternatively, as in **Subheading 3.4.**
2. Use calibration serum samples including at least the following phenotypes: A 3 B 1, A 6 B 1, B 2 B 1, or preferably A 4, B 2 B 1. But inclusion of further allotypes (such as A 12, B 5, B 3, B 92) if available according to RM values listed in **Table 2** will improve the precision of measurements.
3. Ten determinations should be done on each unknown sample with the set of calibration samples on the same plate.
4. Densitometric evaluation of absolute migration differences and calculation of the RM values may be carried out with any laser or video scanning densitometer.
5. Use the mean calculated distance between A 3 and B 1 (A3 → B1) as the RM 100 reference.
6. Calculate the linearity for all calibrations with a curve fitting program using a PC or by plotting all calibrations on a millimeter chart paper.
7. Determine the RM value of unknown C4 variants with the same curve fitting program or by assigning the absolute migration distance between C4B 1 (=0) and the unknown variants in millimeter to the RM value on the linear standard of the plot (**Fig. 2**).
8. *See* **Notes 9** and **10**.

Table 2
C4 Allele Characteristics

Allele Designation	RM Mean	RM Range	Lysis Type	MAB Reactivity
A Q0	silent allele			
A 8	197.2	—	A	n.d.
A 7	183.9	—	A	A
A 6[a]	142.4	138.7–151.3	A	A
A 58	136.0	—	A	A
A 55	127.8	127.5–128.0	A	A
A 5	119.4	116.1–121.1	A	A
A 45	116.9	—	A	A
A 4[a]	106.5	104.4–109.8	A	A
A 3	100	94.0–105.6[b]	A	A
A 2	82.6	—	A	A
A 12	64.0	59.6–67.8	A	A/B
A 1	54.2	53.6–54.8	A	A/B
A 91	47.8	45.3–49.8	A	B
B Q0	silent allele			
B 6	72.6	72.2–73.0	B	B/A
B 5[a]	70.1	68.2–72.9	B	A/B
B 45	59.2	—	B	A
B 4	46.6	46.3–47.2	B	A/B
B 35[a]	43.8	41.4–46.4	B	B
B 3	39.2	37.0–42.5	B	A/B
B 22	33.2	—	B	B
B 2*	31.6	29.5–34.2	B	B
B 13	12.9	12.7–13.2	B	A
B 12	9.5	8.7–10.0	B	A
B 11	7.7	—	B	A
B 1	0.0	—	B	B
B 92	−20.7	−(19.7–21.1)	B	B
B 94	(−40.7)[n]	—	B	B
B 95	n.d.	—	B	n.d.
B 96	(−92.6)[n]	—	B	B

As compiled from **refs. *9,11,27,29***

n.d., not determined.

[a]Reference alleles and RM values as determined during the 7th Workshop ***(27)*** differ from those published from the 6th Complement Genetics Workshop ***(9,11)***.

[b]Range for RM 100 reference was calculated from combined evaluation of five C4A3, B1 phenotypes.

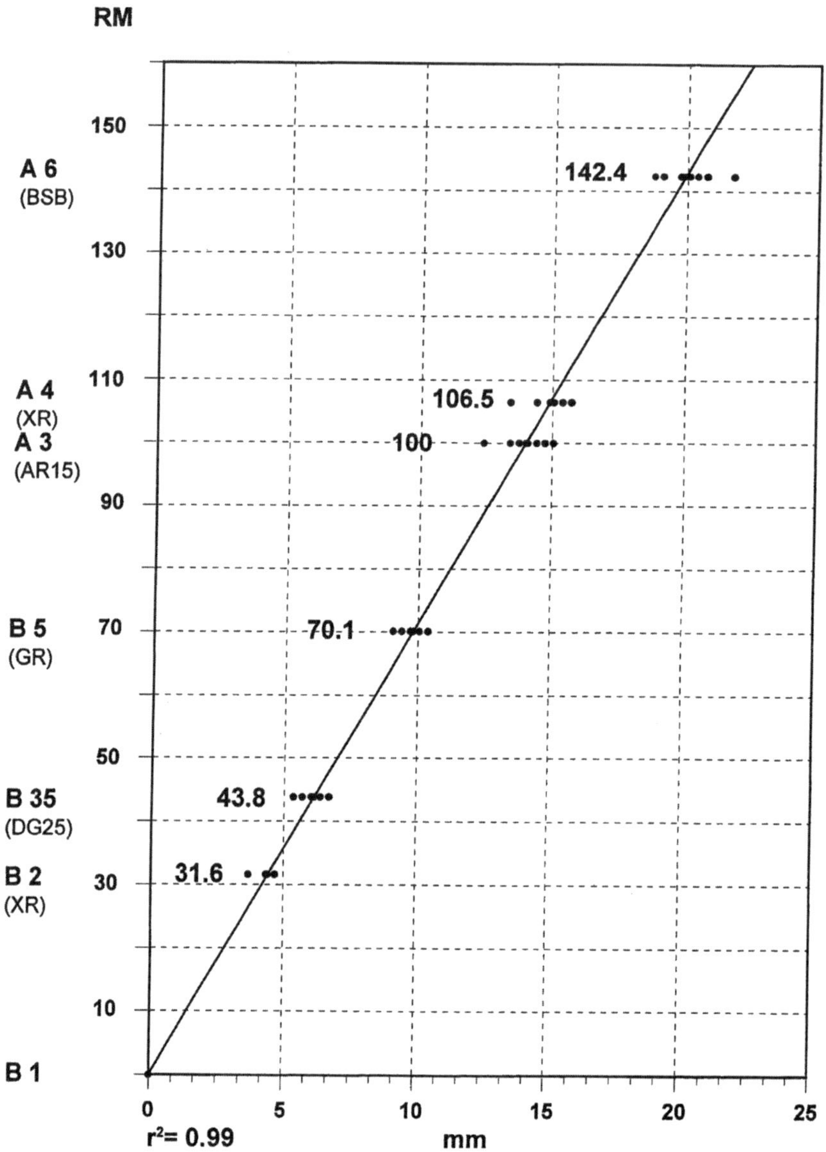

Fig. 2. RM values of C4 reference allotypes against absolute migration distances in mm from video scanning densiometry (10-fold determinations, bold figures denote mean RM values; in parenthesis sample designations of original data in **ref.** ***27***).

3.6. C4 α- and C4 β-Chain SDS-PAGE

1. Grease the Teflon spacers lightly and clamp the gel cast together, making sure that the slot former fits between the glass plates (but not too tightly); place on paper towels and level. Prepare 10% running gel, (for two gels): acrylamide 30% in 25 mL distilled water (prepare fresh daily) (TOXIC!); bis-acrylamide 1% in 4.5 mL distilled water (prepare fresh daily); 14.5 mL running gel buffer (RGB), 0.75 mL 10% SDS, 30 mL H_2O, 0.075 mL TEMED, 0.45 mL 10% APS.
2. Pour into gel cast up to 12.5 cm and carefully overlay with water or butanol with syringe. After overnight polymerization, water or butanol is poured off and the stacking gel (5%) is mixed: 30% acrylamide in 2.5 mL distilled water (prepare fresh daily) (TOXIC!); bis acrylamide 1% in 1.5 mL distilled water (prepare fresh daily); 1.89 mL SGB, 0.15 mL 10% SDS, 8.54 mL H_2O, 0.015 mL TEMED, 0.90 mL 10% APS.
3. Pour all into gel cast and slowly insert slot former; the acrylamide solution will spill over, and gloves *must* be worn. There should be approx 1 cm stacking gel below the ends of the slot formers, and polymerize 30 min in front of UV lamp or overnight without lamp.
4. Prepare samples:
   a. Mix together in a small plastic tube: 120 µL serum or plasma sample, 60 µL anti-C4, 27 µL 4 *M* KCl, 9 µL 0.25 EDTA, 2 µL 0.14 *M* phenylmethylsulfonylfluoride (PMSF). One gel holds 20 samples, including C4A 3, B 1 control.
   b. Incubate overnight at 4°C (or at 37°C in a water bath for 4 h).
   c. Centrifuge 5000*g* for 5 min at 4°C.
   d. Carefully remove supernatant and add 0.5 mL 0.01*M* EDTA, and mix well.
   e. Repeat **steps c** and **d** 2X.
   f. Remove supernatant from final wash and add 40 µL 0.01 *M* EDTA and resuspend precipitate on mixer *completely*.
   g. Add 20 µL disintegration buffer (buffer no. 8.5, **Table 1**), mix and incubate for 15 min at 70°C.
5. When the samples have been prepared, remove the bottom Teflon spacer and assemble the gel cast in the electrophoresis chamber, add electrode buffer (buffer no. 16, **Table 1**), pull out the slot former slowly, and carefully remove air bubbles in the wells with a syringe filled with electrophoresis buffer. The wells are now ready for sample

loading: apply 30 μL (for C4 α-chains) or 8.5 μL (for C4 β-chains) per sample by underlaying the buffer in the well using a Hamilton syringe. Wash all removed items carefully with water.

6. The running conditions are: 360–450 V, 100 mA, until blue marker is 1 cm from end of gel (about 4 h for α-chains; about 5 h for β-chains).
7. For staining, remove the gel from cast and mark one corner; stain for 60 min at 56°C in 0.7% Serva Blue R, destain with several changes of destaining solution (methanol:H_2O:acetic acid = 10:10:1). The gels are stored in 7% acetic acid.
8. *See* **Notes 11–13**.

3.7. Southern Blot Analysis

1. Use 10 μg high molecular weight genomic DNA per individual. Set up a restriction enzyme digestion in a 1.5-mL Eppendorf tube combining the appropriate amount of DNA, 50 U *Taq*I restriction enzyme (not more than 10% of total vol), 10% (v/v) 10X concentrated restriction buffer, and distilled water. The total vol depends on the concentration of genomic DNA. If it is 0.5 μg/μL or more, a total vol of 50 μL is sufficient (e.g., 20 μL DNA, 5 μL (5 U/μL) *Taq*I, 5 μL 10x buffer, and 20 μL water). Add 100 μL of paraffin oil to prevent evaporation, and incubate overnight in a 65°C waterbath. Carefully remove the paraffin oil to prevent the DNA sample from floating to the buffer surface during gel loading.
2. If the DNA is too diluted, a larger vol has to be selected, e.g., 100 or 200 μL, and the DNA has to be concentrated after digestion by a single phenol/chloroform extraction (one vol), followed by precipitation with 2 vols of ice-cold absolute ethanol, incubation for 2 h at –20°C, 15 min centrifugation, and resuspension in 30 μL of 1X Tris-EDTA buffer, pH 7.5.
3. Pour a 20 x 20 cm 0.8% agarose gel in 1X TBE which is sufficient to load 20 samples, and submerge gel in the electrophoresis tank. Add 20% 5X loading buffer to the restricted DNA samples, mix carefully by pipeting the samples up and down, and load slowly into the gel wells, using a separate tip for each sample to prevent contamination. Run gel overnight at 40–50V constant voltage. Stop electrophoresis when the blue dye is 1 to 2 cm from the end of the gel. Add ethidium bromide to a final concentration of 20 mg/L about 30 min before the end of the run to visualize the restriction fragments in order to monitor complete digestion and good separation of the DNA samples.

4. Incubate gel for 15 min in 500 mL 0.25 *M* HCl, and then 2X for 30 min in 500 mL alkaline buffer on a slowly moving shaker platform. Mark the top right corner of a 20 x 20 cm nylon membrane, and carry out standard Southern blotting using either a passive diffusion transfer in a buffer tank with 16X SSC overnight, or a vacuum blot for 2 h. After completion, wash the membrane in 4X SSC with 0.1% SDS to remove any agarose particles, and then fix the transferred DNA fragments by baking the dry membrane for 2 h at 80°C, or by cross-linking using a standardized UV light source.
5. For labeling of the C4 and CYP21 probes, hybridization, and detection, the readers are referred to standard protocols for ^{32}P radioisotope labeling or to manufacturer's protocols for nonradioactive labeling methods. Chemiluminescent detection is preferable in case of nonradioactive detection, as it is more sensitive and, similar to autoradiography, allows multiple exposures from the same membrane to optimize the intensity of the RFLP image on the X-ray film. Both the C4 and CYP21 probes can be used simultaneously in a single hybridization, as the respective restriction fragments are well separated. Posthybridization washing should be carried out at 65°C in several steps using 0.2% SSC, 0.1% SDS for 30 min in the final stringent wash. The interpretation of Taq I fragment patterns is illustrated in **Fig. 3** (*see* **Notes 14–16**).

3.8. PCR-Based Rodgers and Chido Typing

1. PCR conditions: 75–100 ng genomic DNA, 25 pmol of each primer, as well as the control primers, 200 μ*M* deoxynucleotide triphosphates A,G,C,TTP (dNTPs), and 2 U *Taq* DNA polymerase in 50 μL vol using the buffer provided by the *Taq* manufacturer. Use $MgCl_2$ concentrations as below.
2. Cycle conditions: initial denaturation at 94°C for 3 min, then as follows: denaturing at 94°C for 20 s, annealing according to **Table 3** for 45 s, extension at 72°C for 1 min, for the number of cycles shown in **Table 3**, and then the final extension at 72°C for 10 min.
3. Primer combinations, annealing temperatures, PCR cycle numbers, and Mg^{++} concentrations are summarized in **Table 3**.
4. To obtain complete genotype information, the standard and the alternative primer combinations, including the nested PCR primers, have to be carried out initially for each sample for the common epitopes of each isotype (**Table 4**) (*see* **Notes 17–19**).

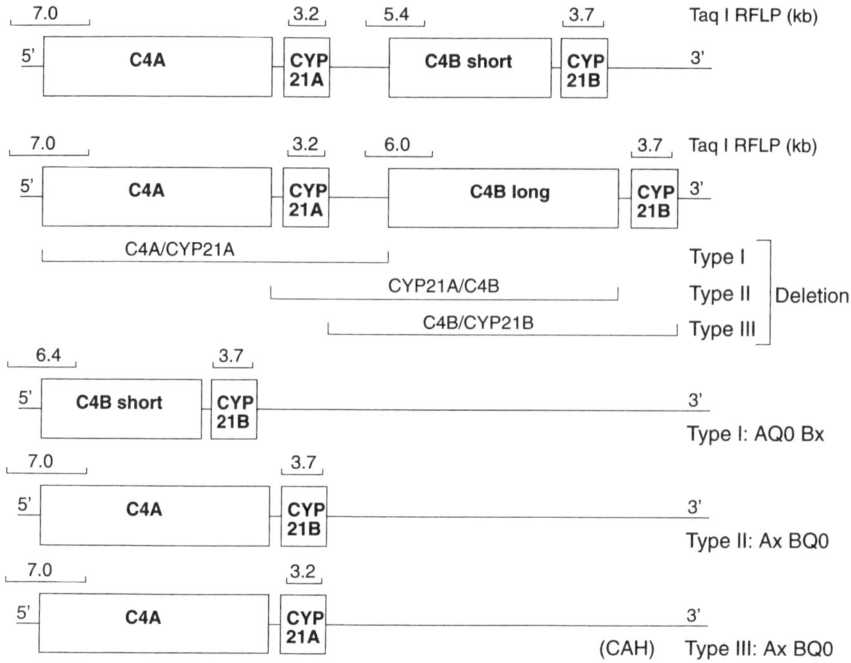

Fig. 3. Organization of the tandemly arranged C4/CYP21 gene pairs in regular haplotypes (top), as well as in haplotypes with a typical deletion of a C4 and CYP21 gene (deletion types I–III; bottom). The location and size of *Taq*I restriction fragments specific for long and short C4 genes, a C4A gene deletion, and CYP21A and B are indicated. Deletion type III is characterized by congenital adrenal hyperplasia (CAH) in its homozygous state.

4. Notes

For a final decision on the characterization of variant C4 alleles, deletions/nonexpressions or duplications a combined evaluation of results from all methods dealt with in this chapter may be necessary in many instances, including C4 α-chain determinations and standard immunochemical quantification of C4 protein concentrations, as described in any textbook of immunological methods. In the published nomenclature statement and the reports on C4 reference

Table 3
PCR Conditions for Rg/Ch Typing

Primer pair	Standard (S) or alternative (A) epitope (*see* **Table 4**)	Annealing at °C	No. cycles	$MgCl_2$
2.1. Chido 5 typing:				
A-up/Ch –5	S	66	35	1.5 m*M*
A-up/Ch 5	A			1.5 m*M*
B-up/Ch 5	S			1.5 m*M*
B-up/Ch –5	A			1.5 m*M*
2.2.1. Rg/Ch nested PCR first round:				
A-down/Rg 1	S	67	20	1.25 m*M*
A-down/Ch 1	A			1.25 m*M*
B-down/Ch 1	S			1.25 m*M*
B-down/Rg 1	A			1.25 m*M*
2.2.2. Five microliters of PCR product from first round for second round PCR:				
Rg 1/Rg 3	S	68	15	1.25 m*M*
Rg 1/Ch 6	A			1.25 m*M*
Ch 1/Ch 6	S			1.5 m*M*
Ch 1/Rg 3	A			1.5 m*M*

Table 4
Primer Combinations for Rg/Ch Typing

Isotype	Table 3	Standard PCR	Alternative PCR
C4A	2.1.	A-up/Ch–5	A-up/Ch5
	2.2.1.	A-down/Rg1	A-down/Ch1
	2.2.2.	Rg1/Rg3, and Rg1/Ch6	Ch1/Ch6 and Ch1/Rg3 *
C4B	2.1.	B-up/Ch5	B-up/Ch–5
	2.2.1.	B-down/Ch1	B-down/Rg1
	2.2.2.	Ch1/Ch6 and Ch1/Rg3	Rg1/Rg3, and Rg1/Ch6 *

Primers 7b/8 should be included in each single tube reaction as positive PCR controls. PCR fragments can be separated in a 1% agarose gel stained with ethidium bromide.

Second round reactions marked with * denote rare combinations, which should only be tested if the respective first round reaction was positive.

typings, many aspects of allotyping are discussed, and numerous examples are given *(9,15,27)*. In the following, comments on special technological difficulties and interpretations are given.

1. Precision of C4 protein separation will be sensitive to a number of electrophoretic conditions. These are: (*i*) reproducible cooling at the same temperature with as little convection as possible; (*ii*) purity of neuraminidase and agarose; and (*iii*) conductivity and pH of electrode and gel buffers.
2. All normal and C4-deficient or depleted GPS for hemolytic overlay will inactivate (deteriorate) if not carefully stored at –70°C.
3. Hydrazine treatment of normal GPS for C4 depletion (R4) will be fastidious but superior in reproducibility to C4-deficient GPS of varying origin, if larger quantities of pooled sera are used and stored at –70°C in aliquots. The optimum concentration of hydrazine needs to be determined by prior pilot lysis tests; repeated treatment of the same serum pool may be necessary if depletion is not sufficient. It must be kept in mind that hydrazine treatment is not C4-specific and will also affect activity of other complement components.
4. In pouring the hemolytic gel into the cast, the fluid mixture of reagents must not be warmer than 50°C; but care should be taken to avoid uneven solidification or air bubbles by prewarming the cast.
5. If lytic bands are not clearly visible after an approx 75-min incubation at 37°C, no further improvement is to be expected.
6. The washing of overlay gels with distilled water after glutardialdehyde fixation will sharpen lytic bands.
7. Aside the immunoblot methods widely in use, no further specialized techniques for C4 immunoblots are necessary. If monoclonal C4A- and C4B-specific antibodies are available, they should be preferred for the distinction of C4 isotypes. However, in this case, the existence of rare allotypes with hybrid or reversed antigenicity has to be considered for the correct interpretation of results.
8. HV-AGE technologically is the most demanding of the C4 protein separation methods. Only experience will improve separations, and care should be taken to use all reagents well standardized. Invariably, bands will migrate in "garlands," making precise RM value determinations difficult. However, the advantage of high-resolution HV-AGE are the larger separation distances for the distinction of closely migrating C4 variants.

9. If repeated trials fail with high-resolution HV-AGE for RM value determinations, very reliable and reproducible results may also be obtained with the well-standardized immunofixation HV-AGE as described in **Subheading 3.1.**
10. On the RM plot linear values as opposed to exponential or logarithmic relationships should be sought for the calibrating variants. A range of less than ±5% from the mean of ten determinations should be accepted (*see* **Table 2**). Electrophoretically faster A variants or cathodally migrating B variants may have a larger range than slower variants. Due to the limitation of the method, sigma values usually are not used as validation criteria. In addition, very closely migrating variants may be distinguished on side-by-side comparison of single and 1-to-1 admixed serum samples (*see* **Fig. 4** for all presently described C4 allotypes).
11. The reading of C4 α- and C4 β- chain types after SDS-PAGE separation usually is done by visual interpretation of the relative intensities of C4 α- A and B or C4 β- H and L bands, using designations from 1 (weak) to 6 (very strong) and 0 (no band detectable).
12. Precautions: in preparing PAGE, special caution should be taken; several chemicals, are dangerous in handling (PMSF: extremely caustic; acrylamide-monomer: neurotoxic, carcinogenic, mutagenic); wear gloves and protect eyes!
13. Due to the very close migration distances between C4 α- and C4β-chains, attempts for automated densitometric evaluation have been unsuccessful until now. Interpretation of α-chain results and identification of rare variants can be facilitated by combining this technique with immunoblotting detection using poly- or monoclonal antibodies (*see* **Subheading 3.3.**) *(19,28)*.
14. As illustrated in **Fig. 3**, a *Taq*I 7.0-kb fragment represents a long C4A gene, a 6.0-kb fragment represents a long C4B gene, and 5.4-kb fragment a short C4B gene. A 6.4-kb fragment represents a short C4B gene in case of a deletion of the flanking C4A/CYP21A gene pair, as is usually the case in the Caucasian HLA haplotype A1, Cw7, B8, C4AQ0 B1, DR3. Similarly, a 3.7-kb fragment represents the functionally active CYP21B gene, and a 3.2-kb fragment represents the CYP21A pseudogene. A rare short C4A gene also has the 6.4-kb band, but can only be identified correctly in the context of C4 allotyping at the protein level, and by the presence of an even ratio of CYP21 fragments (*see* **Note 15**).

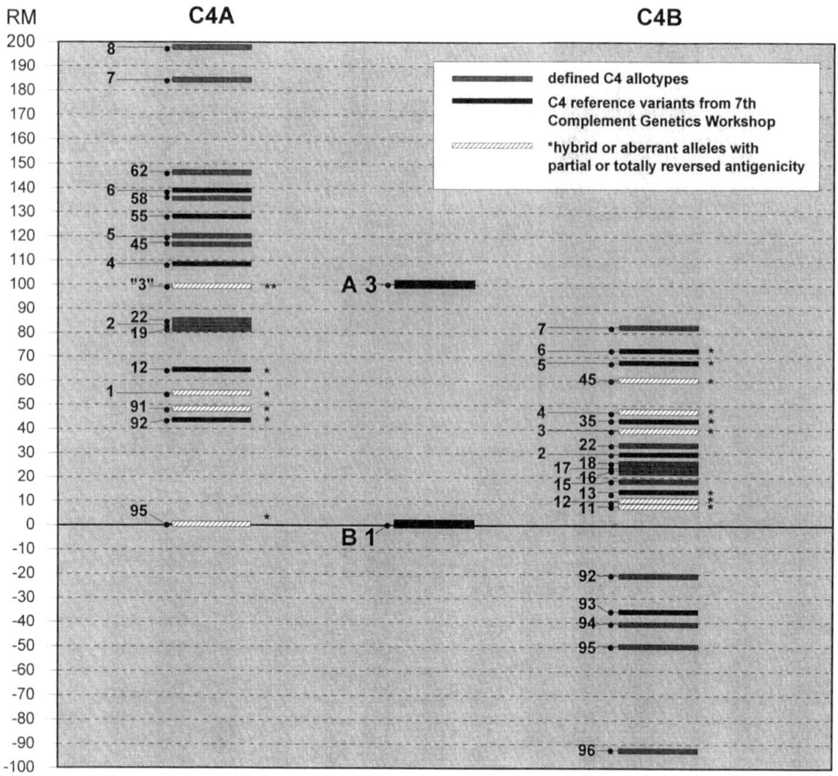

Fig. 4. RM of presently described C4 variants (actual RM positions marked by black dots). **, hybrid allele with C4A and B mAb reactivity *(29)*.

15. Since RFLP results are based on direct visualization of genomic fragments, the intensity of bands is correlated with the gene dosage at a given locus. Thus, it is possible to detect a heterozygous gene deletion, as well as gene duplications, which are both common at the C4/CYP21 loci. It has to be considered that intensities should reflect a ratio of integer gene numbers. A heterozygous deletion of the CYP21A-C4B genes on one chromosome should be represented by a ratio of 2:1 for the 7 kb C4A compared to the 6 (or 5.4) kb C4B fragments, as well as the 3.7 kb CYP21B compared to the 3.2 kb CYP21A fragments. A normal genotype with two C4A genes and one long and one short C4B gene should give a ratio of 2:1:1 for the respective fragments 7.0:6.0:5.4, and a 1:1 ratio for the CYP21A and B fragments.

16. Gene duplications are characterized by ratios of 2:2:1 or 2:3 for the C4 fragments, and 2:3 or 3:2 for the CYP21 fragments. Matters are more complicated in cases when C4B deletions and duplications occur in one individual, as such cases have usually to be confirmed by family studies.
17. The PCR typing strategy is based on the theoretically possible combinations of determinants in a given C4 gene. Furthermore, not all polymorphic residues can be detected at the protein level by either human alloantisera or monoclonal antibodies. Thus several determinants (and primers) have a "–" prefix, e.g., the C4A standard epitope "Ch -5" cannot be detected serologically, but represents a "D" residue typically present in C4A genes. Also, Rg3 (as alternate to Ch6) is a hypothetical determinant not detectable by serology, which has been postulated earlier in the structural model ***(14)***.
18. Principally, for a given C4 gene (A or B, as defined by residues 1101-6 representing Ch -4 and 4, respectively, and detected by using the primers A-up/down and B-up/down in each reaction), there are three additional polymorphic sites to be defined (*see* **Fig. 1**): Ch -5/Ch5, Rg3/Ch6, and Rg1/Ch1. As it is well established that there are a number of unusual C4 genes exhibiting partial or complete reversed antigenicity (i.e., Chido-positive C4A and Rodgers-positive C4B genes ***[15,29]***), all theoretically possible combinations of these three sites have to be tested using the standard and alternative primers combinations as listed in **Table 4.** This results in the definition of eight Rg/Ch subtypes for each C4A or C4B gene based on these combinations, which form the basis of the C4 subtype designations as shown in **Table 5**. Finally, in addition to the simple sequence epitopes described above, there are four conformational epitopes defined by the presence of two distinct sites in the same molecule (*see* **Fig. 1**). The conformational epitopes are Rg2 (from Rg1 plus Rg3), Ch2 (from Ch4 plus Ch5), Ch3 (from Ch1 plus Ch6), and WH (from Rg1 plus Ch6). Whereas Ch2 and Ch3 can only be found in C4B genes, as they include Ch4, which defines the C4B isotype, Rg2 and WH are possible in both C4A and B genes.
19. According to the C4 nomenclature proposal ***(21)***, the new four-digit nomenclature is a combination of the protein allotype and the PCR subtype, e.g., the standard A3 allotype would be represented by the designation "A0301," and the B1 allotype by "B0101." The rare allotype with reversed antigenicity linked to HLA-B47,DR7 is designated "A9108."

Table 5
Designation of Rodgers and Chido Subtypes

A. C4A Rg/Ch Subtype	C4A	01	02	03	04	05	06	07	08
I. Standard	Ch–5	•	•	•	•				
sequence	Rg+3	•		•		•		•	
epitope[a]	Rg+1	•	•			•	•		
II. Alternative	Ch+5					•	•	•	•
sequence	Ch+6		•		•		•		•
epitope[a]	Ch+1			•	•			•	•
III. Deducted	WH		•				•		
conformational	Rg+2	•				•			
epitope[a]	Ch+3				•				•

B. C4B Rg/Ch Subtype	C4B	01	02	03	04	05	06	07	08
I. Standard	Ch+5	•	•	•	•				
sequence	Ch+6	•		•		•		•	
epitope[a]	Ch+1	•	•			•	•		
II. Alternative	Ch–5					•	•	•	•
sequence	Rg+3		•		•		•		•
epitope[a]	Rg+1			•	•			•	•
III. Deducted	Ch+3	•				•			
conformational	WH			•				•	
epitope[a]	Ch+2	•	•	•	•				
	Rg+2				•				•

•, indicates presence of this residue/epitope based on presence of PCR product using the respective primer.

[a]Section I lists the typical antigen combinations for standard C4A and C4B alleles, i.e., "01" (first column) for all Rg+, Ch– C4A alleles, and all Rg–, Ch+ C4B alleles (in sequential order, *see* **Fig. 3**). Note that for C4A, the Ch–4 sequence is always present, and for C4B the Ch+4 sequence, as the isotype designation is based on these residues.

[b]Section II lists the alternative epitopes which have to be present if the corresponding standard epitope is not detected by PCR.

[c]Section III summarizes the conformational epitopes that can be deducted from the combination of sequence epitopes in a given subtype (e.g., WH is present when Rg1 and Ch6 are found in the same allele, i.e., subtypes "02" or "06" in C4A, and "03" or "07" in C4B).

References

1. Carroll, M. C., Campbell, R. D., Bentley, D. R., and Porter, R. R. (1984) A molecular map of the human major histocompatibility complex class III region linking complement genes C4, C2 and factor B. *Nature* **307**, 237–241.
2. Yu, C. Y. (1991) The complete exon-intron structure of a human complement component C4A gene: DNA sequences, polymorphism, linkage to the 21-hydroxylase gene. *J. Immunol.* **146**, 1057–1066.
3. Prentice, H., Schneider, P. M., and Strominger, J. L. (1986) C4B gene polymorphism detected in a human cosmid clone. *Immunogenetics* **23**, 274–276.
4. Yu, C. Y. (1998) Molecular genetics of the human MHC complement gene cluster. *Exp. Clin. Immunogenet.* **15**, 213–230.

4a. Schneider, P. M., Witzel-Schlomp, K., Rittner, C., Zhang, L. (2001) The endogenous retroviral insertion in the human complement C4 gene modulates the expression of homologous genes by antisense inhibition. *Immunogenetics* **53**, 1-9.

5. Schneider, P. M., Carroll, M. C., Alper, C. A., et al. (1986) Polymorphism of the human complement C4 and steroid 21-hydroxylase genes: restriction fragment length polymorphisms revealing structural deletions, homoduplications and size variants. *J. Clin Invest.* **78**, 650–657.
6. Schneider, P. M., Hartung, K., Seuchter, S. A., et al. (1992) Association of MHC class I, II, and III genes with systemic lupus erythematosus—results of a German collaborative study, in *HLA 1991, Vol. 2* (Tsuji, K., Aizawa, M., and Sasazuki, T., eds.), Oxford University Press, New York, pp. 525–528.
7. Yu, C. Y. and Campbell, R. D. (1987) Definitive RFLPs to distinguish between the human complement C4A/C4B isotypes and the major Rodgers/Chido determinants: application to the study of C4 null alleles. *Immunogenetics* **25**, 383–390.
8. Schneider, P. M. (1990) C4 DNA RFLP reference typing report. *Complement Inflamm.* **7**, 218–224.
9. Mauff, G., Alper, C. A., Dawkins, R. L., et al. (1990) C4 nomenclature statement. *Complement Inflamm.* **7**, 261–268.
10. Carroll, M. C., Fathallah, D. M., Bergamaschini, L., Alicot, E. M., and Isenman, D. E. (1990) Substitution of a single amino acid (aspartic acid for histidine) converts the functional activity of human complement C4B to C4A. *Proc. Natl. Acad. Sci. USA* **87**, 6868–6872.

11. Mauff, G., Brenden, M., and Braun-Stilwell, M. (1990) Relative electrophoretic migration distances for the classification of C4 allotypes. *Complement Inflamm.* **7**, 277–281.
12. Isenman, D. E. and Young, J. R. (1984) The molecular basis for the difference in immune hemolysis activity of the Chido and Rodgers isotypes of human complement component C4. *J. Immunol.* **132**, 3019–3027.
13. Law, S. K. A., Dodds, A. W., and Porter, R. R. (1984) A comparison of the properties of two classes, C4A and C4B, of human complement component C4. *EMBO J.* **3**, 1819–1823.
14. Yu, C. Y., Campbell, R. D., and Porter, R. R. (1988) A structural model for the location of the Rodgers and the Chido antigenic determinants and their correlation with the human complement component C4A/C4B isotypes. *Immunogenetics* **27**, 399–405.
15. Mauff, G., Brenden, M., Braun-Stilwell, M., et al. (1990) C4 reference typing report. *Complement Inflamm.* **7**, 193–212.
16. Awdeh, Z. L. and Alper, C. A. (1980) Inherited structural polymorphism of the fourth component of human complement. *Proc. Natl. Acad. Sci. USA* **77**, 3576–3580.
17. Sim, E. and Cross, S. J. (1986) Phenotyping of human complement C4, a class III HLA antigen. *Biochem. J.* **239**, 763–767.
18. Doxiadis, G. and Grosse-Wilde, H. (1990) C4 allotyping by prolonged gel electrophoresis and immunoblotting using monoclonal and polyclonal antibodies. *Complement Inflamm.* **7**, 269–276.
19. Roos, M. H., Mollenhauer, E., Demant, P., and Rittner, C. (1982) A molecular basis for the two locus model of human complement component C4. *Nature* **298**, 854–856.
20. Barba, G., Braun-Heimer, L., Rittner, C., and Schneider, P. M. (1994) A new PCR-based typing of the Rodgers and Chido antigenic determinants of the fourth component of human complement. *Eur. J. Immunogenet.* **21**, 325–339.
21. Schneider, P. M., Stradmann-Bellinghausen, B., and Rittner, C. (1996) Genetic polymorphism of the fourth component of human complement: population study and proposal for a revised nomenclature based on genomic PCR typing of Rodgers and Chido determinants. *Eur. J. Immunogenet.* **23**, 335–344.
22. Steuer, M., Mauff, G., Adam, C., et al. (1989) An estimate on the frequency of duplicated haplotypes and silent alleles of human C4

protein polymorphism. I. Investigations in healty Caucasoid families. *Tissue Antigens* **33**, 501–510.

23. Belt, K. T., Carroll, M. C., and Porter, R. R. (1984) The structural basis of the multiple forms of human complement component C4. *Cell* **36**, 907–914.
24. Barba, G., Rittner, C., and Schneider, P. M. (1993) Genetic basis of human complement C4A deficiency. Detection of a point mutation leading to nonexpression. *J. Clin. Invest.* **91**, 1681–1686.
25. Carroll, M. C., Palsdottir, A., Belt, K. T., and Porter, R. R. (1985) Deletion of complement C4 and steroid 21-hydroxylase genes in the HLA class III region. *EMBO J.* **4**, 2547–2552.
26. Chu, X., Braun-Heimer, L., Rittner, C., and Schneider, P. M. (1992) Identification of the recombination site within the steriod 21-hydroxylase gene (CYP21) of the HLA-B47, DR7 haplotype. *Exp. Clin. Immunogenet.* **9**, 80–85.
27. Mauff, G., Luther, B., Schneider, P. M., et al. (1998) Reference typing report for complement component C4. *Exp. Clin. Immunogenet.* **15**, 249–260.
28. Rittner, C. and Stradmann-Bellinghausen, B. (1990) C4 alpha-chain reference typing report. *Complement Inflamm.* **7**, 225–229.
29. Weg-Remers, S., Brenden, M., Schwarz, E. et al. (1997) Major histocompatibility complex (MHC) class III genetics in two Amerindian tribes from southern Brazil. The Kaingang and the Guarani. *Hum. Genet.* **100,** 548–556.
30. Giles, C. M., Uring-Lambert, B., Goetz, J., et al. (1988) Antigenic determinants expressed by human C4 allotypes; a study of 325 families provides evidence for the structural antigenic model. *Immunogenetics* **27**, 442–448.

15

Typing of Tumor Necrosis Factor Alleles

Anthony Gerard Wilson

1. Introduction

The tumor necrosis factor (TNF) locus contains the three closely related cytokine genes: *TNF*, and *lymphotoxin* α and β. The gene cluster lies at the telomeric end of the class III region of the major histocompatibility complex (MHC) approximately 1 megabase telomeric to the human leukocyte antigen (HLA)-DR locus and 250 kb centromeric to HLA-B. In view of the potent biological activites of these molecules and the location of their genes within the MHC, the possibility that polymorphism within the TNF locus could influence the onset of, or the outcome from, a large number of autoimmune or infectious diseases, has lead to intensive genetic investigation.

Five polymorphic microsatellites spanning the TNF locus have been characterized (**Fig. 1**). Approximately 3.5 kb telomeric of LTα lie the closely linked TNFα and *LTα*. The first intron of this gene contains the biallelic TNFc marker, while roughly 10 kb centromeric of TNFα lie the closely related TNFd and TNFe markers *(1,2)*. Because of the large number of alleles and high heterozygosity rates (**Table 1**), these are very useful in genetic studies of the TNF locus, however, there is no evidence that they have any direct effects on gene function.

From: *Methods in Molecular Biology, vol. 210: MHC Protocols*
Edited by: S. H. Powis and Robert W. Vaughan © Humana Press Inc., Totowa, NJ

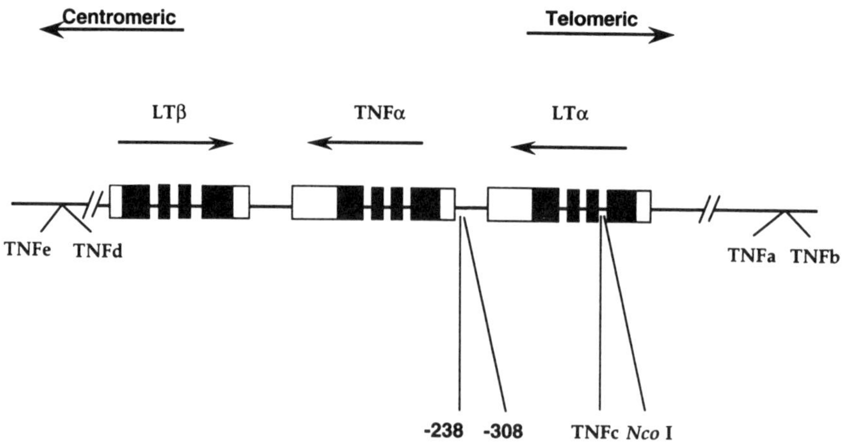

Fig. 1. Polymorphic markers in the TNF locus.

Table 1
Microsatellite Polymorphic Markers in the TNF Locus

Marker	TNFa	TNFb	TNFc	TNFd	TNFe
GDB accession no.	182382	182377	182383		
No. of alleles	13	7	2	7	4
Size of alleles (bp)	97–121	125–131	159–161	124–136	98–102
Heterozygosity index	0.73	0.86	0.35	0.54	0.21

Several restriction fragment length polymorphisms (RFLPs) within the *LTα* gene have also been characterized, including a rare *Eco*RI polymorphism in the 3' untranslated region and an *Nco*I polymorphism in the first intron, which is in strong linkage disequilibrium with a coding polymorphism at position 26 of the mature protein *(3)*. More recently, sequence analysis has revealed a number of biallelic single base substitutions in the TNF promoter (**Table 2**) *(4,5)*, one of which has direct effects on gene expression *(6)*. Analysis of the five microsatellites and the *Nco*I polymorphism have defined 35 distinct TNF haplotypes *(7)*.

Table 2
Single Base Transition Markers in the TNF Locus

Marker	TNFA-308	TNFA-238	LTα
GDB accession no.	GDB:162544	GDB:636492	GDB:156766
Allele frequencies	G 0.84 A 0.16	G 0.94 A 0.06	1 0.42 2 0.58
Heterozygosity index	0.27	0.11	0.49
Methods of screening	SSCP or *Nco*I digestion of PCR product	*Ava*II digestion of PCR product	*Nco*I digestion of PCR product

Typing of these markers is conveniently performed by polymerase chain reaction (PCR). Those that involve single base changes can be easily characterized by endonuclease digestion of PCR products *(5,8)* or single-stranded conformational polymorphism (SSCP) analysis *(9)*. The exact method chosen depends on personal preferences and the facilities available (*see* **Note 1**).

2. Materials

2.1. Polymerase Chain Reactions

2.1.1. Microsatellite Typing

1. First stage PCR mixture: 10 m*M* Tris-HCl, pH 8.3, 50 m*M* KCl; 10 m*M* $MgCl_2$, 20 μ*M* of each oligonucleotide primer, 250 μ*M* of each dNTP, 200 ng of genomic DNA, 0.5 U of *Taq* DNA polymerase in 50 μL total volume (*see* **Note 2**).
2. Second stage PCR typing requires [α-^{33}P]dATP, 3000 Ci/mmol.

2.1.2. Single Base Transition Typing

1. TNFA-308 and TNFB. PCR mixture: 10 m*M* Tris-HCl, 1.5 m*M* $MgCl_2$, 50 m*M* KCl, and 0.1% (v/v) Triton® X-100, 0.2 μ*M* of each oligonucleotide primer, 200 μ*M* of each dNTP, 100 ng of genomic DNA, 1.25 U of *Taq* DNA polymerase in 50 μL total vol.

2. TNFA-238. PCR mixture: 10 m*M* Tris-HCl, 1.5 m*M* $MgCl_2$, 50 m*M* KCl, and 0.1% (v/v) Triton X-100, 0.25 μ*M* of each oligonucleotide primer, 200 μ*M* of each dNTP, 100 ng of genomic DNA, 1.25 U of *Taq* DNA polymerase in 50 μL total vol.

2.2. Equipment

1. Thermal cycler
2. Pipets: 0.5–2 mL, 0.5–20 mL, 10–100 mL.
3. Agarose gel and acrylamide gel electrophoresis reagents and equipment.
4. DNA sequence analyzer and software (if available).

3. Methods

3.1. Microsatellite Typing

1. All PCR typing consist of two stages, the second stage leading to the incorporation of [α-^{33}P]dATP into the product.
2. The initial PCR cycles are: one cycle at 94°C for 3 min and 30 cycles of 94°C for 1 min, 60°C for 1 min, and 72°C for 1 min using the following oligonucleotides: TNFa and b, IR1 and IR2; TNFd and e, IR11 and IR14; and TNFc, IR6 and IR7 (**Table 3**).
3. PCR products should be visualized on a 2% agarose gel in 1X Tris-borate EDTA (TBE) to check for successful amplification.
4. The initial product (2 μL) is used as template in the second stage.
5. Conditions for use in the second stage are identical, except that 25 μ*M* of dATP and 5 μCi of [α-^{33}P]dATP, 3000 Ci/mmol are used.
6. Typing of TNFc is different, in that it involves 5 further cycles in the presence of [α-^{33}P]dATP, but does not require the addition of further oligonucleotide primers.
7. Following completion of the second stage PCR, 2 μL of PCR product is incubated with 2 U Klenow fragment and 100 m*M* dATP at room temperature for 15 min, to remove nontemplate-derived nucleotides, and the reaction should be stopped using 3 μL of formamide containing stop solution.
8. The products are then separated by electrophoresis on a 5% polyacrylamide gel.
9. Unambiguous assignment of TNF microsatellite alleles can be achieved by typing HLA homozygous cell lines at the same time *(7)*.

Table 3
Oligonucleotide Primers Used in Typing of the TNF Locus

TNF Locus		Primers
TNFA-308		5'-AGGCAATAGGTTTTGAGGGCCAT-3'
		5'-TCCTCCCTGCTCCGATTCCG-3'
TNFA-238		5'-GAAGCCCCTCCCAGTTCTAGTTC-3'
		5'-CACTCCCCATCCTCCCTGGTC-3'
LTα (*Nco*I)		5'-CCGTGCTTCGTGCTTTGGACTA-3'
		5'-AGAGCTGGTGGGACATGTCTG-3'
TNFa, b	IR1	5'-GCACTCCAGCCTAGGCCACAGA-3'
	IR2	5'-GCCTCTAGATTTCATCCAGGCACA-3'
	IR4	5'-CCTCTCTCCCCTGCAACACACA-3'
	IR5	5'-GTGTGTGTTGCAGGGGAGAGAG-3'
TNFc	IR6	5'-GGTTTCTCTGACTGCATCTTGTCC-3'
	IR7	5'-TCATGGGGAGAACCTGCAGAGAA-3'
TNFd, e	IR11	5'-AGATCCTTCCCTGTGAGTTCTGCT-3'
	IR12	5'-CATAGTGGGACTCTGTCTCCAAAG-3'
	IR13	5'-GTGCCTGGTTCTGGAGCCTCTC-3'
	IR14	5'-TGAGACAGAGGATAGGAGAGACAG-3'

10. As can be seen in **Table 2**, the sizes of TNFa, b, and c alleles do not overlap, and they can therefore be loaded together into a single well. However TNFd and e alleles should be run separately from them.
11. If available, an automated DNA sequence analyzer (e.g., Model 373; Applied Biosystems, Foster City, CA, USA) and suitable software can greatly increase typing throughput. Furthermore, the use of appropriately labeled oligonucleotides makes it possible to type all 5 microsatellites in a single well.

3.2. TNFA-308 Typing

1. This polymorphism can be easily screened by restriction digestion of the PCR product as a result of the creation of an *Nco*I site on amplification of allele 2 *(**5**)*, or by SSCP analysis *(**9**)* of the same PCR product (**Table 3**) (*see* **Note 3**).
2. Cycling conditions are as follows: one cycle at 94°C for 3 min, and 30 cycles of 94°C for 1 min, 60°C for 1 min, and 72°C for 1 min.

3. Restriction digestion of the PCR product by *Nco*I leads to the generation of 87-bp and 20-bp fragments from the A allele (TNF2) and a 117-bp fragment from G (TNF1).
4. The fragments are easily resolved by electrophoresis on a 9% polyacrylamide gel.
5. Typing by SSCP is performed using alkaline denaturation of the products of PCR: 2.5 µL of 1 *M* NaOH is added to 50 µL of PCR product, after heating this mixture at 42°C for 5 min, 6 L of 100% formamide is then added. The samples are then electrophoresed in a 9% polyacrylaimide gel for 16 h at 4 V/cm. DNA band patterns are visualized by UV transillumination after staining with ethidium bromide. Band patterns of the two alleles are illustrated in Wilson et al. ***(9)***.

3.3. LTα Typing

1. Typing of this polymorphism is most easily performed by NcoI digestion of a 740 bp PCR product ***(8)*** (*see* **Note 3**).
2. Cycling conditions are as follows: one cycle at 95°C for 6 min, and 35 cycles of 94°C for 1 min, 65°C for 1 min, and 72°C for 1 min.
3. Restriction digestion of the PCR product yields 555- and 185-bp fragments for TNFB*2 and a single 740-bp fragment for TNFB*1.
4. These fragments can be resolved by electrophoresis through a 1% agarose gel.

3.4. TNFA-238 Typing

1. Typing of this polymorphism is most easily performed by AvaII digestion of a 211-bp PCR product. One of the PCR primers has a base change that creates an *Ava*II site when amplifying allele G (*see* **Note 3**).
2. Cycling conditions are as follows: one cycle at 94°C for 3 min, and 35 cycles of 94°C for 1 min, 61°C for 1 min, and 72°C for 1 min, followed by 5 min at 72°C.
3. Restriction digestion of the PCR with *Ava*II will produce a constant 77 bp. In addition, allele G will give addition bands of 63 + 49 + 21 bp, and allele A will give an additional 70 + 63 bp bands.
4. These fragments can be resolved by electrophoresis through a 12% polyacrylamide gel.

4. Notes

1. Detailed information on these markers is available on the Human Genome Projects Genome Database (GDB) on the World Wide Web at (http://gdbwww.gdb.org/).
2. Negative controls, using ultrapure water as a template, should be used in all PCRs.
3. Restriction digests should include a positive control in each tube. This can either be incorporated into the PCR product, as in the typing of TNFA-238, or a second PCR product should be generated that always contains an appropriate restiction site. SSCP does not have this problem and may, therefore, be preferred.
4. Other techniques, such as dot blotting of PCR products and the Taqman® allelic discrimination system, have also been described for several of these markers *(**5,10,11**)*.

References

1. Nedospasov, S. A., Udalova, I. A., Kuprash, D. V., and Turetskaya. R. L. (1991) DNA sequence polymorphism at the human tumor necrosis factor (TNF) locus. Numerous TNF/lymphotoxin alleles tagged by two closely linked microsatellites in the upstream region of the lymphotoxin (TNF-β) gene. *J. Immunol.* **147**, 1053–1059.
2. Jongeneel, C. V., Briant, L., Udalova, I. A., Sevin, A., Nedospasov, S. A., and Cambon-Thomsen, A. (1991) Extensive genetic polymorphism in the human tumor necrosis factor region and relation to extended HLA haplotypes. *Proc. Natl. Acad. Sci. USA* **88**, 9717–9721.
3. Messer, G., Spengler, U., Jung, M. C., et al. (1991) Polymorphic structure of the tumor necrosis factor (TNF) locus: an *Nco*I polymorphism in the first intron of the human TNF-β gene correlates with a variant amino acid in position 26 and a reduced level of TNF-β production. *J. Exp. Med.* **173**, 209–219.
4. Wilson, A. G., di Giovine, F. S., Blakemore, A. I. F., and Duff, G. W. (1992) Single base polymorphism in the human tumor necrosis factor alpha (TNFα) gene detectable by NcoI restriction of PCR product. *Hum. Mol. Genet.* **1**, 353.
5. D'Alfonso, S. and Momigliano Richiardi, P. (1994) A polymorphic variation in a putative regulation box of the TNFA promoter region. *Immunogenetics* **39**, 150–154.

6. Wilson, A. G., Symons, J. A., McDowell, T. L., McDevitt, H. O., and Duff, G. W. (1997) Effects of a polymorphism in the human tumor necrosis factor alpha promoter on transcriptional activity. *Proc. Natl. Acad. Sci. USA* **94**, 3195–3199.
7. Udalova, I. A., Nedospasov, S. A., Webb, G. C., Chaplin, D. D., and Turetskaya, R. L. (1993) Highly informative typing of the human TNF locus using six adjacent polymorphic markers. *Genomics* **16**, 180–186.
8. Cabrera, M., Shaw, M. A., Sharpes, C., et al. (1995) Polymorphism in tumor necrosis factor genes associated with Mucocutaneous Leishmaniasis. *J. Exp. Med.* **182**, 1259–1264.
9. Wilson, A. G., de Vries, N., Pociot, F., di Giovine, F. S., van de Putte, L. B. A., and Duff, G. W. (1993) An allelic polymorphism within the human tumor necrosis factor alpha promoter region is strongly associated with the HLA A1, B8 and DR3 alleles. *J. Exp. Med.* **177**, 577–560.
10. Brinkman, B. M. N., Huizinga, T. W. J., Kurban, S. S., et al. (1997) Tumor necrosis factor α gene polymorphisms in rheumatoid arthritis: association with susceptibility to, or severity of, disease? *Br. J. Rheumatol.* **36**, 516–521.
11. Millard, T. P., Kondeatis, E., Cox, A., et al. (2001) A candidate gene analysis of three related photosensitivity disorders: cutaneous lupus erythematosus, polymorphic light eruption and actinic prurigo. *Br. J. Dermatol.* **145,** 229–236.

16

Molecular Typing of the MHC Class I Chain-Related Gene Locus

R. W. M. Collins, Henry A. F. Stephens and Robert W. Vaughan

1. Introduction

The human major histocompatibility complex (MHC) class I chain-related genes A and B (*MICA* and *MICB*) are located within the human leukocyte antigen (HLA) class I region of chromosome 6 *(1)*. Their organization, expression, and products differ considerably from classical HLA class I genes *(2)*. MIC proteins are considered to be markers of "stress" in the epithelia *(3)*, and they act as ligands for cells expressing a common activatory natural killer cell receptor (NKG2D) *(4)*. MIC proteins are also highly polymorphic *(5)*, although the functional implications of this polymorphism remain unclear *(6)*.

MIC genes are transcribed in fibroblast and epithelial cell lines and, reputedly, in most tissues with epithelial cell types *(2)*. MIC products are constitutively expressed on the cell surface of freshly isolated gastric epithelium, endothelial cells, and fibroblasts, but are not present on T and B cells *(7–9)*. MIC surface expression would appear to be considerably deregulated or overexpressed in transformed cells of various types, particularly in those of an epithelial origin *(3)*, and there is evidence for heat and viral induced up-regu-

From: *Methods in Molecular Biology, vol. 210: MHC Protocols*
Edited by: S. H. Powis and Robert W. Vaughan © Humana Press Inc., Totowa, NJ

lation of MIC cell surface expression *(10,11)*. Kidney allografts undergoing both acute and chronic rejection episodes have been shown to express MIC molecules *(12)*. There is also evidence of MIC protein-specific antibodies being detected in recipients of kidney allografts *(13)*.

The *MICA* locus encodes membrane-bound polypeptides of 383 amino acids, with a relative molecular mass of 43 kDa (1). The predicted domain structure is similar to that of classical HLA class I chains, including three external domains (α 1–3), a transmembrane, and a cytoplasmic domain *(1)*. Amino acid sequence alignments with classical HLA class I genes shows between 18–30% homology across the putative α 1–3 domains. MIC products, unlike their classical class HLA class I counterparts, do not bind β-2 microglobulin *(3,7)*, are independent of any transporter-associated protein (TAP) *(3)*, and attempts to acid-elute any bound peptides have been unsuccessful *(10)*. However, human MIC molecules do act as ligands for NK cells expressing an activatory lectin-like NK cell receptor in conjunction with a transmembrane signaling adaptor protein (DAP10) *(4,14)*.

Mapping studies of the human MHC genes on chromosome 6 have identified seven *MIC* genes (*MICA-MICG*), of which only *MICA* and *MICB* encode expressed transcripts, while *MICC*, *MICD*, *MICE*, *MICF* and *MICG* are pseudogenes *(2)*. *MICA* and *MICB* are located approximately 46.4 and 141.2 kb centromeric to HLA-B, respectively *(2)*. MIC genes are highly conserved and are present in humans, primates, dogs, goats, cows and pigs, but are not found in mice or rats *(1,2)*.

There are now some 51 recognized human MICA alleles *(5)*. There are numerous polymorphisms within exons 2–4 of MICA (which encode the three extracellular domains of the molecule), resulting mainly in single amino acid substitutions generating dimorphic positions, of which many are nonconservative changes *(5)*. The top surface of the MICA α 1 and 2 domains, which contains the majority of polymorphic positions *(6)*, is now known to make direct contact with NKG2D *(15)*.

The transmembrane domain of MICA encoded by exon 5, also shows some unusual features *(**5**)*. There are at least eight specific variants now described, of which six essentially differ in the number of polyalanine repeats inserted at position 296 *(**5**)*. Another MICA transmembrane variant is a truncated molecule, which has arisen from a premature stop codon most commonly seen in the MICA*008 allele. A further MICA transmembrane variant, which is detected in the MICA*017 allele, has also been shown to contain a large polylysine repeat followed by a truncation *(**5**)*.

Although a variety of methods have been developed to analyze *MIC* gene polymorphisms *(**16–19**)*, there is, at present, a relative paucity of information available on the population frequencies of *MIC* alleles in different ethnic groups. Haplotypes composed of stable combinations of MICA and HLA-B alleles that extend right through into the class II region have been described *(**20**)*.

Here, we describe three methods, a high resolution method for the external domain *MICA* alleles, the polymerase chain reaction (PCR) amplification method of Ota et al. *(**21**)* for analyzing the number of transmembrane repeats and a nucleotide sequencing method for *MICA* based on Fodil et al. *(**16**)*.

2. Materials

2.1. PCR Amplification

1. Genomic DNA.
2. PCR Buffer: PCR Buffer IV (Advanced Biotechnologies, UK).
3. 5' and 3' amplimer (*see* **Subheading 3.** for specific oligonucleotide sequences).
4. *Taq* DNA polymerase.
5. dNTP mixture.
6. $MgCl_2$.
7. Thermal cycler.
8. QIAquick™ PCR Purification Kit (Qiagen, Germany).

2.2. Agarose Gel Electrophoresis

1. Agarose.
2. 5X Tris-borate EDTA (TBE) electrophoresis buffer: 108 g Trizma base (Sigma, St. Louis, MO, USA), 27.5 g boric acid, 20 mL 0.5 *M* ethylinediaminetetraacetic acid (EDTA), pH 8.9, made up to 1 L with water. Use at 0.5X TBE for agarose gel electrophoresis.
3. 100–1000 bp DNA molecular weight marker ladder (Advanced Biotechnologies).
4. UV imager.
5. Ethidium bromide.

2.3. Microsatellite Analysis

1. Automated DNA sequencer (we have used a Model 373A Sequencer; Applied Biosystems, Foster City, CA, USA).
2. GeneScan 672 to automatically identify microsatellite alleles.

2.4. DNA Sequencing

1. ABI Prism™ BigDye Terminator Cycle Sequencing Ready Reaction Kit (Applied Biosystems).
2. ABI Prism 377 Sequencer.
3. MatchTools software (Applied Biosystems).

3. Methods

3.1. High Resolution MICA Phototyping by PCR-SSP

We have previously developed a PCR-based phototyping scheme, which was capable of defining the first 16 recognized alleles of the *MICA* locus, using sequence-specific primers (SSPs) that detected polymorphisms within exons 2–4 *(18)*. Here, we have expanded the original scheme to identify all 51 of the currently recognized *MICA* alleles *(5)*. Our scheme is specifically designed to identify all nucleotide polymorphisms that result in nonsynonymous amino acid changes within the MICA α 1–3 external domains. Certain heterozygous combinations of alleles, such as MICA*002 and MICA*020, or MICA*007 and MICA*026, share the same external amino acid sequences, but differ by the number of alanine repeats in their transmembrane regions *(5)*. Such minor variations of GCT repeats in exon 5 cannot be detected in this scheme.

3.1.1. PCR Amplification

1. As an internal control, we use primers that recognize conserved homologous sequences in the human growth hormone (HGH) gene: sense 5'-CAgTgCCTTCCCAACCATTCCCTTA-3' and antisense 5'-ATCCACTCACggATTTCTgTTgTgT-3'.
2. The sequences of the *MICA* amplimers are given in **Table 1** and their combinations are as given in **Table 2**.
3. Prepare the following PCR amplification reaction mixture: 25 ng of test DNA, 1.0 μL of 10X PCR Buffer IV, 0.43 μL of 25.0 mm $MgCl_2$, 1.0 μL of 2 m*M* dNTP mixture, 0.5 μL of a 2.0 μ*M* stock of sense and antisense internal control primers, 5.0 μL of a 1.0 μ*M* stock of the MICA amplimer combinations, and 0.4 U of *Taq* DNA polymerase (all nonoligonucleotide reagents from Advanced Biotechnologies). The final vol of all PCR was 10 μL.
4. PCR amplifications are carried out in a GeneAmp® PCR system 9600 (Applied Biosystems). The cycling conditions for *MICA* amplification are 96°C for 5 min followed by 30 cycles of 94°C for 20 s, 65°C for 50 s, and 72°C for 30 s, before cooling to 4°C.

3.1.2. Agarose Gel Electrophoresis

1. MICA allele and group-specific PCR products or amplicons are identified by gel electrophoresis in 1.0% agarose gels containing 1 μg/L ethidium bromide, which are run for 30 min at 200 V in 0.5X TBE buffer, pH 8.0.
2. A permanent digital image of each gel is obtained using UV illumination and a UVP Imagestore 7500.
3. The relative size of the *MICA*-specific amplicons is determined by comparison against the migration of a 100–1000 bp DNA molecular weight marker ladder. Successful PCR amplifications are verified with the HGH internal control primers, which generate a 480-bp product.

Table 1
***MICA* Amplimer Sequences**

Sense Primers	Exon; Posn	Code	Antisense primers	Exon; Posn	Code
CAgAgCCCCACAgTCTTCg	2;86	86G*	TCTCTCTgTCCCATgTCTTg	2;237	237G'
CAgAgCCCCACAgTCTTCC	2;86	86C	CTCACAgACCCTAATCTCCC	3;341	341C'
TAACCTCACggTgCTgTCCT	2;109	109T	gTAgAAATgCTgggAgCTCT	3;384	384T'
CCTCACggTgCTgTCCg	2;109	109G	TCCATTCCTCAgTCTCCAC	3;433	433C'
TgTgCAgTCAgggTTTCTTg	2;139	139TG	ggCACTgTCCATTCCTCAgA	3;439	439A'
GTgCAgTCAgggTTTCTCA	2;139	139CA	TggggCATTgTCCATTCCTT	3;442	442T'
TgTgCAgTCAgggTTTCTCg	2;139	139CG*	ggggCATTgTCCATTCCTC	3;442	442C'
CAgggTTTCTCgCTgAggg	2;146	146G*	TCTggAggACTggggCAC	3;454	454C'*
TCAgCCCTTCCTgCgCTA	2;176	176A*	CTTCCTTCAAgAAATTCCTgAT	3;493	493T'
CATgggACAgAgAgACCAgA	2;261	261A	CTgCATgCATAgCgTgATAgA	3;536	536A'
CAgAAgATgTCCTgggAAAT	2;306	306T	CTgCATgCATAgCgTgATAgT	3;536	536T'*
CCAgCATTTCTACTACgATA	3;409	409A	gCATgCATAgCgTgATAgC	3;536	536C'
AAggCTTgCATTCCCTCCA	3;410	410A	CTCAggACTACgCCggATTT	3;586	586T'
GCTCTTCCTCTCCCAAAACg	3;433	433G	TCAggACTACgCCggATTC	3;586	586C'
CTCTCCCAAAACCTggAgA	3;439	439A*	TTCTCCTCAggACTACgCT	3;592	592T'
GgAgACTgAggAATggACAA	3;454	454A	CCTgTTCTCCTCAggACTAT	3;595	595T'
ATTTCTTgAAggAAgATgCCA	3;520	520A*	CCTgTTCTCCTCAggACTAC	3;595	595C'
TCTTgAAggAAgATgCCg	3;520	520G	ggCCAgCgTCCgTACCTg	3;611	611G'
AggAACTACggCgATATCTAA	3;586	586A	gTTgCCCTCTgAggCCTCA	4;642	642A'
GgAACTACggCgATATCTAg	3;586	586G	AgCCCTgCATgTCACggTA	4;663	663A'

(cont.)

Table 1
MICA* Amplimer Sequences *(cont.)

Sense Primers	Exon; Posn	Code	Antisense primers	Exon; Posn	Code
ACggCgATATCTAgAATCCA	3;592	592A	AgCCCTgCATgTCACggTg	4;663	663G'
ACggCgATATCTAgAATCCg	3;592	592G	CCTgACgCCAggTCAgTA	4;707	707A'
GCgTAgTCCTgAggAgAAg	3;611	611G	CCCATCCTgACgCCAgg	4;713	713G'*
GCCTCAgAgggCAACATC	4;663	663C*	CCCCATCCTgACgCCAgC	4;713	713C'
CTTCTATCCCCggAATATCAC	4;707	707C*	gTgTCgTggCTCAAAgATAg	4;730	730G'
GCAgggCTTCTggCTTCTA	4;761	761A	gTgTCgTggCTCAAAgATAC	4;730	730C'*
			gTgTgTgATATTCCAgggAC	4;761	761C'
			CTCTCCTCggCAAATCCTA	4;810	810A'
			AACCTCTgCTCCTCTCCTTC	4;820	820C'
			CCTCTgCTCCTCTCCTTg	4;820	820G'*
			AACCTCTgCTCCTCTCCTC	4;821	821C
			AACCTCTgCTCCTCTCCTT	4;821	821T'*
			CTCACCAgAgggCACAgC	4;880	880C'
			ATAACAAAAATAgCAgCAgCC	5;953	953C'
			TTCTTCTTACAACAACAgACg	5;1002	1002G'

Amplimers marked * also share homology with alleles encoded by the *MICB* locus. All amplimers were slected to have strand melting temperatures within the range of 58–62°C, and allele and group-specific nucleotides on the terminal 3' base. The given positions of amplimers is based on the ImMunoGeneTics (IMGT)/HLA database alignment of *MICA* found at http://www.ebi.ac.uk/imgt/hla/.

Table 2
Sense and Antisense Amplimer Combinations Used to Discriminate *MICA*001–046* Alleles[a]

Mix	Target Allele	Amplimer Codes		Other Alleles Recognized	Amplicon Size
	*MICA**	Sense	Antisense	*MICA**	(bp)
1	001	139CA	442T’		616
2	002/020	761A	1002G’	017	449
3		146T	1002G’	017,046	1854
4		410A	1002G’	041,046	1317
5	004	611G	713G’	044	724
6		109T	433C’	006,009	637
7	005	707C	821C’	013	154
8		176A	663A’	025,028,031,032	1384
9	006	176A	595T’		731
10[a]	00701/026	109T	642A’	045	1432
11		642GT	820G’	00202	215
12	008	663C	953C’		428
13	009	176A	433C’	004,006,044	568
14		433G	611G’	006	216
15		433G	595C’	004,044	202
16	010	86C	454C’		679
17[b]	011	520G	880C’		983
18	012	139CA	536A’	021	711

(cont.)

Table 2
Sense and Antisense Amplimer Combinations Used to Discriminate *MICA*001–046* Alleles[a] *(cont.)*

Mix	Target Allele	Amplimer Codes		Other Alleles Recognized	Amplicon Size
	*MICA**	Sense	Antisense	*MICA**	(bp)
19		306T	536A'	032	545
20	013	109G	821C'	022,035	1608
21	014	409A	586C'		216
22	015	409A	586T'		217
23	016	586G	730G'		770
24	017	139CG	341C'		516
25	018	261A	536T'	001	590
26		139CA	442C'	012,021	615
27	019	592A	821C'	010,016,022,033	854
28		86G	592T'	004,006,009,016,022,032,033,036,044	818
29		592A	730C'	004,006,009,010,022,032,033,036,044	764
30		439A	592T'	004,006,009,010,016,022,032,036,044	191
31		139TG	663G'	004,006,008,009,010,016,024,027,033,042	1423
32	021	109T	237G'		168
33	022	109G	592T'	036,044	793
34	023	586A	953C'		1094
35	024	586G	821T'	004,006,009,014,028,036,044	860
36		592G	713G'	005,008,027,034,035,037,038,039,042	744

(cont.)

Table 2
Sense and Antisense Amplimer Combinations Used to Discriminate *MICA*001–046* Alleles[a] *(cont.)*

Mix	Target Allele	Amplimer Codes		Other Alleles Recognized	Amplicon Size
	*MICA**	Sense	Antisense	*MICA**	(bp)
37	025	86C	586T'		813
38	027/008	139TG	707A'	010,016,019,033,042	1466
39		586G	707A'	010,016,019,022,033	745
40	028	586G	713C'	013,014	751
41	029	109T	493T'		700
42	030	520A	880C'		985
43	031	176A	442T'		578
44	032	176A	536A'		673
45	033	176A	439A'		575
46	034	520G	713G'		814
47	035	586A	707A'	037,039,042	746
48		109G	409C'	002,011,013,017,020,022,023,030,034,035,036, 041,044,046	609
49	036	109G	384T'		586
50	037	454A	821C'	035,039,042	993
51		109T	730C'	001,004-010,012,018,019,021,024-029, 031-033,038,040,042,043,045	1521

(cont.)

Table 2
Sense and Antisense Amplimer Combinations Used to Discriminate *MICA*001–046* Alleles[a] *(cont.)*

Mix	Target Allele	Amplimer Codes		Other Alleles Recognized	Amplicon Size
	*MICA**	Sense	Antisense	*MICA**	(bp)
52		139CG	586T'	002,007,011,015,017,020,023,026,029,030,034, 035,038-041,043,045,046	761
53	038	454A	663G'	034,035,037,039,042	834
54		109T	820G'	001,004,005,006,007,009,010,012,016,018,019, 021,024-029,031-033,037-040,042,043	1610
55		433C	821T'	001,002,007,011,012,014,015,017,018,020,021, 023-026,028-032,034,036,038,040,041,043, 045,046	1012
56	039	586A	730G'		771
57	040	139CG	442T'		617
58	041	146G	586T'		753
59	042	586A	810A'		850
60	043	86G	536C'		762
61	044	139CG	433C'		607
62	045	586A	820C'		861
63	046	520A	761C'		868

The addition of betaine[a,b] (final concentration 1 *M*) and increased $MgCl_2$[b] (final concentration of 10 m*M*) enhanced reactions (*see* **Note 1**).

3.1.3. MICA Typing Scheme

The typing scheme represented in **Table 2** is able to identify all MICA alleles at the amino acid level within the external domain. A unique pattern of reactions for each allele allows identification in any combination.

3.2. Typing for the MICA Transmembrane Microsatellite Repeat Polymorphism

This method is from Ota et al. *(21)*. More details on microsatellite analysis is provided in Chapter 17 of this book.

3.2.1. PCR Amplification

1. The following PCR primers flanking the transmembrane region of the *MICA* gene are used: MICA5F, 5'-CCTTTTTTTCAGGG AAAGTGC-3'; MICA5R, 5'-CCTTACCATCTCCAGAAACTGC-3'. The MICA5R primer corresponds to the intron 4 and exon 5 boundary region, while MICA5F is located in intron 5.
2. Prepare the following PCR amplification reaction mixture: 10–50 ng of DNA, 10 m*M* Tris-HCL, pH 8.3, 50 m*M* KCl, 5 m*M* $MgCl_2$, 1 U *Taq* DNA polymerase, 10 pmol of each of the primers MICA5F and MICA5R, and 200 mm deoxyribonucleoside triphosphates. The final PCR vol is 25 µL.
3. The reaction conditions consist of 94°C for 2 min, followed by 30 cycles of 1 min at 94°C, 1 min annealing at 55°C, and 2 min at 72°C.

3.2.2. Microsatellite Analysis

1. To determine the number of triplet repeats, the forward primer is end-labeled with 6-FAM amidite (Applied Biosystems), and the amplified products are separated on 6% polyacrylamide gels containing 8 *M* urea in an automated DNA sequencer.
2. The number of microsatellite repeats is estimated automatically using a GeneScan 672.
3. The amplified sizes of the alleles are: 179 bp for the 4 repeat, 182 bp for the 5 repeat, 183 bp for the 5 + 1 nucleotide repeat, 185 bp for the 6 repeat, 194 bp for the 9 repeat, and 197 bp for the 10 repeat.

3.3. Nucleotide Sequencing of MICA

Direct sequencing of a PCR-amplified *MICA* gene segment is based on the method of Fodil et al. ***(16)***. The nucleotide sequence for the human MHC Class I *MICA* gene has been reported by Bahram and Spies ***(22)*** and comprises 11,722 bp. Exon 2, exon 3 and exon 4, which encode the three extracellular domains α1, α2 and α3 respectively, are located between bases 6950–8631.

3.3.1. PCR Amplification

1. Oligonucleotide primers: forward primer 5'-CgT TCT TgT CCC TTT gCC CgT gTg C-3' (residues 6823–6847), reverse primer 5'-gAT gCT gCC CCC ATT CCC TTC CCA A-3' (residues 8999–9023).
2. Prepare the following PCR amplification reaction mixture: 27.5 µL deionized water, 1.0 µL forward primer (10 m*M*), 1.0 µL reverse primer (10 m*M*), 9.7 µL 2 m*M* dNTP mixture, 5.0 µL test DNA (50 ng/mL), 5.0 µL Expand™ buffer mixt 1 (Roche Molecular Biochemicals), 8 µL Expand enzyme (Roche Molecular Biochemicals). The final reaction volume is 50 µL.
3. PCR amplifications are carried out in PTC 200 thermal cycler (NJ Research, Waltham, MA, USA) using the following cycling conditions: 1 cycle of 94°C for 2 min, 35 cycles of 94°C for 10 s, 65°C for 45 s, and 68°C for 3 min, 1 cycle of 68°C for 7 min, and 1 cycle of 20°C forever.
4. The expected product is a 2201-bp MICA gene fragment
5. Purify the amplified product using the QIAquick PCR Purification Kit and estimate the DNA concentration by running an aliquot on a 1% agarose gel together with a ladder containing a known amount of DNA. Dilute the purified product to a concentration of approx 20 ng/mL.

3.3.2. DNA Sequencing

Sequencing of exons 2, 3, and 4 of the *MICA* gene is performed using the ABI Prism BigDye Terminator Cycle Sequencing Ready Reaction Kit.

1. 8.0 µL one BigDye Terminator Cycle Sequence Ready Reaction Mix.

2. 5.0 μL one forward or reverse exon specific primer (0.75 m*M*). The following primers are used to sequence each exon in both the forward and reverse directions:

   Exon 2

   Forward: 5'-CAT CAA TgT gAA gTT ACT TCC-3' (residues 6924–6948).

   Reverse: 5'-TgC TCT Ctg CCC CTA ACT TTT CTg g-3' (residues 7246–7270).

   Exon 3

   Forward: 5'-TTC ggg AAT ggA gAA gTC AC-3' (residues 7443–7462).

   Reverse: 5'-AAT Tgg Agg gAg Agg AgA gC-3' (residues 7786–7805).

   Exon 4

   Forward: 5'-gTT CCT CTC CCC TCC TTA gA-3' (residues 8294–8313).

   Reverse: 5'-ACT TgT TAT ACA Ctg ggC AgA-3' (residues 8724–8744).

3. Seven microliters purified PCR product (*see* **Subheading 3.3.1.**).
4. The final vol is 20 μL.
5. Cycle sequencing reactions are carried out in PTC 200 thermal cycler using the following cycling conditions: 25 cycles of 96°C for 10 s, 50°C for 5 s, 60°C for 4 min, and 1 cycle of 4°C forever. Ramp speed should be no greater than 1°C per s.
6. The extension products are purified to remove any excess BigDye Terminators using either ethanol precipitation, carried out according to the instructions in the ABI Prism BigDye Terminator Cycle Sequencing Ready Reaction Kit protocol; or for a more rapid purification, AGTC Spin Columns (Edge Biosystems Inc.) can be used followed by evaporation at 80°C for about 90 min.
7. Five milliliters of gel loading buffer (5 parts deionized formamide and 1 part 25 m*M* EDTA with blue dextran) is added to each dried-down product and mixed well before denaturing at 95°C for 2 min. After denaturation, the tubes are plunged into ice until they were ready to be loaded.
8. The samples are loaded and separated overnight on a 4% acrylamide gel in an ABI Prism 377 Sequencer and analyzed using the current

version of PE Applied Biosystems Sequence Analysis software. The resultant sequences of exons 2, 3 and 4 can be compared to a library of known MICA alleles using MatchTools software.

4. Notes

1. Please note the following concerning the matching scheme provided in **Table 2**.
   a. *MICA*002* and *MICA*020* share the same external amino acid sequence, but differ by one alanine repeat in the transmembrane region.
   b. *MICA*007* and *MICA*026* have an identical amino acid sequence in the external domain, but differ by two alanine repeats in the transmembrane region.
   c. *MICA*027* and *MICA*008* share the same external amino acid, but again differ in the transmembrane region.
   d. *MICA*008* has a G insertion, which transcribes a premature stop codon.
   e. *MICA*002* and *MICA*008* are identified using primers located in exon 5, which encodes the transmembrane region.
   f. The nucleotide sequence of exon 5 of the following alleles has not yet been determined: *MICA*005*, *013*, *014*, *021*, *022*, *024*, *025*, *030–032*, *034–040*, *042*, and *044*. Until these are defined, there is a possibility of these alleles reacting with primer mixes 2, 3, 4, 12, and 34.

References

1. Bahram, S., Bresnahan, M., Geraghty, D. E., and Spies T. (1994) A second lineage of mammalian major histocompatibility complex class I genes. *Proc. Natl. Acad. Sci. USA* **91**, 6259–6263.
2. Bahram, S. (2000) MIC genes: from genetics to biology. *Adv. Immunol.* **76**, 1–60.
3. Groh, V., Bahram, S., Bauer, S., Herman, A., Beauchamp, M., and Spies, T. (1996) Cell stress-related human major histocompatibility complex class I gene expressed in gastrointestinal epithelium. *Proc. Natl. Acad. Sci. USA* **93**, 12445–12450.
4. Bauer, S., Groh, V., Wu, J., et al. (1999) Activation of NK cells and T

cells by NKG2D, a receptor for stress-inducible MICA. *Science* **285**, 727–729.

5. Robinson, J., Perez-Rodriguez, M., Waller, M. J., et al. (2001) MICA Sequences 2000. *Immunogenetics* **53**, 150–169.
6. Stephens, H. A. F. (2001) MICA and MICB genes: can the enigma of their polymorphism be resolved? *Trends Immunol.* **22**, 378–385.
7. Zwirner, N. W., Fernandez-Vina, M.A., and Stastny, P. (1998) MICA, a new polymorphic HLA-related antigen, is expressed mainly by keratinocytes, endothelial cells, and monocytes. *Immunogenetics* **47**, 139–148.
8. Groh, V., Rhinehart, R., Secrist, H., Bauer, s., Gabstein, K. H., Spies, T. (1999) Broad tumor-associated expression and recognition by tumor-derived gamma-delta T cells of MICA and MICB. *Proc. Natl. Acad. Sci. USA* **96**, 6879–6884.
9. Zwirner, N. W., Dole, K., Stastny, P., et al. (1999) Differential surface expression of MICA by endothelial cells, fibroblasts, keratinocytes and monocytes. *Hum. Immunol.* **60**, 323–330.
10. Groh, V., Steinle,A., Bauer, S., and Spies, T. (1998) Recognition of stress-induced MHC molecules by intestinal epithelial gamma-delta T cells. *Science* **279**, 1737–1740.
11. Groh, V., Rhinehart, R., Randolph-Habecker, J., Topp, M. S., Riddell, S. R., and Spies, T. (2001) Costimulation of CD8 alpha-beta T cells by NKG2D via engagement by MIC induced on virus-infected cells. *Nat. Immunol.* **2**, 255–260.
12. Hankey, K. G., Drachenberg, C. B., and Papdimetriou, J. C. (2000) Expression of MICA/MICB in renal and pancreatic allografts. *Transplantation* **73**, 304–306.
13. Zwirner, N. W., Marcos, C. Y., Mirbaha, F., Zou, Y., and Stastny, P. (2000) Identification of MICA as a new polymorphic alloantigen recognized by antibodies in sera of organ transplant recipients. *Hum. Immunol.* **61**, 917–924.
14. Wu, J., Song, Y., Bakker, A. B., et al. (1999) An activating immunoreceptor complex formed by NKG2D and DAP10. *Science* **285**, 730–732.
15. Li, P., Morris, D. L., Willcox, B. E., Steinle, A., Spies, T., and Strong, R. K. (2001) Complex structure of the activating immunoreceptor NKG2D and its MHC class I-like ligand MICA. *Nat. Immunol.* **2**, 443–451.

16. Fodil, N., Laloux, L., Wanner, V., et al. (1996) Allelic repertoire of the human MHC class I MICA gene. *Immunogenetics* **44**, 351–357.
17. Tsuchiya, N., Shiota, M., Moriyama, S., et al. (1998) MICA allele typing of HLA-B27 positive Japanese patients with seronegative spondylarthropathies and healthy individuals. Differential linkage disequilibrium with HLA-B27 subtypes. *Arthritis Rheum.* **41**, 68–73.
18. Stephens, H. A. F., Vaughan, R. W., Collins, R., Kondentis, E., Theron, J., and Payne, A. (1999) Towards a molecular phototyping system for allelic variants of MICA, encoded by polymorphisms in exons 2, 3, and 4 of MHC class I chain-related genes. *Tissue Antigens* **53**, 167–174.
19. Mendoza-Rincon, J., Arguello, J. R., Perez-Rodriguez, M., et al. (1999) Characterization of the MICA polymorphism by sequence-specific oligonucleotide probing. *Immunogenetics* **49**, 471–478.
20. Katsuyama, Y., Ota, M., Ando, H., et al. (1999) Sequence based typing for genetic polymorphisms in exons 2, 3, and 4 of the MICA gene. *Tissue Antigens* **54**, 178–184.
21. Ota, M., Katsuyama, Y., Mizuki, N., et al. (1997) Tri-nucleotide repeat polymorphism within exon 5 of the MICA gene (MHC class I chain-related gene A): allele frequency data in the nine population groups Japanese, Northern Han, Hui, Uygur, Kazakhstan, Iranian, Saudi Arabian, Greek and Italian. *Tissue Antigens* **49**, 448–454.
22. Bahram, S. and Spies, T. (1996) Nucleotide sequence of a human MHC class I MICB cDNA. *Immunogenetics* **43**, 230–233.

IV

MICROSATELLITES

17

HLA Microsatellite Analysis

Mary Carrington

1. Introduction

The human genome contains 50,000–100,000 microsatellite repeats of $(CA)_n$ occurring every 30–60 kb in euchromatic regions of DNA *(1)*. Their characteristic length polymorphism is probably due to DNA slippage during replication *(2)*. While the normal function of microsatellite loci remains unknown, they have become extremely valuable markers for genetic mapping *(3,4)*, construction of evolutionary trees *(5)*, individual identification in forensic cases *(6)*, and relatedness testing *(7)*.

Many microsatellites within and flanking the human major histocompatibility complex (MHC) have already been identified, most of which are $(CA)_n$ repeats *(8)*. Reference information has been published *(9)* for 12 of these microsatellite loci using the well-characterized 10th International Histocompatibility Workshop homozygous typing cell lines (HTCs), which have previously been typed at more than 20 genes within the class I, II, and III regions. Further, a number of these microsatellite markers have been used for various purposes, such as mapping of the hemochromatosis gene *(10,11)*, determination of recombination frequencies across the MHC *(12)*, identification of recombinant chromosomes within the class II region *(13)*, and human leukocyte antigen (HLA)

From: *Methods in Molecular Biology, vol. 210: MHC Protocols*
Edited by: S. H. Powis and Robert W. Vaughan © Humana Press Inc., Totowa, NJ

matching of related bone marrow donors and recipients *(14)*. Recent data also indicate that these markers are useful in mapping regions of disease association within the MHC due to the strong linkage disequilibrium between HLA genes and their close neighboring microsatellite markers *(15)*.

Typing of microsatellites is rapid and accurate, providing additional incentive for their use as polymorphic markers. The technique involves amplification of the locus using primers specific for sequences flanking the repeats. As long as sequence information is available, primers may be designed to amplify products ranging from 100 to several hundred base pairs. These markers have been typed by amplification of the marker using polymerase chain reaction (PCR) in the presence of radioactive nuclides, running the products on 40-cm denaturing gels to distinguish length polymorphism, and analyzing the banding patterns on film exposed to the gels. More efficient techniques, such as that described in this chapter, have been developed which eliminate the use of radioactive material and allow much more rapid collection of data with greater accuracy. A map illustrating the location of the microsatellites used in this protocol is provided in **Fig. 1**.

2. Materials

2.1. DNA Amplification

1. Primers: the sequences representing the nine sets of primers used are listed in **Table 1**. Each 5' primer is fluorescently tagged with 6-FAM, HEX, or NED (*see* **Table 1**) by Applied Biosystems Division (Foster City, CA, USA). Size range of the products and the fluorescent label were chosen so that up to 3 loci with alleles in the same size range could be run in a single lane, each labeled with a different color. Stock primer concentration is 10 μ*M*. Loci are amplified in three multiplex reactions as follows: group 1, D6S273, D6S291, TAP1CA, D6S276, G51152; group 2, D6S265, MIB, RING3CA; group 3 MOGCA (*see* **Note 1**).
2. DNA: 50–100 ng DNA/2 μL.
3. dNTP: 2 m*M* dNTP stock (10X).

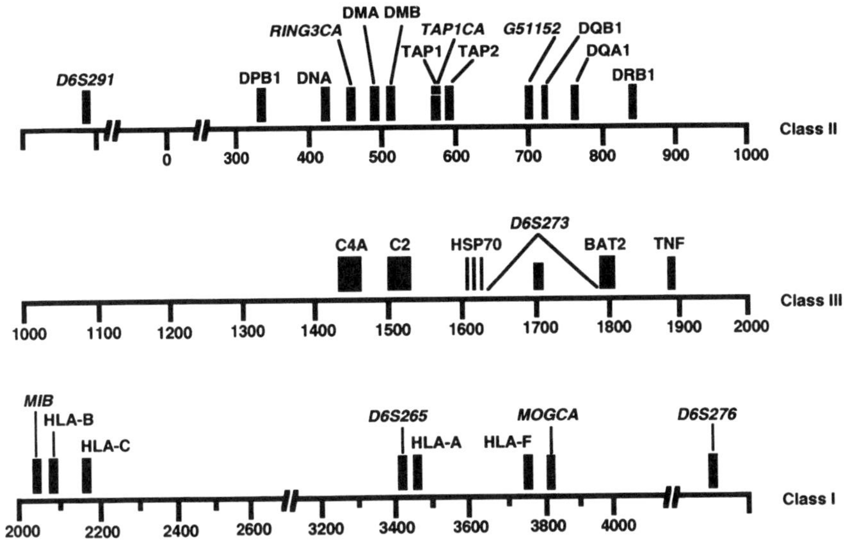

Fig. 1. Map of the HLA class I, II, and III regions showing the relative location of the 9 microsatellites (shown in italics) employed herein.

4. 10X PCR buffer: 100 m*M* Tris-HCl, pH 8.3, and 500 m*M* KCl.
5. 20 m*M* $MgCl_2$ (10X).
6. Polymerase: Amplitaq Gold™ (Applied Biosystems) (*see* **Note 2**).

2.2. Electrophoresis

1. Loading/standard buffer: fluorescently labeled molecular weight marker (GeneScan®-350 ROX, Applied Biosystems), blue dextran (comes with the 350 ROX), and formamide.
2. 10X TBE running buffer: 0.9 *M* Tris base, 0.9 *M* boric acid, 0.02 *M* EDTA, pH 8.0.
3. Electrophoresis gel: Long Ranger™ gel solution (FMC Bioproducts, Rockland, ME, USA), urea, 10X running buffer (*see* **step 2**), 10% ammonium persulfate, and N,N,N',N'-tetramethylethylenediamine (TEMED).
4. AG 501-X8 resin (Bio-Rad, Hercules, CA, USA).
5. Nalgene 115-mL 0.2-µm filters (Nalgene, Rochester, NY, USA).

Table 1
Primers Used for Microsatellite Typing

Locus	5' Primer Sequence	3' Primer Sequence
D6S276	5'-tcaatcaaatcatccccagaag-3'	5'-ccttctttgcagactgtcacc-3'
MOGCA	5'-gaaatgtgagaataaaggaga-3'	5'-gataaaggggaactactaca-3'
D6S265	5'-agtcaccctactgtgctatc-3'	5'-atcgaggtaaacagcagaaag-3'
MIB	5'-gcttcacccgatcagtagaagac-3'	5'-gcatggtgtcagagatagtcaggtc-3'
D6S273	5'-ggagaagttgagtatttctgc-3'	5'-accaaacttcaaattttcgg-3'
G51152	5'-ggtaaaattcctgactggcc-3'	5'-gacagctcttcttaacctgc-3'
TAP1CA	5'-gctttgatctcccccctc-3'	5'-ggacaatattttgctcctgagg-3'
RING3CA	5'-tgcttatagggagactaccg-3'	5'-gaggtaatgtcacaggatggg-3'
D6S291	5'-ggcattcaggcatgcctggc-3'	5'-ggggatgacgaattattcactaact-3'

2.3. Equipment

1. Thermal cycler: conditions given in **Subheading 3.** for amplification are appropriate for a Model 9700 thermal cycler (Applied Biosystems), although they can be adapted to other types of thermal cyclers.
2. ABI model 377 DNA sequencer (Applied Biosystems): 36-cm well-to-read plates, 0.2-mm spacers, 50-well 0.2-mm square-tooth comb.
3. ABI Prism™ GeneScan and Genotyper DNA fragment analysis software: these software allow analysis and interpretation of nucleic acid fragment size and quantitation by converting it into user-defined results, which can then be transferred to a database for storage and analysis.

3. Methods

3.1. DNA Amplification

1. The multiplex amplification reactions are carried out in a total of 20 µL. Thus, the cocktail for the five Group 1 loci will require the addition of less water than that for the three Group 2 loci, and the one Group 3 locus. Distilled deionized water (dDW) should be added to bring the final vol of each amplification to 20 µL. Prepare the PCR mixture by combining the following: 0.45–0.9 µL of each primer (*see* **Note 3**), 50–100 ng DNA, 2 µL 10X dNTP, 2 µL 10X PCR buffer, 2 µL 10X $MgCl_2$, 0.2 µL (1 U) Amplitaq Gold®.
2. Amplify under the following conditions: 8 min at 94°C (denaturation of DNA and activation of Amplitaq Gold), followed by 30 cycles (of 15 s at 94°C, 15 s at 55°C, and 30 s at 72°C), with a final 30 min extension step at 72°C (*see* **Note 4**) .
3. Following amplification, dilute the samples by mixing 4 µL of Group 1 product, 2 µL Group 2 product, 4 µL Group 3 product, and 10 µL dDW. These amounts may need to be adjusted to optimize data collection.
4. Mix 2.5 µL of diluted products with 3.5 µL loading/standard buffer.
5. Immediately prior to loading the gel, denature the amplified products at 96°C for 2 min and place samples on ice.

3.2. Electrophoresis

1. Weigh 36 g molecular biology grade urea in a 250-mL beaker.
2. Add 54 mL dDW and 10 mL 50% Long Ranger Gel Solution.
3. Microwave (high) the mixture for 10 s only.
4. Add 5 g AG 501-X8 resin and a stir bar. Cover with parafilm and mix on a stir plate for about 30 min.
5. Filter the gel solution in a 115-mL 0.2-μm filter unit and degas the solution for 5 min from the time all of the gel solution has passed through the filter.
6. Pour solution into a 100-mL cylinder and add 10 mL 10X TBE.
7. Pour solution into a 250-mL beaker and add 500 μL fresh 10% ammonium persulfate and 50 μL TEMED. Mix gently without forming bubbles.
8. Draw solution into a 100-cc syringe, eject bubbles, and insert into coupler on gel pouring apparatus, dispensing until the solution has reached the top of the plates (*see* **Note 5**).
9. Insert comb, cover ends of plates with Saran® Wrap, and place 2 metal clips over the comb. Allow polymerization for 2 h (*see* **Note 6**).
10. Load 1.5 μL of sample per well and run the gel for 2 h at 3000 V using the GS 36D–2400 run module.
11. Follow the GeneScan and Genotyper DNA fragment analysis software to interpret the data.

4. Notes

1. Additional microsatellite markers within the MHC can be amplified and grouped as described for the nine markers used in the protocol described in **Subheading 3.** However, no more than 9–12 markers can be run per lane without affecting the clarity of the data.
2. Amplitaq Gold is a modified *Taq* DNA polymerase that does not become enzymatically active until it has been exposed to a high temperature soak (94°–95°C). This confers a "hot start" to the PCR, which greatly improves the efficiency and specificity of the multiplex PCR.
3. Although each primer is kept at a stock concentration of 10 μM, the amount used in the amplification reaction depends on the quality of the amplification of each individual marker, because some markers

amplify more robustly than others. Thus, if the signal for one locus is much higher than that for the other loci, then the primer concentration for the first locus should be reduced. A reasonable volume range for all primers is 0.45–0.9 µL.

4. *Taq* DNA polymerase can catalyze nontemplated addition of a nucleotide (usually adenosine) to the 3' end of the amplified product. Because this addition is not uniform, some products have the additional adenosine and others do not. In order to generate a more homogenous population of products, the final extension step at 72°C is lengthened to 30 min. This increases the likelihood that adenosine will be added to the amplified products.
5. Air bubbles in the gel can cause problems in reading the bands. To avoid bubble formation, try to remove all air from the syringe prior to pouring the gel, and pour the gel slowly.
6. The gel can be poured 1 day in advance of electrophoresis.

References

1. Hamada, H. and Kakunaga, T. (1982) Potential Z-DNA forming sequences are highly dispersed in the human genome. *Nature* **298**, 396–398.
2. Schlötterer, C. and Tautz, D. (1992) Slippage synthesis of simple sequence DNA. *Nucleic Acids Res.* **20**, 211–215.
3. Hearne, C. M., Gosh, S., and Todd, J. A. (1992) Microsatellites for linkage analysis of genetic traits. *Trends Genet.* **8**, 288–294.
4. Bruford, M. W. and Wayne, R. K. (1993) Microsatellites and their application to population genetical studies. *Curr. Opin. Genet. Dev.* **3**, 939–943.
5. Bowcock, A. M., Ruiz-Linares, A., Tomfohrde, J., Minch, E., Kidd, J. R., and Cavalli-Sforza, L. L. (1994) High resolution of human evolutionary trees with polymorphic microsatellites. *Nature* **368**, 455–457.
6. Hagelberg, E., Gray, I. C., and Jeffreys, A. J. (1991) Identification of the skeletal remains of a murder victim by DNA analysis. *Nature* **352**, 427–429.
7. Jeffreys, A. J., Allen, M. J., Hagelberg, E., and Sonnberg, A. (1992) Identification of the skeletal remains of Joseph Mengele by DNA analysis. *Forensic Sci. Int.* **56**, 65–76.

8. Foissac, A., Crouau-Roy, B., Fauré, S., Thomsen, M., and Cambon-Thomsen, A. (1997) Microsatellites in the HLA region: an overview. *Tissue Antigens* **49**, 197–214.
9. Martin, M. P., Harding, A., Chadwick, R., et al. (1998) Characterization of 12 microsatellite loci of the human MHC in a panel of reference cell lines. *Immunogenetics* **47**, 503.
10. Feder, J. N., Gnirke, A., Thomas, W., et al. (1996) A novel MHC class I-like gene is mutated in patients with hereditary haemochromatosis. *Nat. Genet.* **13**, 399–408.
11. Raha-Chowdhury, R., Bowen, D., Stone, C., et al. (1995) New polymorphic microsatellite markers place the haemochromatosis gene telomeric to D6S105. *Hum. Mol. Genet.* **4**, 1869–1874.
12. Martin, M., Mann, D., and Carrington, M. (1995) Recombination rates across the HLA complex: use of microsatellites as a rapid screen for recombinant chromosomes. *Hum. Mol. Genet.* **4**, 423–428.
13. Cullen, M., Noble, J., Erlich, H., et al. (1997) Characterization of recombination in the HLA class II region. *Amer. J. Hum. Genet.* **60**, 397–407
14. Carrington, M. and Wade, J. (1996) Selection of transplant donors based on MHC microsatellite data. *Hum. Immunol.* **51**, 106–109.
15. Carrington, M., Marti, D., Wade, J., et al. (1999) Microsatellite markers in complex disease: mapping disease-associated regions within the human major histocompatibility complex. In: *Microsatellites–Evolution and Applications* (Goldstein, D. B., and Schlötterer, C., eds.), Oxford University Press, Inc., New York, pp. 225–237.

Index

From: *Methods in Molecular Biology, vol. 210:MHC Protocols*
Edited by: S. H. Powis and R. W. Vaughan © Humana Press Inc., Totowa, NJ

T

W, X